Integrated Environmental Solutions: Approaches in Microbiology, Biotechnology, and Engineering

Edited By

Rajneesh Kumar
Department of Chemical Engineering and Technology
Indian Institute of Technology (BHU), Varanasi
Uttar Pradesh, India

Ram Sharan Singh
Department of Chemical Engineering and Technology
Indian Institute of Technology (BHU), Varanasi
Uttar Pradesh, India

&

Maulin P. Shah
Department of Research Impact and Outcome Research
Research and Development Cell, Lovely Professional University
Phagwara, Punjab, India

Integrated Environmental Solutions: Approaches in Microbiology, Biotechnology, and Engineering

Editors: Rajneesh Kumar, Ram Sharan Singh & Maulin P. Shah

ISBN (Online): 979-8-89881-375-8

ISBN (Print): 979-8-89881-376-5

ISBN (Paperback): 979-8-89881-377-2

© 2026, Bentham Books imprint.

Published by Bentham Science Publishers Pte. Ltd. Singapore, in collaboration with Eureka Conferences, USA. All Rights Reserved.

First published in 2026.

BENTHAM SCIENCE PUBLISHERS LTD.
End User License Agreement (for non-institutional, personal use)

This is an agreement between you and Bentham Science Publishers Ltd. Please read this License Agreement carefully before using the ebook/echapter/ejournal (**"Work"**). Your use of the Work constitutes your agreement to the terms and conditions set forth in this License Agreement. If you do not agree to these terms and conditions then you should not use the Work.

Bentham Science Publishers agrees to grant you a non-exclusive, non-transferable limited license to use the Work subject to and in accordance with the following terms and conditions. This License Agreement is for non-library, personal use only. For a library / institutional / multi user license in respect of the Work, please contact: permission@benthamscience.org.

Usage Rules:

1. All rights reserved: The Work is the subject of copyright and Bentham Science Publishers either owns the Work (and the copyright in it) or is licensed to distribute the Work. You shall not copy, reproduce, modify, remove, delete, augment, add to, publish, transmit, sell, resell, create derivative works from, or in any way exploit the Work or make the Work available for others to do any of the same, in any form or by any means, in whole or in part, in each case without the prior written permission of Bentham Science Publishers, unless stated otherwise in this License Agreement.
2. You may download a copy of the Work on one occasion to one personal computer (including tablet, laptop, desktop, or other such devices). You may make one back-up copy of the Work to avoid losing it.
3. The unauthorised use or distribution of copyrighted or other proprietary content is illegal and could subject you to liability for substantial money damages. You will be liable for any damage resulting from your misuse of the Work or any violation of this License Agreement, including any infringement by you of copyrights or proprietary rights.

Disclaimer:

Bentham Science Publishers does not guarantee that the information in the Work is error-free, or warrant that it will meet your requirements or that access to the Work will be uninterrupted or error-free. The Work is provided "as is" without warranty of any kind, either express or implied or statutory, including, without limitation, implied warranties of merchantability and fitness for a particular purpose. The entire risk as to the results and performance of the Work is assumed by you. No responsibility is assumed by Bentham Science Publishers, its staff, editors and/or authors for any injury and/or damage to persons or property as a matter of products liability, negligence or otherwise, or from any use or operation of any methods, products instruction, advertisements or ideas contained in the Work.

Limitation of Liability:

In no event will Bentham Science Publishers, its staff, editors and/or authors, be liable for any damages, including, without limitation, special, incidental and/or consequential damages and/or damages for lost data and/or profits arising out of (whether directly or indirectly) the use or inability to use the Work. The entire liability of Bentham Science Publishers shall be limited to the amount actually paid by you for the Work.

General:

1. Any dispute or claim arising out of or in connection with this License Agreement or the Work (including non-contractual disputes or claims) will be governed by and construed in accordance with the laws of Singapore. Each party agrees that the courts of the state of Singapore shall have exclusive jurisdiction to settle any dispute or claim arising out of or in connection with this License Agreement or the Work (including non-contractual disputes or claims).
2. Your rights under this License Agreement will automatically terminate without notice and without the

need for a court order if at any point you breach any terms of this License Agreement. In no event will any delay or failure by Bentham Science Publishers in enforcing your compliance with this License Agreement constitute a waiver of any of its rights.
3. You acknowledge that you have read this License Agreement, and agree to be bound by its terms and conditions. To the extent that any other terms and conditions presented on any website of Bentham Science Publishers conflict with, or are inconsistent with, the terms and conditions set out in this License Agreement, you acknowledge that the terms and conditions set out in this License Agreement shall prevail.

Bentham Science Publishers Pte. Ltd.
No. 9 Raffles Place
Office No. 26-01
Singapore 048619
Singapore
Email: subscriptions@benthamscience.net

CONTENTS

FOREWORD	i
PREFACE	iii
LIST OF CONTRIBUTORS	v

CHAPTER 1 THE BLUEPRINT OF LIFE: BIO-ENGINEERING OF SUSTAINABLE MATERIALS FOR A GREENER FUTURE 1
G. Vijaya Laxmi, Chelemala Katyayani, K. Shivathmika Reddy, Y. Swarna Manjari, Sanjeeb Kumar Mandal and *Bishwambhar Mishra*

INTRODUCTION	1
What are Sustainable Materials, and Why are they Important?	2
The Role of Bioengineering in Sustainability	3
How Bioengineering Differs from Traditional Material Manufacturing	3
BIOENGINEERED SUSTAINABILITY MATERIAL TYPES	4
Biodegradable Materials	5
Plant-derived and Agricultural Waste Materials	5
Sustainable Textiles and Fibers	5
Bio-based Composites and Hybrid Materials	6
MICROBIAL CONTRIBUTIONS TO SUSTAINABLE MATERIAL PRODUCTION	6
Microorganisms in the Synthesis of Biodegradable Materials	6
Harnessing Microbial Metabolism for Waste-to-resource Conversion	7
Genetic Engineering of Microbes for Enhanced Material Properties	8
Integration of Microbiology in Bioprocessing for Material Development	8
BIOTECHNOLOGY APPROACHES FOR SUSTAINABLE MATERIALS	9
Biocatalysis as a Sustainable Approach to Material Synthesis	9
Engineered Enzymes for Biodegradation	9
Enzymatic Biodegradation of Plastics	10
Polyhydroxyalkanoates -PHA-Degrading Enzymes	10
Techniques for Engineered Enzymes	10
Synthetic Biology for the Production of Advanced Bio-based Materials	11
Protein Engineering for Improved Biocatalytic Efficiency	12
BIOREMEDIATION AND ITS ROLE IN SUSTAINABLE MATERIAL DEVELOPMENT	12
Engineering Microbes for Environmental Cleanup	13
Applications of Bioremediation in Recycling and Reuse of Materials	14
Integrating Bioremediation with Material Production Processes	14
FACTORS AFFECTING THE EFFECTIVENESS OF BIOENGINEERED MATERIALS	15
Choice of Microorganisms and Enzymes	16
Properties of Feedstock and Raw Materials	17
Efficiency of Bioprocesses	19
Environmental Impact and Lifecycle Analysis	19
Economic Feasibility and Scalability	20
• *Costs of Production*	20
• *Production Scalability*	20
• *Acceptance and Market Demand*	21
CHALLENGES AND FUTURE DIRECTIONS	21
CONCLUSION	22
REFERENCES	23

CHAPTER 2 SYNTHETIC BIOLOGY FOR ENVIRONMENTAL APPLICATIONS 26
S S Kirthiga and *R Dhinesh*

INTRODUCTION	27

SYNTHETIC BIOLOGY TOOLS AND TECHNIQUES	28
Genetic Circuit Design and Optimization	28
Metabolic Engineering for Environmental Applications	29
Integration of Microbial Systems	30
SCOPE OF SYNTHETIC BIOLOGY FOR ENVIRONMENTAL APPLICATIONS	31
SYNTHETIC BIOLOGY FOR POLLUTION REMEDIATION	32
Engineered Microorganisms in Pollution Control	33
Biodegradation of Toxic Pollutants	34
CARBON CAPTURE AND GREENHOUSE GAS MITIGATION	35
BIOSENSORS FOR ENVIRONMENTAL MONITORING	35
APPLICATIONS IN ENVIRONMENTAL MANAGEMENT	36
Wastewater Treatment	36
Soil Bioremediation	37
Air Quality Management	38
ETHICAL AND RISK CONSIDERATIONS	39
FUTURE DIRECTIONS IN SYNTHETIC BIOLOGY FOR ENVIRONMENTAL APPLICATIONS	39
CONCLUSION	40
REFERENCES	40

CHAPTER 3 MECHANISMS AND ADVANCED PROCESSES IN WASTEWATER TREATMENT TECHNOLOGIES 47
R Dhinesh and *S S Kirthiga*

INTRODUCTION	48
WASTEWATER TREATMENT PROCESSES AND THEIR IMPORTANCE	49
PRELIMINARY AND PRIMARY TREATMENT	51
Screening and Removal of Large Solids	51
Sedimentation and Grit Removal Mechanisms	52
Limitations and Efficiency of Primary Treatment	53
SECONDARY TREATMENT: BIOLOGICAL METHODS	53
Principles of Biological Wastewater Treatment	53
Activated Sludge Systems: Design and Operation	54
Trickling Filters and Sequencing Batch Reactors	55
Biological Nutrient Removal (BNR)	56
TERTIARY TREATMENT TECHNOLOGIES	57
Advanced Filtration Techniques	57
Disinfection Methods	58
Ultraviolet (UV) Irradiation	58
Ozonation Processes	59
Chemical Treatments	59
SLUDGE MANAGEMENT AND RESOURCE UTILIZATION	60
ADVANCED AND EMERGING WASTEWATER TREATMENT TECHNOLOGIES	61
Membrane Bioreactors	61
Electrochemical Treatment Processes	62
CONCLUSION	63
REFERENCES	63

CHAPTER 4 BIOAUGMENTATION AND BIOSTIMULATION 70
Azimul Hasan and *Arun Kumar K.*

INTRODUCTION	71
Principle of Bioaugmentation	71

    Microorganisms used in Bioaugmentation 71
    Factors Influencing Bioaugmentation 73
        Environmental Parameters 73
        Competition with Indigenous Microorganisms 73
        Predation and Abiotic Factors 73
        Recent Advances in Bioaugmentation 74
        Genetic Engineering of Microbial Strains 74
        Encapsulation and Immobilization Techniques 74
        Co-cultures and Synergistic Microbial Communities 74
        Metagenomics and Environmental Genomics 74
BIOSTIMULATION IN ENVIRONMENTAL REMEDIATION: RECENT ADVANCES AND APPROACHES 75
    Principles of Biostimulation 75
    Recent Advances in Biostimulation 75
        Molecular Techniques and Genomics 76
        Systems Biology and Predictive Modeling 76
APPLICATIONS OF BIOSTIMULATION AT CONTAMINATED SITES 76
    Hydrocarbon Contamination 76
    Degradation of Pesticides and Industrial Chemicals 77
    Remediation of Heavy Metal 77
CHALLENGES AND LIMITATIONS OF BIOSTIMULATION 77
    Environmental Factors 77
    Site Specificity 77
    Contaminant Characteristics 78
RECENT INNOVATIONS AND ADVANCEMENTS IN BIOSTIMULATION 78
    Integrated Bioaugmentation and Biostimulation 78
    Genetic Engineering and Synthetic Biology 79
    Synergistic Fungi and Bacteria 80
    Nutrient and Substrate Amendment 80
    Use of Nanoparticles 80
    Biochar Amendment 81
    Oxygen and Hydrogen Peroxide Injection 83
    Electro-bioremediation (Bioelectrochemical Systems) 83
    CONCLUSION 84
    REFERENCES 84

CHAPTER 5 BIOREMEDIATION TECHNIQUES 92
Shruti Khanna Ahuja, Akriti Gupta and *Preeti Mehta Kakkar*
    INTRODUCTION 92
    BIOREMEDIATION TECHNIQUES 94
        In-situ Bioremediation Techniques 94
            Intrinsic Bioremediation 95
            Engineered In-situ bioremediation 95
        Ex-situ Bioremediation 100
            Slurry phase bioremediation 100
            Solid phase bioremediation 101
    FUTURE PERSPECTIVES IN BIOREMEDIATION 102
        Genetically Engineered Microorganisms (GEM) 103
        Bioremediation by Nano Materials 103
        Bioinformatics in Bioremediation 103
    CONCLUSION 104

REFERENCES	104
CHAPTER 6 MICROBIAL COMMUNITIES IN ENVIRONMENTAL ENGINEERING	108

Arunima Singh, Sonali Ranjan, Suparna Bardhan, Anupam Jayas, Yogesh Kumar Vishwakarma and R. S. Singh

INTRODUCTION	108
ENGINEERING APPROACHES TO WASTE TREATMENT USING MICROBIAL COMMUNITIES	111
In-situ Bioremediation	112
Intrinsic In-situ Bioremediation	113
Enhanced In-situ Bioremediation	113
Ex-situ Bioremediation	116
Solid Phase Treatment	116
Solid–Liquid Mix Phase Treatment	117
Liquid Phase Treatment	118
TYPE OF MICROBIAL COMMUNITIES USED IN ENVIRONMENTAL ENGINEERING	118
Bacterial Communities in Different Engineering Approaches	119
Degradation of Pollutants	119
Heavy Metal Reducers	120
Biosorption	120
Wastewater Treatment	120
Pesticide Degradation	121
New Molecular Techniques used in Bacteria to Study Communities	122
Applications	122
Fungal Communities in Different Engineering Approaches	122
Wastewater Treatment	123
Sanitary Landfill	123
Bioremediation	123
Bioelectricity and Fuel Cell	124
Algal Communities in Different Engineering Approaches	124
CO_2 Sequestration	124
Bioindicators	124
Nutrient Recovery	125
Biofuel Production	125
Biofuel Production from Heavy Metal-Contaminated Wastewater	125
MODERN APPROACH TO WASTE TREATMENT	125
Using Genetically Modified Organisms	129
Using a Hybrid System	130
LIMITATIONS	131
Use of Proteomics in Microbes Identification	131
Analytic Computational Tools	131
Risks of Genetically Engineered Organisms	131
CONCLUSION	132
REFERENCES	132
CHAPTER 7 CLIMATE CHANGE AND MICROBIAL PROCESSES	146

Preeti Mehta Kakkar, Aindree Lohumi, Diya Saha and Shruti Khanna Ahuja

INTRODUCTION	146
MICROBIAL DIVERSITY AND CLIMATE CHANGE	149
Microbial Biodiversity in Terrestrial and Aquatic Ecosystems	149
Impacts of Climate Change on Microbial Diversity	150
Temperature Variations	150

- *Rising CO2 Levels* 152
- *Ocean Acidification and Marine Microbiomes* 152
- *Altered Precipitation Patterns* 153
- *Emerging Microbial Pathogens* 154
- **MICROBIAL INFLUENCE ON BIOGEOCHEMICAL CYCLES** 154
  - Carbon Cycle: Production and Consumption of Greenhouse Gases 154
    - *Role of Microorganisms in Carbon Sequestration* 155
  - Nitrogen and Phosphorus Cycles: Agricultural and Environmental Impacts 155
  - Methane Production and Microbial Regulation 155
    - *Why does Microbial Stewardship Matter?* 156
- **CLIMATE CHANGE AND MICROBIAL INTERACTIONS** 156
  - Microbe-Plant Interactions: Agricultural and Forest Ecosystems 156
    - *Soil Microbes and Plant Health* 157
    - *Microbes in Forests* 157
    - *Impact on Agricultural Productivity* 158
  - Microbe-Animal and Human Pathogens: Health Implications 158
  - Microbial Interactions in Extreme Environments 159
    - *Extremophiles and Their Adaptation* 159
- **ADAPTATION AND EVOLUTION OF MICROBIAL COMMUNITIES** 160
  - Mechanisms of Microbial Adaptation to Changing Climates 160
  - Evolutionary Responses and Potential for Resilience 160
- **HUMAN HEALTH IMPLICATIONS** 161
  - Emerging Pathogens and Disease Spread 162
  - Impact on Antibiotic Resistance 162
  - Microbial Influence on Air Quality 163
- **AGRICULTURAL AND ENVIRONMENTAL IMPLICATIONS** 164
  - Soil Microbes and Crop Productivity 164
  - Microbes in Bioremediation: Managing Environmental Pollution 165
  - Role of Microorganisms in Mitigating Climate Change 165
- **FUTURE RESEARCH DIRECTIONS IN MICROBIAL SCIENCE** 167
  - Addressing Climate Change and Exploring New Technologies 167
  - Knowledge Gaps in Microbial Responses to Climate Change 167
  - Potential for Microbial Technologies in Climate Mitigation 168
  - Advances in Genomics and Microbial Ecology 169
- **CONCLUSION** 170
  - Microbes and Climate Change: The Hidden Architects of Our Future 170
  - Synthesis of Microbial Contributions to Climate Dynamics 170
  - Integrating Microbial Science into Climate Change Mitigation Strategies 171
- **REFERENCES** 173

CHAPTER 8 ADVANCED OXIDATION PROCESS AS AN EMERGING TECHNOLOGY FOR THE TREATMENT OF PHARMACEUTICAL WASTEWATER 177
M. Mounica, V.V. Vaishnavi and *M. Vijay Pradhap Singh*
- **INTRODUCTION** 178
- **PHARMACEUTICAL EFFLUENTS** 181
  - Different Sources of Pharmaceutical Products That Cause Pollution To The Environment 182
- **ADVANCED OXIDATION PROCESS** 183
  - Theory of Advanced Oxidation Process 183
  - Status and Emergence of the AOP Process 186
  - Homogeneous Process 186
    - *(O_3/UV) Ozone couples with UV* 186

(H_2O_2/UV) (H_2O_2) coupled with UV	187
(H_2O_2/O_3)H_2O_2 couples with O_3	188
O_3/Fe^{2+}) process	188
ULTRA SOUND IRRADIATION	188
FENTON PROCESS AND PHOTO FENTON PROCESS	188
Electrochemical Oxidation	190
Electro Fenton	191
HETEROGENEOUS PROCESS	191
Heterogeneous Photocatalysis	192
Sonochemical Process	193
INTEGRATED AOP	193
FACTORS AFFECTING AOP	194
CONCLUSION	194
REFERENCES	195

CHAPTER 9 DIGITAL TECHNOLOGIES USED IN ENVIRONMENTAL MANAGEMENT   199

G. Vijaya Laxmi, V. Sai Nikhitha, K. Sree Saahitthi Reddy, B. Indira, Vanitha Guda, Keshetti Sreekala, Ramesh Ponnala, Sanjeeb Kumar Mandal and *Bishwambhar Mishra*

INTRODUCTION	200
What are Digital Technologies in Environmental Management?	200
Evolution of Digital Technologies in Environmental Conservation	201
Importance of Digital Solutions for Sustainable Environmental Management	202
Data Driven Decision Making	202
Monitoring and Transparency	202
Eco-Friendliness Efficiency	203
Support sustainability goals	203
Scope of Digital Technologies in Various Environmental Sectors	203
TYPES OF DIGITAL TECHNOLOGIES USED IN ENVIRONMENTAL MANAGEMENT	204
Geographic Information System (GIS)	204
Remote Sensing	204
Internet of Things (IoT) and Environmental Sensors	205
Artificial Intelligence and Machine Learning	206
Big Data Analytics	207
Ongoing Environmental Projects Utilizing Big Data Analytics	207
APPLICATIONS OF DIGITAL TECHNOLOGIES IN ENVIRONMENTAL MANAGEMENT	210
Climate Monitoring and Weather Forecasting	210
Biodiversity Conservation	210
Pollution Prevention and Waste Management	210
Water Resource Management	211
Disaster Risk Reduction and Mitigation	211
Sustainable Farming and Land Management Planning	211
Renewable Energy Monitoring and Optimization	212
BENEFITS OF DIGITAL TECHNOLOGIES IN ENVIRONMENTAL MANAGEMENT	212
Better data collection with accuracy	214
Real-time environmental monitoring	215
Predictive modeling and forecasting	215
Improved decision-making and policy formulation	216
Cost-effectiveness in Environmental Conservation Efforts	217

- **LIMITATIONS AND CHALLENGES OF DIGITAL TECHNOLOGIES IN ENVIRONMENTAL MANAGEMENT** ... 218
  - Data Privacy and Security Issues ... 218
  - High Cost of Implementing Digital Technologies ... 218
  - Digital Gap and Access to Technology in Developing Areas ... 219
  - Limited understanding and skill gaps in using advanced technologies ... 219
  - Technological Aging and the Necessity for Continuous Updates ... 220
- **FUTURE DIRECTIONS IN DIGITAL TECHNOLOGIES FOR ENVIRONMENTAL MANAGEMENT** ... 221
  - AI and IoT-Based Intelligent Environmental Systems ... 221
  - The Development of Affordable and Scalable Digital Solutions ... 221
  - Strengthening Public-Private Partnerships in Digital Environmental Initiatives ... 222
  - Digital Engagement Through Public Participation: Future Directions ... 223
  - Open Data and Cooperative Environmental Monitoring ... 223
- **CONCLUSION** ... 224
- **REFERENCES** ... 225

CHAPTER 10 SUSTAINABLE DEVELOPMENT GOALS OF HEALTH AND ENVIRONMENT AND CURRENT STATUS OF INDIA WITH MEASUREMENT STRATEGIES FOR FUTURE ... 228
Rajal Dave and Abhijeet Joshi
- **INTRODUCTION** ... 228
- **SUSTAINABLE DEVELOPMENT GOALS (SDGS) – STATUS OF INDIA** ... 233
- **SUSTAINABLE DEVELOPMENT GOALS RELATED TO HEALTH** ... 234
  - Risk Factors Associated with Health and Well-being ... 235
    - *Nutritional Risk* ... 235
    - *Behaviour Risk Factor* ... 236
  - Measurement Strategies for Infectious Diseases and Non-communicable Diseases ... 237
- **SUSTAINABLE DEVELOPMENT GOALS FOR SUSTAINABLE ENVIRONMENT** ... 239
  - Strategies for Sustainable Environment ... 241
    - *Promote the Genome Editing technology, Organic Farming for high yields, and climate-resilient crop production for a Sustainable Environment* ... 242
    - *Promote the Application of Biopesticides Compared to the use of Conventional Chemical Pesticides* ... 243
    - *Bioremediation Technology for the Removal of Pollutants from the Environment* ... 243
  - Achievements of India Towards Environmental Sustainability ... 244
  - Challenges ... 245
  - Measuring Strategies ... 246
  - Improvement on Indoor and Outdoor Air ... 247
- **DISCUSSION** ... 248
- **CONCLUSION** ... 248
- **LIST OF ABBREVIATIONS** ... 249
- **ACKNOWLEDGMENTS** ... 250
- **REFERENCES** ... 250

CHAPTER 11 TRANSFORMING WASTEWATER INTO RENEWABLE ENERGY: A PATHWAY TO ACHIEVE SUSTAINABILITY AND THE CIRCULAR ECONOMY ... 253
V.V. Vaishnavi, M. Mounica and M. Vijay Pradhap Singh
- **INTRODUCTION** ... 254
- **WASTEWATER** ... 254
  - Wastewater Generation ... 255
  - Composition ... 256

ENERGY RECOVERY FROM WASTEWATER	257
Techniques used	257
Anaerobic Digestion	258
BIOGAS	260
CIRCULAR ECONOMY	262
EMERGING TRENDS	262
CONCLUSION	263
REFERENCES	263
CHAPTER 12 COMMUNITY-BASED ENVIRONMENTAL MANAGEMENT	268

Preeti Mehta Kakkar, Meet Sharma, Shivani Singh, Yashna Tiwari and *Ruchi Jakhmola Mani*

INTRODUCTION	269
HISTORICAL EVOLUTION OF CBEM	271
SUCCESS FACTORS IN CBEM	273
Strong Community Participation	273
Role of Community Involvement	273
Methods for Effective Engagement	273
Case Study	274
Supportive Policy Frameworks	275
Policy Enablers for CBEM	275
Features of a Supportive Framework	275
Examples of Successful Policy Frameworks	275
Collaboration between Local and Global Stakeholders	276
Importance of Multi-Stakeholder Partnerships	276
Models of Collaboration	276
Case Studies	276
CHALLENGES AND GAPS IN CBEM	277
Lack of Long-term Sustainability	277
Insufficient Integration of Scientific Research and Traditional Practices	279
Socio-economic Barriers and Cultural Resistance in CBEM	280
Inadequate Funding and Resources in CBEM	281
EMERGING SOLUTIONS AND INNOVATIONS	282
ROLE OF AI IN CBEM	283
COMMUNITY LED MONITORING SYSTEM	283
FUTURE DIRECTIONS IN CBEM	285
Adopting Advanced Technologies	285
Addressing Challenges	286
Leveraging Technological Advancements	286
Strategies for Strengthening CBEM	286
Promoting Knowledge Exchange	287
CONCLUSION	287
REFERENCES	288
CHAPTER 13 REDUCED GRAPHENE OXIDE-BASED SOLUTIONS FOR WATER PURIFICATION: ADVANCES IN SUSTAINABLE NANOCOMPOSITES	292

Deepak Dahiya, Ashish Sharma, Sweety Dahiya, Pooja Yadav, Kiran Kaushik and *Sudesh Chaudhary*

INTRODUCTION	292
SYNTHESIS AND CHARACTERISTICS	294
RGO Synthesis	294
rGO's Characteristics	300

METHODS OF REMOVING POLLUTANTS USING RGO-BASED NANOCOMPOSITES 303
   Mechanism of Adsorption 303
   Significance of Surface Area and Porosity 304
   Actuation of Functional Groups Containing Oxygen 305
   Augmenting Adsorption via Functionalization 305
   The Kinetics and Isotherms of Adsorption 306
   Catalysis and Photocatalysis Catalytic Pathways in Photocatalytic Degradation of Organic Pollutants Using rGO Nanocomposites Explanation of the Flow: 306
RGO-BASED NANOCOMPOSITES FOR THE ELIMINATION OF POLLUTANTS 308
   Nanocomposites of Metal Oxides (rGO) 309
      rGO-TiO$_2$ Nanocomposites 309
      Nanocomposites in rGO-Fe$_3$O$_4$ 309
      The rGO-ZnO Nanocomposites 310
   Nanocomposites of rGO-polymers 310
      Iron-graphene Oxide-polyaniline Nanocomposites 310
      rGO-Polyvinyl Alcohol Nanocomposites 311
      Specialization of Nanocomposites for Targeted Pollutants 311
   Biopolymer Composites using rGO 312
ENVIRONMENTAL IMPACT AND SUSTAINABILITY 312
CONCLUSION 313
REFERENCES 314
SUBJECT INDEX 317

FOREWORD

The 21st century presents a dual challenge and opportunity: to mitigate the degradation of natural ecosystems while advancing Sustainable Development Goals (SDGs) through science and innovation. Environmental systems today are under unprecedented stress—driven by urbanization, industrial expansion, climate change, and resource overexploitation. We are entering an era where the intersection of biology, technology, and engineering is not just transformative, but it is imperative. The integration of microbiology, biotechnology, and engineering offers a promising approach to developing sustainable environmental solutions.

Integrated Environmental Solutions: Microbiology, Biotechnology, and Engineering Approaches is a compilation that reflects this paradigm shift. The book opens with "The Blueprint of Life: Bio-engineering Nature for a Sustainable Future," a conceptual anchor that sets the stage for synthetic biology's transformative potential. This foundation is expanded in "Synthetic Biology for Environmental Applications," which explores that microorganisms can be engineered for targeted remediation and pollutant sensing and are directly tied to practical challenges and solutions in environmental biotechnology, as discussed in "Wastewater Treatment Processes," "Bioaugmentation and Biostimulation," and "Bioremediation Techniques," where biological processes are optimized to treat contaminated matrices effectively. A comparative understanding is offered in "Comparative Analysis of Traditional and Advanced Bioremediation Techniques," which evaluates the evolution of environmental remediation methods. Meanwhile, "Microbial Communities in Environmental Engineering" dissects the ecological complexity and functional potential of microbial consortia, revealing the biochemical networks that underpin pollutant transformation. The climate lens is sharply focused in "Climate Change and Microbial Processes" and "Microbes and Climate: A Dynamic Interaction." These chapters concentrate on the bidirectional influence between climate systems and microbial activity, providing details on carbon cycling, feedback loops, and resilience building in changing ecosystems. The integration of advanced technologies is discussed in "Advanced Oxidation Process as an Emerging Technology for the Treatment of Pharmaceutical Wastewater," showcasing a physicochemical complement to biological remediation, and in "Digital Technologies in Environmental Management," which showed the role of data-driven systems, Artificial Intelligence (AI), and sensor networks in environmental decision-making. The book is committed to sustainability and the circular economy, particularly through "Transforming Wastewater into Renewable Energy," a chapter that discusses energy recovery, waste minimization, and resource optimization. This is reinforced in the chapter "Sustainable Development Goals in the Health Care Sector," which connects environmental biotechnology with public health and planetary well-being. On the frontier of material science, "Reduced Graphene Oxide-Based Solutions for Water Purification" explores the use of nanocomposites for achieving high-efficiency pollutant removal with minimal environmental footprint. And, perhaps most crucially, the book concludes by shifting the focus outward—toward society—with "Community-Based Environmental Management," which affirms the role of participatory governance and localized knowledge in implementing sustainable practices.

This book is more than an academic resource—it is a call to redesign the systems we rely on to interact with and protect our environment. An advanced scientific framework and practical methodologies make this book an indispensable guide for researchers, academicians, environmental consultants, and sustainability professionals.

I extend my admiration to the editors, Dr. Rajneesh Kumar, Dr. R. S. Singh, and Dr. Maulin P. Shah, who have curated this book with intellectual rigor and visionary clarity. Their work ensures that this book will be a cornerstone reference in the ongoing pursuit of interdisciplinary, innovative, and impactful environmental solutions.

Susana Rodríguez-Couto
Biological Water Treatment
Lappeenranta-Lahti University of Technology (LUT)
Lappeenranta, Finland

PREFACE

Environmental challenges in the 21st century require innovative and interdisciplinary solutions that integrate microbiology, biotechnology, and engineering principles. Pollution, resource depletion, and climate change have intensified the need for sustainable approaches to environmental management. "Integrated Environmental Solutions: Microbiology, Biotechnology, and Engineering Approaches" is a response to these global concerns, offering a comprehensive exploration of how these disciplines converge to address environmental issues effectively. This book provides an in-depth analysis of modern techniques and technologies for cleaning and managing environmental pollutants. It highlights the crucial role of microbial communities in environmental engineering, particularly in biodegradation, bioremediation, and bioaugmentation. These processes demonstrate the ability of microorganisms to break down pollutants and restore ecological balance, showcasing nature's potential to mitigate human-induced contamination.

Biotechnological advancements have revolutionized environmental management, and this book delves into key innovations such as genetically engineered microorganisms for pollutant degradation, biosensors for real-time environmental monitoring, and biofuel production from waste materials. These technologies illustrate how biotechnology is instrumental in reducing environmental footprints and promoting sustainability. Engineering principles play a fundamental role in designing and optimizing environmental solutions. This book examines critical engineering applications, including biofiltration systems, wastewater treatment technologies, and sustainable material development. By integrating microbiology and biotechnology with engineering design, innovative systems can be developed to enhance efficiency and sustainability in pollution control and resource management.

To bridge the gap between theory and practice, the book includes case studies highlighting successful interdisciplinary environmental projects. These real-world examples provide insights into the implementation of integrated solutions and the benefits of collaborative approaches in tackling complex environmental problems. Looking ahead, the book explores emerging trends in environmental science, including the impact of climate change on microbial processes, advances in synthetic biology for environmental applications, and the use of digital technologies such as artificial intelligence and big data analytics for environmental monitoring and management. These forward-looking perspectives aim to inspire future research and innovation in the field.

The intended audience for this book includes academics and researchers in the fields of environmental science, microbiology, biotechnology, and engineering. Environmental practitioners and engineers will find valuable insights into practical applications, while graduate and advanced undergraduate students will gain a strong foundation in interdisciplinary environmental solutions. Policymakers and environmental consultants will also benefit from the book's recommendations for sustainable practices. By integrating diverse scientific perspectives, "Integrated Environmental Solutions: Microbiology, Biotechnology, and Engineering Approaches" serves as a vital resource for understanding and implementing holistic environmental solutions. It is our hope that this book will inspire further research, innovation, and collaboration in the pursuit of a more sustainable and resilient future.

Rajneesh Kumar
Department of Chemical Engineering and Technology
Indian Institute of Technology (BHU), Varanasi
Uttar Pradesh, India

Ram Sharan Singh
Department of Chemical Engineering and Technology
Indian Institute of Technology (BHU), Varanasi
Uttar Pradesh, India

&

Maulin P. Shah
Department of Research Impact and Outcome Research
Research and Development Cell, Lovely Professional University
Phagwara, Punjab, India

List of Contributors

Azimul Hasan	School of Biosciences, Engineering and Technology, VIT Bhopal University, Sehore, Madhya Pradesh, India
Arun Kumar K.	School of Biosciences, Engineering and Technology, VIT Bhopal University, Sehore, Madhya Pradesh, India
Akriti Gupta	Department of Biotechnology, Amity Institute of Biotechnology, Amity University, Noida, Uttar Pradesh, India
Arunima Singh	Department of Chemical Engineering and Technology, Indian Institute of Technology, Banaras Hindu University, Varanasi, Uttar Pradesh, India
Anupam Jayas	Department of Botany, Institute of Science, Banaras Hindu University, Varanasi, Uttar Pradesh, India
Aindree Lohumi	Department of Biotechnology, Amity Institute of Biotechnology, Amity University, Noida, Uttar Pradesh, India
Abhijeet Joshi	Department of Microbiology, Faculty of Science, Atmiya University, Rajkot, Gujarat, India
Ashish Sharma	Department of Chemical Engineering, Deenbandhu Chhotu Ram University of Science and Technology, Murthal, Sonipat, Haryana, India
Bishwambhar Mishra	Department of Biotechnology, Chaitanya Bharathi Institute of Technology, Hyderabad, Telangana, India
B. Indira	Department of Master of Computer Applications, Chaitanya Bharathi Institute of Technology, Hyderabad, Telangana, India
Chelemala Katyayani	Department of Biotechnology, Chaitanya Bharathi Institute of Technology, Hyderabad, Telangana, India
Diya Saha	Department of Biotechnology, Amity Institute of Biotechnology, Amity University, Noida, Uttar Pradesh, India
Deepak Dahiya	Center of Excellence for Energy and Environmental Studies, Deenbandhu Chhotu Ram University of Science and Technology, Murthal, Sonipat, Haryana, India
G. Vijaya Laxmi	Department of Biotechnology, Chaitanya Bharathi Institute of Technology, Hyderabad, Telangana, India
K. Shivathmika Reddy	Department of Biotechnology, Chaitanya Bharathi Institute of Technology, Hyderabad, Telangana, India
K. Sree Saahitthi Reddy	Department of Biotechnology, Chaitanya Bharathi Institute of Technology, Hyderabad, Telangana, India
Keshetti Sreekala	Department of Computer Science and Engineering, Mahatma Gandhi Institute of Technology, Hyderabad, Telangana, India
Kiran Kaushik	Center of Excellence for Energy and Environmental Studies, Deenbandhu Chhotu Ram University of Science and Technology, Murthal, Sonipat, Haryana, India
M. Mounica	Department of Biotechnology, Vivekanandha College of Engineering for Women (Autonomous), Elayampalayam, Tamil Nadu, India

M. Vijay Pradhap Singh	Department of Biotechnology, Vivekanandha College of Engineering for Women (Autonomous), Elayampalayam, Tamil Nadu, India
Meet Sharma	Department of Biotechnology, Amity Institute of Biotechnology, Amity University, Noida, Uttar Pradesh, India
Preeti Mehta Kakkar	Department of Biotechnology, Amity Institute of Biotechnology, Amity University, Noida, Uttar Pradesh, India
Pooja Yadav	Center of Excellence for Energy and Environmental Studies, Deenbandhu Chhotu Ram University of Science and Technology, Murthal, Sonipat, Haryana, India
R Dhinesh	Department of Aquatic Environment Management, Faculty of Fisheries Science, Kerala University of Fisheries and Ocean Studies, Kochi, Kerala, India
R. S. Singh	Department of Chemical Engineering and Technology, Indian Institute of Technology, Banaras Hindu University, Varanasi, Uttar Pradesh, India
Ramesh Ponnala	Department of Computer Science and Engineering, Faculty of Science and Technology (ICFAITech), ICFAI Foundation for Higher Education, Hyderabad, Telangana, India
Rajal Dave	Department of Microbiology, Faculty of Science, Atmiya University, Rajkot, Gujarat, India
Ruchi Jakhmola Mani	Proteomics and Translational Research Lab, Centre for Medical Biotechnology, Amity Institute of Biotechnology, Amity University, Noida, Uttar Pradesh, India
Sanjeeb Kumar Mandal	Department of Biotechnology, Chaitanya Bharathi Institute of Technology, Hyderabad, Telangana, India
S S Kirthiga	Department of Aquatic Environment Management, Faculty of Fisheries Science, Kerala University of Fisheries and Ocean Studies, Kochi, Kerala, India
Shruti Khanna Ahuja	Centre for Medical Biotechnology, Amity Institute of Biotechnology, Amity University, Noida, Uttar Pradesh, India
Sonali Ranjan	Department of Chemical Engineering and Technology, Indian Institute of Technology, Banaras Hindu University, Varanasi, Uttar Pradesh, India
Suparna Bardhan	Department of Botany, Institute of Science, Banaras Hindu University, Varanasi, Uttar Pradesh, India
Shruti Khanna Ahuja	Centre for Medical Biotechnology, Amity Institute of Biotechnology, Amity University, Noida, Uttar Pradesh, India
Sanjeeb Kumar Mandal	Department of Biotechnology, Chaitanya Bharathi Institute of Technology, Hyderabad, Telangana, India
Shivani Singh	Department of Biotechnology, Amity Institute of Biotechnology, Amity University, Noida, Uttar Pradesh, India
Sweety Dahiya	Department of Environmental Sciences, School of Basic and Applied Sciences, SGT University, Gurugram, Haryana, India
Sudesh Chaudhary	Center of Excellence for Energy and Environmental Studies, Deenbandhu Chhotu Ram University of Science and Technology, Murthal, Sonipat, Haryana, India

V.V. Vaishnavi	Department of Biotechnology, Vivekanandha College of Engineering for Women (Autonomous), Elayampalayam, Tamil Nadu, India
V. Sai Nikhitha	Department of Biotechnology, Chaitanya Bharathi Institute of Technology, Hyderabad, Telangana, India
Vanitha Guda	Department of Computer Science and Engineering, Chaitanya Bharathi Institute of Technology, Hyderabad, Telangana, India
Y. Swarna Manjari	Department of Biotechnology, Chaitanya Bharathi Institute of Technology, Hyderabad, Telangana, India
Yogesh Kumar Vishwakarma	Department of Chemical Engineering and Technology, Indian Institute of Technology, Banaras Hindu University, Varanasi, Uttar Pradesh, India
Yashna Tiwari	Department of Biotechnology, Amity Institute of Biotechnology, Amity University, Noida, Uttar Pradesh, India

CHAPTER 1

The Blueprint of Life: Bio-engineering of Sustainable Materials for a Greener Future

G. Vijaya Laxmi[1,*], Chelemala Katyayani[1], K. Shivathmika Reddy[1], Y. Swarna Manjari[1], Sanjeeb Kumar Mandal[1] and Bishwambhar Mishra[1]

[1] *Department of Biotechnology, Chaitanya Bharathi Institute of Technology, Hyderabad, Telangana, India*

Abstract: Bioengineering has emerged as a transformative approach to material development in the pursuit of sustainability. It is the blueprint for bioengineered sustainable materials that have the potential to change industries while ensuring ecological balance in the face of pollution, resource depletion, and climate change. Material derived from renewable resources—such as plants, microorganisms, and agricultural waste—made into biodegradable polymers, bio-based composites, and self-healing materials, while embracing the principles of the circular economy, reduces wastes, conserves energy, and supports environmentally responsible industrial practices. Breakthroughs in biocatalysis, synthetic biology, and bioremediation are cited as key methodologies that enable the development of advanced sustainable materials. The chapter looks into microbial contributions, bioprocess efficiency, and highlights innovations that improve biodegradability and resource use. Bioremediation in environmental cleanup and material recycling also underlines the integration of the approach into sustainable production.

Keywords: Biodegradable, Bioremediation, Biocatalysis, Degradation, Eco-friendly, Micro pollutants, Mycelium, Pollutants, Polymers, Plant-based.

INTRODUCTION

Sustainability is defined in a variety of ways, both conceptually and technically. First, sustainable development is a way of using resources that seeks to satisfy human needs while protecting the environment and reducing resource consumption so that these needs can be satisfied for both the present and the future. The Brundtland Commission coined the widely accepted notion of sustainable development as "meeting the needs of the present without compromising the ability of future generations to meet their own needs" [1].

---

*Corresponding author G. Vijaya Laxmi: Department of Biotechnology, Chaitanya Bharathi Institute of Technology, Hyderabad, Telangana, India; E-mail: drgvlaxmi@gmail.com

Rajneesh Kumar, Ram Sharan Singh & Maulin P. Shah. (Eds.)
All rights reserved-© 2026 Bentham Science Publishers

Looking at sustainability from the perspective of ecology, there is a need for contributions to ensure that a healthy ecosystem and the environment are maintained at all times. Sustainable development has to do with our relationship with the natural environment, on which we depend for food, water, energy, and raw materials. But there is more to it than that. It is also about our relationship with the global economic system in which we source raw materials, manufacture products, and trade. To put it simply, sustainable materials are typically made from renewable resources and can be recycled into other items, thus reducing waste processes and preserving the environment. As the world continues to move towards a more sustainable future, it is important to understand what makes a material "sustainable."

What are Sustainable Materials, and Why are they Important?

Sustainable materials have some qualities that set them apart from others. They are renewable, derived from fast-reproducing resources like plants and water. They are also recyclable, meaning they may be recovered and reprocessed into new items after their first usage, minimizing the requirement for virgin resources. Additionally, the materials are biodegradable. They are easily decomposed by microbes into innocuous components such as water, carbon dioxide, and organic waste. They also have low environmental impact, from extraction to processing, transportation, usage, and disposal. All of these characteristics ensure that sustainable materials are derived from renewable resources, reused through recycling, and degrade safely after use without damaging the environment. This reduces dependency on finite resources, cuts waste, lowers greenhouse gas emissions, and reduces pollution. In broad terms, the increase in environmental sustainability depends heavily on sustainable materials. Their use in a variety of sectors, including electronics, packaging, and construction, shows possibilities for a more sustainable future [1]. Bioplastics include polylactic acid, polyhydroxyalkanoates, and starch-based plastics that are biodegradable and made from renewable sources such as cornstarch and sugarcane. They serve to address the rising issue of pollution from traditional plastics and are used in medical equipment, consumer goods, cutlery, and packaging. Cotton, wool, and linen are examples of sustainable organic fibers produced utilizing environmentally friendly agricultural practices. Biodegradable fibers have a smaller environmental impact than traditional fibers and are utilized in apparel, textiles, and housing. Recycling is important for environmental sustainability. The extraction of virgin metals from ores is both energy-intensive and environmentally hazardous. Metals like aluminum, steel, and copper may be recycled indefinitely without losing their qualities, eliminating the need to mine and process new material [2].

The Role of Bioengineering in Sustainability

Bioengineering is rapidly advancing, creating sustainable solutions that address critical challenges such as environmental degradation, climate change, and resource scarcity. These innovations could revolutionize industries like energy, agriculture, and healthcare. One of the most impactful areas where bioengineering has made strides is in energy production. The world is increasingly relying on biofuels and biomass to transition to renewable energy sources. Bioengineering is key to answering global challenges and reaching sustainable development. Bioengineers keep our practices sustainable in diverse fields by manipulating biological systems and engineering the principles that govern them. Bioengineering related to environmental solutions: As bioengineering becomes more mainstream, it offers a new way of thinking about sustainable solutions. For instance, bioengineers are more focused on the development methods of sustainable agriculture and food production, like increasing crop yield with a lower environmental impact. Further, using biological processes to mitigate pollution and restore ecosystems, they also serve bioremediation and environmental cleanup services. Additionally, the important work of bioengineering towards a greener economy and ecosystem also includes the creation of biofuels and sustainable energy resources. Right now, bioengineering offers big economic benefits for the establishment of sustainable biomaterials. Bioengineered sustainable biomaterials can bring significant economic benefits. Such materials sourced from readily available and renewable resources, cheapen the cost of production. At the same time, businesses utilizing bio-based resources can lower their costs and focus on long-term sustainability. It is time to establish markets for sustainable products such as bioengineered textiles and bioplastics, which will also create jobs and fuel economic growth. Coupled with government incentives for green technology innovation, these industry savings provide a boost to cost efficiency and profitability [3].

How Bioengineering Differs from Traditional Material Manufacturing

The differences between bioengineering and traditional materials manufacturing are often understood through their distinct approaches to sustainability, resource utilization, and environmental impact. Non-renewable resources such as petroleum, metals, and other minerals are often relied upon by traditional manufacturing processes, resulting in new materials at the expense of high-energy use and significant waste emissions. Bioengineering, on the other hand, utilizes biological systems (such as microorganisms, enzymes, or plant-based processes) to develop materials with a lower environmental impact and therefore more sustainable. Bioengineering is the process of creating bio-based materials like sustainable fabrics, biodegradable plastics, and even biofuels using renewable biological resources. These procedures use less energy since they frequently take place at room temperature and pressure, negating the need for harsh heat or chemical treatments that are typical of conventional techniques. Furthermore, bioengineering is opening the door for breakthroughs like biomanufacturing, which uses biological processes to produce sustainable building materials and environmentally friendly packaging. By minimizing waste and allowing end-of--life products to be composted or fully recycled back into the manufacturing chain, this strategy is in line with the circular economy paradigm. On the other hand, traditional material production typically results in a higher amount of non-biodegradable trash, which exacerbates long-term environmental problems. Optimizing biological processes to create more environmentally friendly or biodegradable materials is the main goal of bioengineering. Bioplastics and bio-based chemicals, for instance, are made to break down organically, which lowers pollution. However, traditional manufacturing frequently uses artificial chemicals and procedures that result in long-lasting environmental contaminants [4].

BIOENGINEERED SUSTAINABILITY MATERIAL TYPES

The creation of bioengineered materials that put environmental responsibility first was spurred by the need for sustainable solutions across industries. The negative effects of conventional, resource-intensive materials are intended to be lessened by these materials, which are made from natural and renewable resources. Utilizing cutting-edge biotechnology and engineering methods, bioengineered sustainable materials provide substitutes that not only satisfy performance requirements but also tackle global issues, including pollution, resource depletion, and climate change [5]. Plants, microbes, and agricultural waste are examples of biologically sourced components used to create bioengineered materials, as opposed to traditional materials that frequently rely on petrochemicals and cause long-term environmental harm. Generally speaking, their industrial methods produce less waste and use fewer resources, which promotes a more circular economy. These materials are essential to a sustainable future since they frequently provide improved qualities, including biodegradability, carbon sequestration, and reduced toxicity. Bioengineered sustainable materials are used in a wide range of industries, including construction, automobile production, textiles, and packaging. For example, plant-derived fibers and agricultural waste materials are offering environmentally benign substitutes for synthetic fabrics and composites, while biodegradable plastics created from plant-based polymers aid in lowering plastic pollution. In addition to lessening their negative effects on the environment, these materials encourage innovation in product design and production techniques by turning waste materials into useful resources [5].

Biodegradable Materials

Biodegradable materials are made to decompose spontaneously through biological processes, which lowers pollution and waste buildup. Natural polymers used to make these materials have the ability to break down when microbes are present. Important instances consist of: Biodegradable Plastics: Biodegradable plastics made from renewable resources like corn or sugarcane include polylactic acid (PLA), polyhydroxyalkanoates (PHA), and polybutylene succinate (PBS).Bio-based Packaging: Biodegradable packaging can take the place of traditional plastics, which take hundreds of years to break down. Compostable Materials: These materials may break down in composting environments and are generally used for throwaway products like flatware, plates, and bags. Biodegradable materials have the advantages of being less energy-intensive to produce, lowering landfill waste, and minimizing plastic pollution [6].

Plant-derived and Agricultural Waste Materials

These materials make use of biodegradable and widely accessible renewable resources, such as plants or agricultural waste. The need for petrochemical-based substitutes can be decreased by processing them into a range of sustainable products. Hemp, flax, and bamboo are fibers that find extensive use in everything from construction materials to textiles because of their strength, resilience, and minimal environmental impact. Agricultural Waste (Corn Stalks, Straw, and Rice Husk): Often seen as waste, agricultural by-products are used in bioengineering. These materials can be utilized as building materials or transformed into biofuels or bioplastics. Mushroom Mycelium: The root structure of mushrooms, or mycelium, can be utilized to make biodegradable textiles, building materials, and packaging. It is an eco-friendly substitute because it grows quickly and uses few resources [7].

Sustainable Textiles and Fibers

Fibers that are bio-based, biodegradable, or created with eco-friendly methods are used to make sustainable textiles. These materials are intended to lessen the textile industry's negative environmental effects, which include high water consumption, pesticide use, and textile waste. Organic Cotton: Compared to conventional cotton, organic cotton is more sustainable because it is grown without the use of artificial fertilizers or pesticides. It improves soil health and consumes less water.

Tencel/Lyocell: Made from wood pulp, Tencel is more environmentally friendly than conventional textile fibers since it is made using a closed-loop method that recycles solvents and water. Hemp and linen are sustainable substitutes for synthetic fabrics since they require less water and herbicides to grow.

Algae-based Textiles: A biodegradable and sustainable substitute for traditional textiles, innovative fibers derived from algae are making their debut in the textile business.

Recycled Fibers: Closing the loop on textile waste and lowering the requirement for virgin raw materials are two benefits of using textiles derived from recycled plastics or textiles.

One viable solution to lessen the negative environmental effects of the fashion and textile industries—two of the most polluting industries in the world—is to use sustainable textiles [8].

Bio-based Composites and Hybrid Materials

Renewable bio-based materials, such as fibers, bio-resins, or agricultural waste, are combined with other materials, like metals or plastics, to create bio-based composites. Compared to their single-component predecessors, these hybrid materials provide improved usefulness, durability, and mechanical qualities.

Natural Fiber-Reinforced Composites: These materials reinforce bio-resins with fibers derived from plants, such as flax, jute, hemp, or sisal, making the composite material lighter and more environmentally friendly.

Bio-based resins are more environmentally friendly than petroleum-based resins since they are made from renewable resources like vegetable oils or starch.

Composites made of wood and plastic (WPCs): These composites, which are made from a blend of wood fibers and bio-based polymers (or recycled plastics), are frequently utilized in the decking, automobile, and construction sectors' industries where traditional materials are extremely resource-intensive, such as packaging, building, and automobile manufacture, bio- based composites and hybrid materials are thought to be an efficient way to lower carbon footprints [9].

MICROBIAL CONTRIBUTIONS TO SUSTAINABLE MATERIAL PRODUCTION

Microorganisms in the Synthesis of Biodegradable Materials

Microorganisms are important in the production of biodegradable materials, which can be substituted for plastics. Major biodegradable materials and their applications have been summarized in Table **1**.

Table 1. Major biodegradable materials and their source organisms.

Material	Microorganisms	Applications	References
Polylactic Acid (PLA)	*Lactobacillus species*	Food packaging, biomedical devices	[10]
Bacterial Cellulose	*Acetobacter xylinum*	Wound dressings, scaffolds	[11]
Mycelium-based Materials	Fungi like *Trichoderma reesei*	Packaging, construction	[12]
Xanthan Gum	*Xanthomonas campestris*	Food additives, thickening agents	[13]
Hyaluronic Acid	*Streptococcus species*	Cosmetics, drug delivery systems	[14]
Polyhydroxybutyrate (PHB)	*Ralstonia eutropha*	Biodegradable plastics	[15]
Pullulan	*Aureobasidium pullulans*	Edible films, drug delivery	[16]
Carrageenan	*Kappaphycus alvarezii*	Food stabilizers, gelling agents	[17]

Harnessing Microbial Metabolism for Waste-to-resource Conversion

Microorganisms are essential in the metabolism of waste, as they recycle it into useful materials. This method not only reduces environmental contamination but also adds to the sustainable management of resources, with one key example being Biogas Production *via* Anaerobic Digestion.

Microorganisms decompose organic waste anaerobically, producing biogas—a mixture of methane and carbon dioxide. Bio-gas serves as a renewable energy source for electricity generation, heating, and as a vehicle fuel.

Composting for Soil Enrichment: Microbial activity transforms organic waste into nutrient-rich compost, enhancing soil fertility. Compost is used in agriculture and horticulture to improve soil structure and nutrient content.

Production of Bioplastics: Certain bacteria synthesize PHAs from organic substrates, offering biodegradable plastic alternatives. PHAs are utilized in packaging, medical devices, and agricultural films.

Bioelectrochemical Systems for Wastewater Treatment; Microbial fuel cells employ bacteria to degrade organic pollutants in wastewater, generating electricity in the process. These systems are used for wastewater treatment and energy recovery [18].

Genetic Engineering of Microbes for Enhanced Material Properties

Conventional methods for remediating toxic substances are costly and harmful to the environment. Genetically Engineered Microbes (GEMs) offer a safer, cost-effective alternative for pollution abatement through bioremediation. Using biotechnology, GEMs are designed to over-express potent proteins, enhancing their ability to degrade pollutants like oil spills, camphor, naphthalene, and halobenzenes. Compared to natural strains, GEMs exhibit higher degradative capacity and rapid adaptation to diverse pollutants. This approach holds promise for improving environmental quality and public health [19].

Integration of Microbiology in Bioprocessing for Material Development

Microbial products are essential for developing various therapeutic agents, including antibiotics, anticancer drugs, vaccines, and therapeutic enzymes. Genetic engineering techniques, functional genomics, and synthetic biology unlock previously uncharacterized natural products. The concept of genetic engineering, which involves the artificial manipulation, modification, and recombination of DNA or other nucleic acid molecules to alter organisms, has garnered significant interest over the past few decades. Recent Advancements in genetic and molecular biology have propelled genetic engineering into the forefront of scientific and technological disciplines. Notably, two interconnected themes—microbial biotechnology and genetic engineering—exhibit positive feedback. Microbial biotechnology plays a pivotal role in shaping the field of genetic engineering.

Genetic engineering contributes significantly to the precise development of microbial biotechnology. For instance, the discovery of clustered regularly interspaced short palindromic repeats CRISPR-Cas components in bacteria has revolutionized genome editing. These breakthroughs can enhance the biotechnological capabilities of specific microorganisms, such as improving antibiotic production efficiency. These two disciplines are interdependent and often challenging to differentiate. Together, they have transformed both the industrial sector and the field of medicine. In medicine, microbial biotechnology and genetic engineering extend beyond therapeutic compound development (such as antibiotics and proteins). They also impact diagnosis, prevention, gene expression regulation, and the construction of medical devices using biocompatible biopolymers [20].

BIOTECHNOLOGY APPROACHES FOR SUSTAINABLE MATERIALS

Biocatalysis as a Sustainable Approach to Material Synthesis

The term "biocatalysis" defines catalysis using natural catalysts, enzymes, and cells. This method has increasingly become an accepted "green" strategy for material synthesis due to its environmental friendliness and operational efficiency. Biocatalysts catalyze reactions under mild, environmentally friendly conditions, whereas traditional chemical catalysts often require high temperatures, extreme pH, and hazardous reagents. This reduces energy consumption and minimizes the emission of harmful waste, aligning with the principles of green chemistry. A significant advantage of biocatalysis lies in its exceptional selectivity. Enzymes are highly specific, targeting desired substrates and producing minimal by-products, which reduces the need for extensive purification. This precision is particularly important in pharmaceutical and fine chemical synthesis, enabling the production of optically pure chiral compounds with the desired stereochemistry, thereby making drug manufacturing safer.

Biocatalysis is versatile and can catalyze hydrolysis, oxidation, and reduction reactions. For example, lipases catalyze polymerization in polyesters, which naturally results in the production of materials with reduced environmental impact compared to conventional plastics. In biofuel production, cellulases and amylases hydrolyze biomass into sugars, which are then fermented into ethanol, reducing dependence on fossil fuels and greenhouse gas emissions.

Recent advancements in recombinant DNA technology and protein engineering have expanded the applications of biocatalysis. For example, biodegradable polymers such as PLA, derived from renewable materials like corn starch, are synthesized using enzymatic processes. PLA has diverse applications, including packaging and medical implants.

Biocatalysis exemplifies a sustainable approach that leverages natural catalysts to manufacture products by cleaner means, offering innovative solutions that diminish environmental impacts and support the circular economy. Applications include using enzymes such as lipases and proteases in producing bio-based plastics, detergents, and pharmaceuticals [21].

Engineered Enzymes for Biodegradation

There is an interesting, rapidly developing field in the realm of biotechnology, wherein the use of engineered enzymes, especially for biodegradation, boosts the ramifications for sustainability. Engineered enzymes facilitate the ecological decomposition of materials like plastics, chemicals, and environmental contaminants, leading to waste reduction and the generation of bio-based or biodegradable products. The role of science in sustainability and green chemistry is driving new frontiers in enzyme engineering to tackle complex environmental problems, such as plastic pollution and reliance on non-renewable resources.

Enzymatic Biodegradation of Plastics

Conventional polymer breakdown methods for plastics are ineffective and slow, making them some of the most persistent pollutants in the environment. Finding strategies to accelerate the breakdown of plastics, particularly non-biodegradable polymers like polyethylene (PE), polypropylene (PP), and polystyrene (PS), is a challenge. Thankfully, synthetic polymers can be broken down by specific enzymes into smaller, biodegradable monomers or other less toxic substances. Polyethylene terephthalate (PET), a common plastic found in bottles and clothes, can be broken down by PETase, an enzyme found in the bacterium *Ideonella sakaiensis*. PETase hydrolyzes PET into its monomers, ethylene glycol and terephthalic acid, by breaking down the ester bonds in PET [22].

Polyhydroxyalkanoates -PHA-Degrading Enzymes

These are biodegradable polymers produced by microbes as a means of carbon storage. The enzymatic agents facilitating the breakdown of PHAs can be engineered to improve their degradative capabilities. PHAs are bio-based and environmentally biodegradable, unlike polymers derived from petroleum. Scientists are working hard to enhance these enzymes for quicker and more effective breakdown, which could speed up the transition to biodegradable plastics. Laccases and peroxidases are enzymes capable of degrading aromatic compounds, which include plastics such as polystyrene (PS). Laccases utilize oxygen to oxidize organic compounds, breaking them down into smaller non-toxic fragments. Engineered laccases are now being developed for enhanced polystyrene degradation rates, specifically under environmentally friendly conditions. This could reduce some of the environmentally damaging issues caused by polystyrene, which is predominantly used in packaging and disposable products.

Techniques for Engineered Enzymes

Techniques used to create enzyme mutants with higher activity on certain kinds of plastics include directed evolution and mutagenesis. This approach primarily involves the mutation of the encoding gene of the enzyme, followed by screening for variants with improved catalytic activity. Protein Engineering involves altering the enzyme's active site or regions involved in substrate binding, which can increase stability, substrate affinity, and catalytic performance. For instance, tailoring the PETase enzyme to function at higher temperatures or with more complex PET formulations can significantly accelerate the biodegradation process in industrial applications. Inter-Species Enzyme Fusion is another area of research, where hybrid enzymes are created by fusing enzymes from different species to achieve higher substrate specificity or enhanced catalytic activity [22].

Synthetic Biology for the Production of Advanced Bio-based Materials

Synthetic biology has emerged as the next promising tool to produce complex bio-based materials, making environmentally friendly alternatives to traditional chemical approaches. It works in engineered biological systems made from modified microorganisms, producing new materials with specific properties. For instance, synthetic biology allows microbes to be designed and constructed, synthesizing biopolymers and renewable fibers that are biodegradable and reduce environmental impacts.

Recent advances include metabolic engineering applied to refining microbial pathways to produce high-value substances like spider silk and bio-based adhesives. Biomaterials developed through synthetic biology exhibit superior performance characteristics and high potential for applications in textiles, aerospace, and medicine. Synthetic biology has been applied to create novel intelligent materials that respond to environmental stimuli, offering significant potential in adaptive clothing as well as biomedical devices. Synthetic biology addresses the synthesis of complex materials like biodegradable polymers, biosilks, or bio-adhesives with precise control over microorganisms or cell-free systems. This technology helps diminish reliance on fossil fuels and reduces environmental impact through renewable biological pathways. This opens applications ranging from textiles to biomedicine to smart materials that display responsiveness to environmental changes and enhance functionality with greater sustainability. Synthetic biological approaches are used to create bio-based materials derived from plant biomass, offering superior value. Products Based on Lignin: Lignin is a byproduct associated with biofuel processing and is a complex organic polymer found in the cellular walls of plants. Various efforts are underway to bioengineer microorganisms that can process and degrade lignin into usable metabolites, carbon fibers, or biofuels. The complex polysaccharides found in plant cell walls, such as xylan and hemicellulose, present significant opportunities for their degradation process, with potential for value chains in textile, paper, and biopolymer composite applications. These polysaccharides can be efficiently utilized through synthetic biology for producing microbial strains that degrade hemicellulose and cellulose [23].

Protein Engineering for Improved Biocatalytic Efficiency

Enhancing biocatalytic efficiency is crucial for maximizing enzyme performance in industrial applications. Protein engineering advances biocatalysis by tailoring enzymes for enhanced activity, stability, selectivity, and substrate specificity. Techniques such as directed evolution, rational design, and computational methods are employed to design enzymes that function optimally under challenging conditions with specific substrates. The primary goal is to support environmentally friendly manufacturing across industries like biofuels, pharmaceuticals, and bioremediation.

Protein engineering improves biocatalyst performance in generating eco-friendly, biodegradable alternatives to plastics and toxic packaging materials. By enhancing the activity, stability, selectivity, and substrate specificity of enzymes through directed evolution and rational design, renewable materials such as cellulose, starch, chitin, and lignin can be processed efficiently for bioplastic or biodegradable packaging production. Engineered enzymes like cellulases, amylases, and lipases break down feedstocks into fermentable sugars or monomers, enabling the creation of biodegradable plastics such as PHA or PLA. Protein engineering ensures consistent biocatalyst performance under industrial conditions, reducing waste generation, increasing scalability, and minimizing environmental impact. This leads to improved performance, such as enhanced thermal stability and solvent resistance. Additionally, protein engineering supports enzyme recycling through immobilization techniques. Through these advancements, protein engineering contributes to greener bio-based packaging and promotes a circular economy approach in waste management, helping to reduce plastic pollution [24].

BIOREMEDIATION AND ITS ROLE IN SUSTAINABLE MATERIAL DEVELOPMENT

Bioremediation is an eco-friendly way to destroy or detoxify environmental pollutants with the help of living organisms, including microorganisms, fungi, or plants. Such a natural process has drawn much interest due to its efficacy in the detoxification of highly contaminated ecosystems and the potential for its use in the development of sustainable materials, specifically for biodegradable alternatives to plastics. This would thus entail the conversion of such organic wastes, such as agricultural by-products of biomass, into biodegradable and renewable bioplastics like PLA or PHA through bioremediation by microbes and plants. These are, thus, bioplastics based on renewable resources and are biodegradable, making them a step in the right direction in alleviating the problem of plastic pollution. More importantly, bioremediation has a vantage point in enhancing the bio micro-eco-ratios of newly generated materials through this approach of, for instance, mycoremediation and phytoremediation that utilizes fungi and plants to enhance the breakdown of plastics. Moreover, bioremediation facilitates the realization of the circular economy paradigm by converting waste into valuable products and thereby lessening the dependency on fossil fuels and plastic waste. In contrast to traditional plastics, which are energy-intensive and produce undesirable by-products, bioremediation methods constitute a low-carbon, low-energy means for creating sustainable materials. Thus, bioremediation paves the way for recovery deals on detoxifying the environment and replacing conventional plastics with the biodegradable variety, which goes a long way in convincing both pollution mitigation and ultimately giving rise to sustainable materials systems for generations to come [25].

Engineering Microbes for Environmental Cleanup

Engineered microbes are those that have been genetically modified organisms through deliberate alteration of their genes to enhance their innate capacity for environmental clean-up. These microorganisms are created by identifying specific contaminants like oil, heavy metals, pesticides, or plastics and selecting or designing genetic pathways to enable them to degrade or transform these very contaminants more efficiently. Genetic modifications in microbes for pollutant metabolism include transformation by plasmids, CRISPR-Cas9 technology, and synthetic biology. Engineered microbes find several applications in various fields of environmental cleanup. These microbes may be used for oil degradation for spill-control efforts, detoxifying heavy metals from contaminated water or soil, degrading pesticides from agricultural runoff, and even degrading plastic-like PET under certain conditions. Besides, more uses may be in the area of wastewater treatment, where engineered microbes remove organic pollutants and pathogens, and soil bioremediation, where they bioremediate industrial waste. The advantages of engineered microbes include cost-effectiveness, environmental friendliness, and versatility. Genetically engineered microorganisms are used in bioremediation to reduce the level of environmental pollutants. This long-term, cost-effective approach includes the remediation of hazardous chemicals, xenobiotics, and pesticides. These synthetic microbial communities consist of functionally very specialized species and have been engineered to reside in specific environments. In order for such systems to perform effectively, there are two important aspects that require special attention: a thorough understanding of the dynamics of natural microbial communities and the engineering of microorganisms with genes for metal absorption, chelator synthesis, and survival in various environments [26].

Applications of Bioremediation in Recycling and Reuse of Materials

Bioremediation, the use of microorganisms to degrade or transform waste, is gaining prominence as a sustainable approach to recycling and reuse. In the management of electronic waste (e-waste), certain bacteria and fungi have shown the ability to extract valuable metals like gold, silver, and copper from discarded devices in such a way that the need for environmentally destructive mining practices is reduced. Likewise, bioremediation offers many innovative options to address plastic waste. Microbes, such as *Ideonella sakaiensis*, are capable of breaking down PET into subunits that can be recycled into new plastics. Whereas in textile industries, microbial enzymes are used in the decolorization of dyes and the separation of blended fibers, allowing materials to be recovered and used again in valorizing pollution wrought by synthetic dyes. This is also presented on the grounds of bioremediation, making it successful in recycling organic agricultural waste into biofertilizers and biopesticides geared at restoring soil and cropping yield, thereby resulting in closing a loop in farming systems. Bioremediation is another crucial step forward in construction and demolition waste. Fungi and bacteria can be used to extract minerals from concrete debris, which can be reused in newer construction projects. Current research directions look into using microbial methodologies to recover rare earth elements, a crucial step in waste streams in industries, which is now drawing a lot of attention. It is crucial for electronics, renewable energy systems, and high-tech applications. These applications demonstrate the great potential of bioremediation in transforming waste into viable resources, on the path toward a circular economy, whereby pollution is not being caused [27].

Integrating Bioremediation with Material Production Processes

Integration of bioremediation with materials manufacturing has the potential to transform sustainable manufacturing. Such integration allows industries the opportunity to improve resource efficiency, reduce waste, and lessen environmental impact by integrating microbial systems into production cycles. For example, in the plastic industry, designed microorganisms are directed to break down waste polymers into monomers that can be used subsequently to make new plastics. Besides, this tackles the constantly looming issue of plastic pollution while reducing dependence on fossil fuels. Mining and metallurgy extensively exploit bioremediation to recover rare metals from industrial effluents and low-grade ores. In the case of bioleaching, metals such as copper and gold are extracted in a much eco-friendly manner due to the action of certain types of bacteria like *Acidithiobacillus*, eliminating energy-intensive conventional methods. The use of bioremediation techniques in construction can recycle demolition waste. For example, microorganisms can mineralize concrete to recover resources for the production of eco-friendly building materials. Bioremediation is also utilized in agriculture for the production of fertilizers. Through the action of bacteria, organic waste is converted into nutrient-rich bio-fertilizers that will increase soil fertility and replace chemical fertilizers. Bioremediation allows firms to create systems in which waste is a resource instead of an end product, in line with the ideas promoted in the circular economy. This, in turn, leads to innovation, reduced environmental damage, and lays a foundation for the industry to attain sustainability in the long run. For example, in the plastic industry, designed microorganisms are directed to break down waste polymers into monomers that can be used subsequently to make new plastics [28]. Besides, this tackles the constantly looming issue of plastic pollution while reducing dependence on fossil fuels. Mining and metallurgy extensively exploit bioremediation to recover rare metals from industrial effluents and low-grade ores. In the case of bioleaching, metals such as copper and gold are extracted in a much eco-friendly manner due to the action of certain types of bacteria like *Acidithiobacillus*, eliminating energy-intensive conventional methods. The use of bioremediation techniques in construction can recycle demolition waste. For example, microorganisms can mineralize concrete to recover resources for the production of eco-friendly building materials.

FACTORS AFFECTING THE EFFECTIVENESS OF BIOENGINEERED MATERIALS

The creation of bioengineered materials—new compounds made from renewable biological sources—has been spurred by the increased focus on sustainability in material science and production. The resource and environmental problems caused by conventional petrochemical-based products are increasingly being addressed by these materials. By providing environmentally friendly substitutes that are biodegradable, renewable, and use fewer resources, bioengineered materials have the potential to completely transform sectors, including packaging, textiles, automotive, and construction.

However, a number of variables that affect these materials' efficacy must be taken into consideration for their successful development and broad use. Every stage of the lifecycle of bioengineered materials—from the biological processes that go into their creation to the sustainability of the raw materials utilized—is vital in defining their effects on the environment and the economy.

It is crucial to comprehend and maximize a number of crucial elements in order to guarantee that bioengineered materials fulfill sustainability and performance requirements. These include the characteristics of the raw materials (or feedstocks) that serve as the basis for these products, the effectiveness of the production techniques utilized to produce them, and the choice of microbes and enzymes that power the biotechnological processes. Furthermore, to make sure that bioengineered materials can be produced and used globally, it is essential to analyze their complete lifecycle and environmental impact in addition to their economic viability and scalability. These elements will play a bigger role in determining the direction of bioengineered materials as bioengineering develops. The success of these materials in the market will be largely determined by their capacity to strike a balance between cost-effectiveness and environmental sustainability while preserving high-quality performance. In this regard, successfully addressing these concerns helps enterprises get closer to reaching their sustainability objectives while also quickening the shift to more sustainable materials [29].

Choice of Microorganisms and Enzymes

Many bioengineered materials are produced using biotechnological techniques that heavily rely on microorganisms and enzymes. They are crucial to the transformation of raw resources (such as plant-based feedstocks or agricultural waste) into bio-based goods like biopolymers, biofuels, and biodegradable plastics. The particular microbes and enzymes chosen for these processes have a significant impact on how effective these bioengineered materials are.

Microorganisms: The synthesis of biopolymers, fermentation, and biodegradation are all facilitated by microorganisms like bacteria, fungi, yeasts, and algae. They play a part in bioengineering by breaking down organic matter for bioconversion into useful products and turning plant-based sugars into bioplastics. In order to create biodegradable plastics like PHA from renewable resources like agricultural waste, bacteria like *Escherichia coli* can be genetically modified.

Strain Selection: Due to their varying metabolic routes and tolerances, particular microbial strains must be chosen carefully. To scale production and preserve consistency in the material qualities, the strain must be optimized for high efficiency, strong growth, and tolerance to environmental stress.

Genetic Engineering: Microorganisms can be genetically modified to improve their ability to produce desired products or to process feedstocks that would otherwise be challenging to use. For instance, by enhancing metabolic pathways, genetically modified bacterial strains have been utilized to create high-yielding bioplastics or biofuels.

Proteins called enzymes catalyze biochemical reactions and are essential for polymerization, depolymerization, and other changes needed to produce bioengineered materials. They may be found in nature or created to enhance.

In order to create bioplastics and bio composites, polymerization enzymes join monomers—the building elements of polymers—into lengthy strands. Lipases, which are employed in the synthesis of the biodegradable polymer PHBs, are one example.

Cellulase enzymes break down cellulose into fermentable sugars, which microorganisms then transform into bio-based products like ethanol, bioplastics, or biocomposites in processes that involve lignocellulosic materials (such as plant biomass).

Selecting enzymes with high specificity for the required reaction and stability under industrial-scale processing conditions (such as temperature, pH, and chemical environment) is essential to the process's success [30].

Properties of Feedstock and Raw Materials

The creation and functionality of bioengineered materials depend heavily on the characteristics of the raw materials and feedstocks. The quality, composition, and sustainability of the feedstock— the raw material used to create bioengineered products—have a direct impact on the production process's scalability, efficiency, and environmental impact. The selection of raw materials affects the bio-based products' functionality, durability, and end-of-life properties in addition to their technological viability.

When assessing feedstocks for bioengineered materials, important factors to take into account are:

• **Feedstock Type and Composition**

Plant-based biomass, agricultural wastes, algae, and even waste materials from other industries are the main feedstock types used in the manufacture of bioengineered materials. The structure and chemical makeup of various feedstocks influence how readily they can be processed into useful bio-based products.

Materials such as wood, straw, cotton, hemp, and sugarcane are examples of plant-based biomass. The cellulose, hemicellulose, and lignin found in plant biomass are plentiful and renewable resources that can be converted into bio-based goods, including textiles, bioplastics, and biocomposites.

One of the essential structural elements of plant cell walls is cellulose, a glucose polymer. Bio-based materials such as films, fibers, and bioplastics are frequently made with it.

Lignocellulose: A complex substance made up of lignin, cellulose, and hemicellulose. Wood, corn stover, or sugarcane bagasse are examples of lignocellulosic biomass, which is a viable feedstock for biofuels, biochemicals, and Biocomposites. However, the presence of lignin, which is difficult to break down, makes processing more complex.

• **Feedstocks' Chemical and Physical Characteristics**

The ease of processing and conversion of feedstocks into bioengineered materials depends on their physical and chemical characteristics. These characteristics include the density, reactivity, composition, and molecular structure of the feedstock.

Cellulose and Hemicellulose Content: Strong, long-lasting materials, such as bio-based composites and films, are produced with a high cellulose content. But because hemicellulose hydrolyzes more readily than cellulose, it can be a more effective substrate for fermentation or the synthesis of biopolymers.

Plant cell walls include lignin, a complex, aromatic polymer that gives them structural stiffness. Although lignin has some uses, such as the production of biofuels or bio-based compounds, its resistance to breakdown presents difficulties for procedures like fermentation.

• **Feedstock Availability and Sustainability**

When choosing the raw materials for bioengineered goods, sustainability is a crucial consideration. The feedstock needs to be plentiful, renewable, and sourced in a method that doesn't seriously damage the environment or deplete natural resources.

Renewability: To ensure a steady supply of raw materials, feedstocks made from renewable resources—such as algae, plant-based biomass, or agricultural residues—are crucial. Unlike materials derived from fossil fuels, these materials can be restored over time through farming or natural processes.

Land and Water Use: A feedstock's sustainability is also influenced by how much land and water it uses. For instance, some feedstocks, such as corn or sugarcane, demand a lot of land and water, which could interfere with the production of food or cause deforestation [31].

Efficiency of Bioprocesses

For bioprocesses, which use biological systems or living creatures to produce bioengineered materials, to be profitable, they must be effective. The yield, resource use, energy consumption, and overall sustainability of material manufacturing are all influenced by how well these processes work.

Conversion Efficiency: It is essential that microbes and enzymes be able to effectively transform basic materials into the intended bioengineered products. Increased yields, less waste, and more sustainable production are all results of high conversion efficiency. For instance, microbes must transform carbohydrates or oils into polymers with only a few byproducts in order to produce bioplastics. Similarly, one of the main challenges is to maximize the conversion of plant-based feedstocks into fermentable sugars *via* enzymatic processes (such as cellulose hydrolysis). Process Optimization: To increase the pace of microbial growth and material synthesis, production processes are optimized by regulating variables including temperature, pH, oxygen levels, and nutrition delivery. Additionally, in order to retain efficiency, process scale-up from laboratory to industrial levels frequently necessitates considerable changes.

Fermentation Systems: To guarantee the steady production of bioengineered products, continuous fermentation systems are frequently used. To control nutrient levels, reduce contamination, and preserve high microbial productivity, these systems must be tuned.

Energy and Resource Efficiency: Production procedures for bioengineered materials must minimize the use of energy and resources. To further lessen the production's environmental impact and carbon footprint, low-energy methods are preferred, such as using solar energy or waste heat from industrial operations [32].

Environmental Impact and Lifecycle Analysis

Bioengineered materials' effects on the environment must be assessed at every stage of their lifetime, from the extraction of raw materials to production, use, and disposal. Lifecycle analysis (LCA), which measures the material's influence on variables including energy use, water use, and waste creation, is frequently used to evaluate this procedure.

- Carbon Footprint: The carbon footprint of bioengineered materials is a crucial factor to take into account. The carbon footprint of materials made from renewable feedstocks, such as plant-based biomass, is often smaller than that of materials made from petroleum. Nonetheless, carbon emissions may still be a

result of the energy required for their manufacture, shipping, and processing.
- End-of-Life Disposal: The environmental impact of bioengineered materials is also greatly influenced by their capacity to decompose or recycle. Biodegradable plastics and biopolymers are examples of materials that reduce pollution and landfill trash because they are made to break down naturally or compost.
- Waste development: Sustainability may be greatly impacted by the industrial process's development of waste, such as leftover feedstock or byproducts. Waste-to-resource techniques that enable the reuse of byproducts are ideal for the production of bioengineered materials [33, 34].

Economic Feasibility and Scalability

For bioengineered materials to be widely used and succeed in the long run, they must be scalable and economically viable. The cost of production, the availability of raw resources, and consumer demand are some of the variables that affect these materials' economic viability. For bioengineered materials to be a competitive alternative in a variety of industries, they must be able to be produced on a large scale at prices comparable to those of traditional materials.

• *Costs of Production*

Raw Material Costs: The price of a whole. Material costs can be decreased by using inexpensive, renewable feedstocks, such as agricultural waste.

Processing Costs: Compared to conventional procedures, bioengineering processes like fermentation or enzymatic treatments are frequently more complicated and energy-intensive, which can increase manufacturing costs. But increasing effectiveness through technological advancements can help reduce these costs.

• *Production Scalability*

There are logistical and technical obstacles to overcome when moving from pilot-scale production to large-scale manufacturing. Greater volumes necessitate additional energy, infrastructure, and raw materials, all of which need to be efficiently optimized.

Process Optimization: In order to cut waste and boost productivity, manufacturing processes must be optimized, and constant product quality must be guaranteed as production ramps up.

• *Acceptance and Market Demand*

Consumer Preferences: The market for bioengineered materials is expanding due to the growing demand for sustainable products. Nonetheless, increasing the market share of these items depends on customer education.

The market for bio-based alternatives can be boosted by government laws and incentives that encourage sustainability, such as requirements for biodegradable products or prohibitions on single-use plastics [33].

CHALLENGES AND FUTURE DIRECTIONS

One of the main challenges faced by large-scale bioengineering is acquiring reliable biological feedstock—these are inevitably prone to seasonality and geography. Furthermore, maintaining production efficiency while minimizing waste is yet another technical bottleneck confronting the industry. Making changes to organisms for the industrial sector is painstakingly hard, as the emphasis is on stability of the modified organism, and for industrial scaling of processes, this often leads to increased water and energy usage. This is coupled with the extraction and purification of materials, which not only drives up the costs but also adds energy usage challenges when competing against the more conventional petrochemical sector. The implementation of such bioengineering modifications also faces resistance from the general public and authorities and requires multidimensional problem-solving approaches. The challenges encountered through mass commercialization have been illustrated in Fig. (**1**).

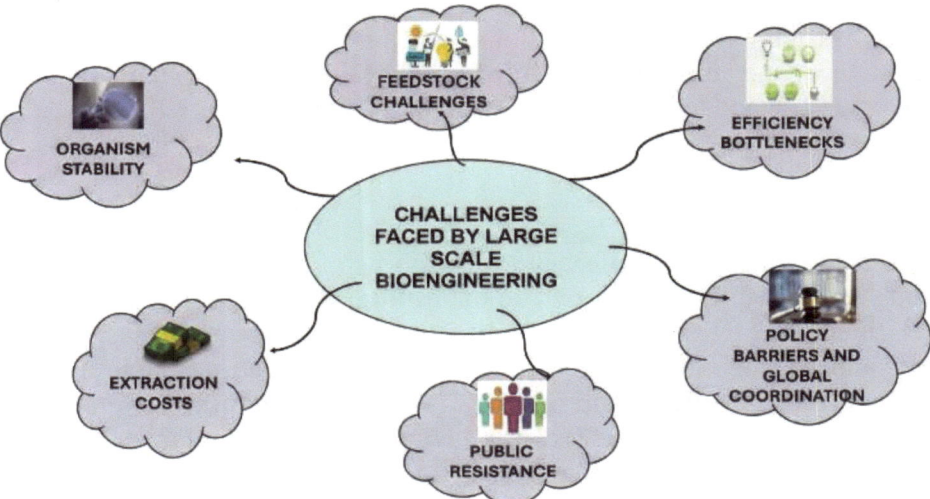

Fig. (1). Challenges faced by large-scale bioengineering.

Understanding the persistence or degradation of biosynthetic materials in the environment requires a focus on their construction and breakdown mechanisms under varying circumstances. However, many materials may decompose and break down into innocuous components, intermediate breakdown products, and genetic materials in contact with an ecosystem, which warrant thorough examination. Life cycle assessment can help make sure that these materials achieve the purpose of being eco-friendly [34]. Successful marketing entails explaining the purpose and advantages of bioengineered materials to the general public, and more importantly, tackling the critics who mistrust the technology due to concerns over genetic engineering. Developing educational programs assisted by government support, clear and open labels aim to increase confidence and even further raise the response to them. Purchase price, performance indicators, and quality are also members of the family.

Exogenous factors include particular trade policies that concentrate on bioproduction and bio business, undefined homogenized nomenclature and frameworks, tardiness or even laziness in drafting and approving bio-products safety assessments, and weak justification motivators oriented on bio-based materials. Fossil fuel policies distort countries' perspectives. Synchronized worldwide policies, keystone labels, and constructive economics are needed for further diffusion.

Biotechnology provides us with the development of new sustainable materials. Living machines that are programmed to make polymers, green organisms that create and sew algae-based AI fabrics for any shapes and sizes, metabolic engineering, and more get the job done and facilitate mass production. Across industries, protein nanostructures and biosynthetic dyes are lightweight, biodegradable, and environmentally friendly substitutes. Such innovations, however, are expected to be game-changing and have the potential of global reach [1, 35].

CONCLUSION

Renewable energy needs to meet the growing resource requirements as well as mankind's unavoidable implications on the ecological system. The emerging field of bioengineering offers environmentally friendly solutions, such as bioplastics, biofuels, and self-healing concrete. They further enhance production of useful bioproducts such as PLA and PHA and support bioremediation by using microorganisms and other genetic modifications. The green-shaped molecules produced can serve multiple purposes across industries, including novel biomedicine and new environmentally friendly materials across manufacturing and construction waste management. The productivity of bioengineered materials hinges on the renewable materials available, efficient bioprocessing methods, and environmental as well as economic considerations. When cost, performance, and sustainability are all achieved, such products have the potential to gain acceptance in other sectors like the construction, packaging, and textile industries. These range from eco-friendly construction materials and bioplastics to textiles and clothes made from algae-based materials and bio-composites. The bioengineered materials are an innovation as well as a way for environmental protection solutions, bringing forth reduced pollution, reduced carbon footprints, and an evolution of a circular economy. With a blend of genetic engineering, optimization of enzymes, and microbial technologies, these materials are practical and address global issues of pollution, resource depletion, and, most importantly, they are cost-effective. Offering such materials is the first step towards a greener society. It ensures a combination of cost-effectiveness, performance, and care for the environment.

REFERENCES

[1] Sales MB, Borges PT, Ribeiro Filho MN, *et al.* Sustainable Feedstocks and Challenges in Biodiesel Production: An Advanced Bibliometric Analysis. Bioengineering (Basel) 2022; 9(10): 539.
[http://dx.doi.org/10.3390/bioengineering9100539] [PMID: 36290507]

[2] Reichert CL, Bugnicourt E, Coltelli MB, *et al.* Bio-Based Packaging: Materials, Modifications, Industrial Applications and Sustainability. Polymers (Basel) 2020; 12(7): 1558.
[http://dx.doi.org/10.3390/polym12071558] [PMID: 32674366]

[3] Taweesan A, Kanabkaew T, Surinkul N, Polprasert C. Integrating clustering algorithms and machine learning to optimize regional snapshot municipal solid waste management for achieving sustainable development goals. Environ Adv 2025; 19: 100607.
[http://dx.doi.org/10.1016/j.envadv.2024.100607]

[4] Zhang YHP, Sun J, Ma Y. Biomanufacturing: history and perspective. J Ind Microbiol Biotechnol 2017; 44(4-5): 773-84.
[http://dx.doi.org/10.1007/s10295-016-1863-2] [PMID: 27837351]

[5] Atmakuri A, Palevicius A, Vilkauskas A, Janusas G. Review of Hybrid Fiber Based Composites with Nano Particles—Material Properties and Applications. Polymers (Basel) 2020; 12(9): 2088.
[http://dx.doi.org/10.3390/polym12092088] [PMID: 32937898]

[6] Abd-Elgawad MMM, Askary TH. Factors affecting success of biological agents used in controlling the plant-parasitic nematodes. Egypt J Biol Pest Control 2020; 30(1): 17.
[http://dx.doi.org/10.1186/s41938-020-00215-2]

[7] Londono R, Badylak SF. Factors which affect the host response to biomaterials. In: Badylak SF, Ed. Host response to biomaterials: the impact of host response on biomaterial selection. London: Academic Press 2015; pp. 1-12.
[http://dx.doi.org/10.1016/B978-0-12-800196-7.00001-3]

[8] Basit A, Rehman A, Iqbal K, Abid HA. Optimization of the Tencel/Cotton and Polyester/Recycled Polyester Blended Knitted Fabrics to Replace CVC Fabric. J Nat Fibers 2023; 20(1): 2167032.
[http://dx.doi.org/10.1080/15440478.2023.2167032]

[9] Chauhan A. Biodegradable plastics: a broad outlook. J Bioremed Biodeg 2013; 4(7): e141.
[http://dx.doi.org/10.4172/2155-6199.1000e141]

[10] Garlotta D. A Literature Review of Poly(Lactic Acid). J Polym Environ 2001; 9(2): 63-84.

[http://dx.doi.org/10.1023/A:1020200822435]

[11] Shah N, Ul-Islam M, Khattak WA, Park JK. Overview of bacterial cellulose composites: A multipurpose advanced material. Carbohydr Polym 2013; 98(2): 1585-98.
[http://dx.doi.org/10.1016/j.carbpol.2013.08.018] [PMID: 24053844]

[12] Jones M, Mautner A, Luenco S, Bismarck A, John S. Engineered mycelium composite construction materials from fungal biorefineries: A critical review. Mater Des 2020; 187: 108397.
[http://dx.doi.org/10.1016/j.matdes.2019.108397]

[13] García-Ochoa F, Santos VE, Casas JA, Gómez E. Xanthan gum: production, recovery, and properties. Biotechnol Adv 2000; 18(7): 549-79.
[http://dx.doi.org/10.1016/S0734-9750(00)00050-1] [PMID: 14538095]

[14] Liu L, Liu Y, Li J, Du G, Chen J. Microbial production of hyaluronic acid: current state, challenges, and perspectives. Microb Cell Fact 2011; 10(1): 99.
[http://dx.doi.org/10.1186/1475-2859-10-99] [PMID: 22088095]

[15] Madison LL, Huisman GW. Metabolic engineering of poly(3-hydroxyalkanoates): from DNA to plastic. Microbiol Mol Biol Rev 1999; 63(1): 21-53.
[http://dx.doi.org/10.1128/MMBR.63.1.21-53.1999] [PMID: 10066830]

[16] Yelithao K, Surayot U, Lee C, *et al.* Studies on structural properties and immune-enhancing activities of glycomannans from *Schizophyllum commune*. Carbohydr Polym 2019; 218: 37-45.
[http://dx.doi.org/10.1016/j.carbpol.2019.04.057] [PMID: 31221341]

[17] Campo VL, Kawano DF, Silva DB Jr, Carvalho I. Carrageenans: Biological properties, chemical modifications and structural analysis – A review. Carbohydr Polym 2009; 77(2): 167-80.
[http://dx.doi.org/10.1016/j.carbpol.2009.01.020]

[18] Czaja W, Krystynowicz A, Bielecki S, Brown RM Jr. Microbial cellulose—the natural power to heal wounds. Biomaterials 2006; 27(2): 145-51.
[http://dx.doi.org/10.1016/j.biomaterials.2005.07.035]

[19] Doudna JA, Charpentier E. The new frontier of genome engineering with CRISPR-Cas9. Science 1979; 2014: 346.
[PMID: 25430774]

[20] Doran PM. Bioprocess Development: An Interdisciplinary Challenge. In: Bioprocess Engineering Principles. 2nd ed. London: Academic Press; 2013. p. 3-11
[http://dx.doi.org/10.1016/B978-0-12-220851-5.00001-0]

[21] Sethi MK, Chakraborty P, Shukla R. Biocatalysis: An Industrial Perspective. In: de Gonzalo G, Domínguez de María P, eds. Biocatalysis: An Industrial Perspective. Cambridge (UK): The Royal Society of Chemistry; 2017. p. 44-76.
[http://dx.doi.org/10.1039/9781782629993-00044]

[22] Joho Y, Royan S, Caputo AT, *et al.* Enhancing PET Degrading Enzymes: A Combinatory Approach. ChemBioChem 2024; 25(10): e202400084.
[http://dx.doi.org/10.1002/cbic.202400084] [PMID: 38584134]

[23] Iram A, Dong Y, Ignea C. Synthetic biology advances towards a bio-based society in the era of artificial intelligence. Curr Opin Biotechnol 2024; 87: 103143.
[http://dx.doi.org/10.1016/j.copbio.2024.103143] [PMID: 38781699]

[24] Turner N, Schneider M. Editorial overview: Biocatalysis — molecular, structural and synthetic advances. Curr Opin Chem Biol. 2000; 4(1): 65-7.
[http://dx.doi.org/10.1016/S1367-5931(99)00053-8]

[25] Li X, Wu Y, Tan Z. An overview on bioremediation technologies for soil pollution in E-waste dismantling areas. J Environ Chem Eng 2022; 10(3): 107839.
[http://dx.doi.org/10.1016/j.jece.2022.107839]

[26] Zhang G, Qi F, Jia H, Zou C, Li C. Advances in bioprocessing for efficient bio manufacture. RSC Advances 2015; 5(65): 52444-51.
[http://dx.doi.org/10.1039/C5RA07699D]

[27] Omokhagbor Adams G, Tawari Fufeyin P, Eruke Okoro S, Ehinomen I. Bioremediation, Biostimulation and Bioaugmention: A Review. Int J Environ Bioremediat Biodegrad 2020; 3(1): 28-39.

[28] Bala S, Garg D, Thirumalesh BV, et al. Recent Strategies for Bioremediation of Emerging Pollutants: A Review for a Green and Sustainable Environment. Toxics 2022; 10(8): 484.
[http://dx.doi.org/10.3390/toxics10080484] [PMID: 36006163]

[29] Agrawal R, Kumar A, Mohammed MKA, Singh S. Biomaterial types, properties, medical applications, and other factors: a recent review. J Zhejiang Univ Sci A 2023; 24(11): 1027-42.
[http://dx.doi.org/10.1631/jzus.A2200403]

[30] Rodrigo-Navarro A, Sankaran S, Dalby MJ, del Campo A, Salmeron-Sanchez M. Engineered living biomaterials. Nat Rev Mater 2021; 6(12): 1175-90.
[http://dx.doi.org/10.1038/s41578-021-00350-8]

[31] New silicon feedstock material. Refocus 2006; 7(5): 16.
[http://dx.doi.org/10.1016/S1471-0846(06)70680-1]

[32] Deepak A, Sharma V, Kumar D. Life cycle assessment of biomedical waste management for reduced environmental impacts. J Clean Prod 2022; 349: 131376.
[http://dx.doi.org/10.1016/j.jclepro.2022.131376]

[33] Tsatsakis AM, Nawaz MA, Kouretas D, et al. Environmental impacts of genetically modified plants: A review. Environ Res 2017; 156: 818-33.
[http://dx.doi.org/10.1016/j.envres.2017.03.011] [PMID: 28347490]

[34] Lang JT. Elements of public trust in the American food system: Experts, organizations, and genetically modified food. Food Policy 2013; 41: 145-54.
[http://dx.doi.org/10.1016/j.foodpol.2013.05.008]

[35] Nath S. Biotechnology and biofuels: paving the way towards a sustainable and equitable energy for the future. Discover Energy 2024; 4(1): 8.
[http://dx.doi.org/10.1007/s43937-024-00032-w]

CHAPTER 2

Synthetic Biology for Environmental Applications

S S Kirthiga[1,*] **and R Dhinesh**[1]

[1] *Department of Aquatic Environment Management, Faculty of Fisheries Science, Kerala University of Fisheries and Ocean Studies, Kochi, Kerala, India*

Abstract: Synthetic biology, an interdisciplinary field combining biology, engineering, and computational science, serves as a tool for the advancement of environmental management by synthesizing new organisms with enhanced abilities to combat pollution, resource depletion, and ecological damage. This chapter explores the role of synthetic biology in environmental applications, focusing on the use of engineered microorganisms for pollution remediation, carbon capture, and resource recovery. By designing and optimizing genetic circuits, researchers can create microorganisms with unique metabolic pathways that efficiently degrade toxic pollutants, such as heavy metals, pesticides, and hydrocarbons, or capture greenhouse gases like carbon dioxide and methane. Synthetic biology also enables the creation of biosensors that detect environmental contaminants with high sensitivity and specificity, providing early warning systems for environmental threats. The chapter further discusses the integration of microbial systems into bio-based production processes and the principles as well as tools of synthetic biology, such as gene circuits, metabolic engineering, and the application of these innovations to tackle environmental challenges in wastewater treatment, soil bioremediation, and air quality management. The use of synthetic biology to create organisms capable of converting waste into biofuels, bioplastics, or valuable chemicals, reducing reliance on fossil fuels and minimizing waste generation, was also reviewed in detail. Additionally, the potential risks and ethical considerations associated with the release of Genetically Modified Organisms (GMOs) into natural ecosystems are addressed, emphasizing the need for safety protocols and regulatory frameworks. The way towards the future directions in synthetic biology, such as the development of synthetic ecosystems and the use of artificial intelligence to design more complex genetic networks, will render synthetic biology a powerful tool for environmental protection and pollution mitigation.

Keywords: Biology, Environment, Genetic engineering, Mitigation, Organisms, Pollution, Remediation, Synthetic biology.

---

[*] **Corresponding author S S Kirthiga:** Department of Aquatic Environment Management, Faculty of Fisheries Science, Kerala University of Fisheries and Ocean Studies, Kochi, Kerala, India; E-mail: sskkirthiga@gmail.com

Rajneesh Kumar, Ram Sharan Singh & Maulin P. Shah. (Eds.)
All rights reserved-© 2026 Bentham Science Publishers

INTRODUCTION

Synthetic biology is a rapidly evolving interdisciplinary field that combines principles from biology, engineering, and computer science to design and construct novel biological systems [1]. The genetic modification of natural organisms or creating entirely new ones, synthetic biology aims to address challenges across diverse sectors, including healthcare, agriculture, and industry [2]. In the environmental context, this innovative field holds immense potential to mitigate critical issues such as pollution, climate change, and ecosystem degradation, offering sustainable and scalable solutions [3]. The environmental applications of synthetic biology are driven by the urgent need for novel approaches to combat ecological crises. With rising levels of pollutants, resource depletion, and biodiversity loss, traditional methods of remediation and restoration are proving inadequate. Synthetic biology—with its ability to engineer microorganisms, plants, and Biosystems—provides a transformative avenue for tackling these challenges.

One of the most promising aspects of synthetic biology is its capacity to design microorganisms that can detect and neutralize pollutants. These engineered organisms can break down hazardous substances such as plastics, heavy metals, and hydrocarbons into harmless components [4]. Similarly, biosensors developed through synthetic biology can monitor environmental conditions in real time, enabling proactive management of ecosystems [5]. These applications demonstrate the field's potential to redefine environmental monitoring and remediation practices. Another critical area where synthetic biology is making an impact is in climate change mitigation. Through the genetic modification of photosynthetic organisms, scientists are enhancing carbon capture and storage capabilities [6]. Synthetic biology also enables the development of bio-based solutions, such as engineered methanotrophs that consume methane, a potent greenhouse gas. These innovations offer promising strategies for reducing atmospheric carbon and mitigating global warming. Resource recovery and waste management also stand to benefit significantly from synthetic biology. Engineered microbes are being utilized to extract valuable elements from waste streams and e-waste, while bio-based plastics and biodegradable materials are being developed to reduce plastic pollution [7]. Furthermore, synthetic biology is being integrated into wastewater treatment systems to recover nutrients and remove micropollutants, contributing to a circular economy approach [8].

While the potential of synthetic biology is vast, it is not without challenges. Issues such as biosafety, regulatory frameworks, and public acceptance need to be addressed to ensure the responsible deployment of synthetic organisms in the environment [9]. Despite these hurdles, advancements in synthetic biology continue to open new frontiers, offering transformative solutions for environmental sustainability. As the field matures, it holds the promise of reshaping our relationship with the natural world, enabling us to restore and protect ecosystems while meeting the needs of a growing global population. This chapter delves deeper into the diverse applications, innovations, and future prospects of synthetic biology in addressing global environmental challenges.

SYNTHETIC BIOLOGY TOOLS AND TECHNIQUES

Synthetic biology tools and techniques are central to the design, construction, and optimization of biological systems that can address complex environmental challenges. With a strong focus on engineering microorganisms, synthetic biology offers a suite of powerful methodologies that allow researchers to reprogram cells, create novel biosynthetic pathways, and design synthetic genetic circuits. These techniques facilitate the development of microbes that can perform functions such as bioremediation of pollutants, carbon sequestration, and production of biofuels. These tools provide high precision in manipulating genetic material, allowing for the targeted optimization of microbial systems. Moreover, advances in bioinformatics, systems biology, and high-throughput screening have further enhanced the potential of synthetic biology in environmental applications. The intersection of these advanced techniques with an understanding of cellular metabolism, gene networks, and environmental dynamics has created a new frontier in the use of engineered organisms for sustainability [10].

Genetic Circuit Design and Optimization

Genetic circuit design and optimization are pivotal components of synthetic biology, enabling the construction of complex, dynamic biological systems, and the process and outputs involved in the synthetic biology are demonstrated in Fig. (1). By designing genetic circuits, researchers can program microorganisms to perform specific tasks in response to environmental inputs, allowing for fine-tuned control of metabolic pathways [11]. These circuits often consist of synthetic DNA elements such as promoters, transcription factors, ribosome binding sites, and terminators, which are assembled to regulate the flow of genetic information and protein synthesis [12, 13].

One of the primary goals of genetic circuit design is to create systems that can respond predictably and efficiently to external stimuli, such as pollutants, temperature, or light, making them ideal for applications in environmental monitoring and pollution control. To optimize genetic circuits, synthetic biologists use a combination of experimental and computational techniques through iterative testing and feedback, and researchers refine genetic elements to enhance their performance, stability, and reliability [14]. Advances in computational modeling

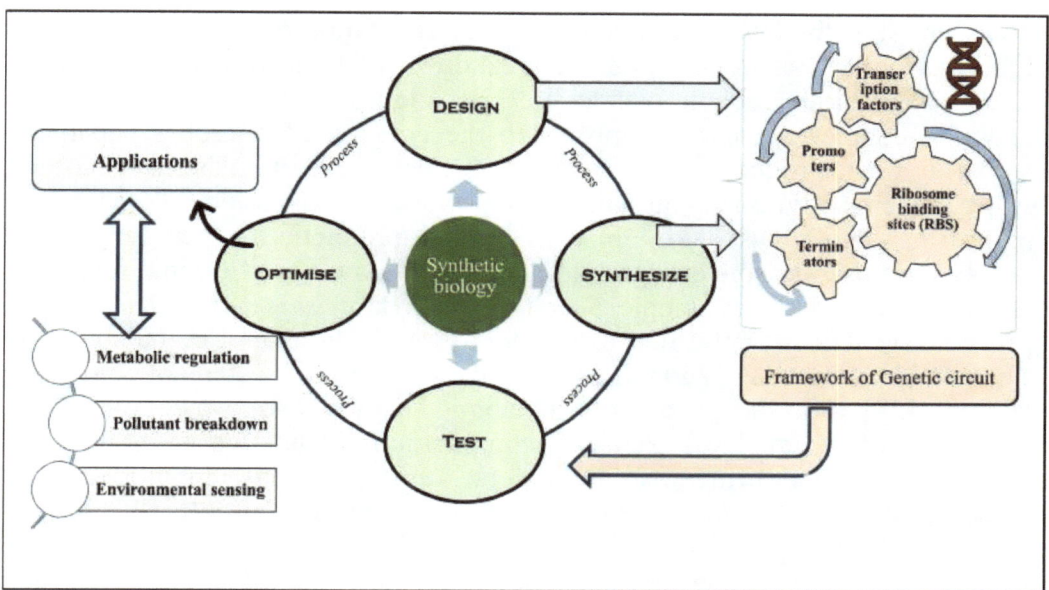

Fig. (1). Schematic representation of the process and output of synthetic biology.

and machine learning have revolutionized the optimization process by predicting the behavior of genetic circuits before they are physically constructed [15]. Optimization can also involve the fine-tuning of circuit parameters such as gene expression levels and metabolic fluxes to improve efficiency, yield, and stability of engineered biological systems [14]. Furthermore, synthetic biologists now routinely integrate circuits with feedback control mechanisms that enable adaptive responses to changing environmental conditions, thereby enhancing the robustness of engineered systems [16]. These advances in genetic circuit design and optimization have enabled a wide array of environmental applications, from the bioremediation of heavy metals to the capture of greenhouse gases.

Metabolic Engineering for Environmental Applications

Metabolic engineering, a core technique within synthetic biology, involves the modification of an organism's metabolic network to enhance or redirect its biochemical pathways for specific applications [17]. In environmental contexts, metabolic engineering can be utilized to design microorganisms that are capable of degrading pollutants, synthesizing valuable chemicals from waste, or capturing carbon dioxide [10]. A critical aspect of metabolic engineering in environmental applications is the identification and optimization of metabolic pathways that allow microbes to metabolize harmful compounds effectively [18]. By manipulating key enzymes, introducing new biosynthetic pathways, or deleting non-essential pathways, synthetic biologists can tailor organisms to perform highly specialized tasks, such as the breakdown of toxic heavy metals or the production of biodegradable plastics [19]. In addition to pollutant degradation, metabolic engineering can be applied to the creation of microbes capable of converting waste products into valuable chemicals, such as bioplastics or biofuels, further reducing environmental pollution and waste accumulation [20]. Metabolic engineering also plays a key role in biofuel production, where engineered microbes are used to convert agricultural waste, biomass, or CO_2 into renewable energy sources [21]. For example, *Saccharomyces cerevisiae* and *Escherichia coli* have been genetically modified to efficiently produce ethanol or butanol from lignocellulosic materials, addressing the growing need for sustainable bioenergy solutions [22]. The design of new metabolic pathways for waste-to-chemical conversion often involves extensive computational modeling to predict and enhance microbial efficiency in diverse environmental conditions [21]. Furthermore, the use of synthetic biology to create custom-tailored microbial systems holds great promise for improving the efficiency of environmental clean-up efforts, such as heavy metal remediation and the treatment of organic pollutants.

Integration of Microbial Systems

The integration of microbial systems is a powerful strategy in synthetic biology, particularly for environmental applications where complex biological processes must be coordinated to achieve desired outcomes. In this context, microbial systems are designed to work together synergistically, often in multispecies consortia, to address specific environmental challenges. These integrated systems can be engineered to perform tasks such as the degradation of pollutants, bioremediation of hazardous waste, or the conversion of waste into valuable products. The integration of microbial systems involves both the engineering of individual microbes with specific traits and the assembly of these microbes into functional communities that can collaborate effectively. For example, in bioremediation applications, consortia of microorganisms can be engineered to degrade a wide range of pollutants in an environment. Different species within the consortium may carry out complementary metabolic processes, each specializing in different stages of pollutant degradation. One example of this is the use of engineered microbial communities to break down oil spills, where one microbe may specialize in breaking down hydrocarbons, while another detoxifies by-products produced during the degradation process. These systems are highly adaptable and can be tuned to respond to varying pollutant concentrations or environmental conditions. Additionally, synthetic biology enables the integration of microbial systems into larger environmental management processes. This includes the development of integrated bioreactors that combine waste treatment, carbon capture, and biofuel production in a single system. The integration of synthetic microbial systems with industrial processes can significantly reduce waste and emissions while increasing resource recovery. This approach is exemplified in integrated biorefinery systems, where engineered microbes are employed in sequence to process organic waste, producing biofuels, bioplastics, and other valuable chemicals while simultaneously remediating pollutants and capturing greenhouse gases [23]. Furthermore, the integration of microbial systems with environmental sensors and monitoring tools enables real-time feedback and control, which is essential for optimizing the performance of engineered systems. Such integrated systems are critical for enhancing the efficiency and sustainability of waste-to-resource conversion processes and bioremediation efforts, providing solutions for a wide range of environmental problems [24]. The ability to design and integrate complex microbial systems with high levels of coordination and flexibility marks a significant advancement in the field of synthetic biology, opening new possibilities for environmental protection and sustainability.

SCOPE OF SYNTHETIC BIOLOGY FOR ENVIRONMENTAL APPLICATIONS

The scope of synthetic biology in environmental applications is expansive, offering innovative solutions to address critical ecological challenges, and the overview alongside the scope of synthetic biology is represented in Fig. (**2**). One of the most impactful areas is pollution control, where synthetic biology enables the design of microorganisms that can detect, degrade, or neutralize harmful substances in the environment. These engineered organisms are capable of breaking down pollutants such as hydrocarbons, heavy metals, pesticides, and plastics, transforming them into less harmful byproducts [25]. This bioremediation potential is a game-changer for managing contamination in water, soil, and air, particularly in industrial and urban areas. Another significant area of application is in climate change mitigation. Synthetic biology is advancing carbon capture technologies by enhancing the efficiency of photosynthetic organisms like algae and cyanobacteria [26]. These engineered organisms can absorb higher amounts of carbon dioxide, helping to offset emissions and reduce atmospheric greenhouse gases. Additionally, synthetic biology offers solutions for managing methane emissions by creating organisms that consume methane as a carbon source [27]. These strategies not only contribute to reducing global warming but also open new avenues for sustainable energy production through biofuels derived from engineered biomass. Synthetic biology also plays a vital role in environmental monitoring through the development of biosensors. These sensors are engineered to detect specific pollutants, toxins, or changes in environmental conditions in real time [28]. In addition to mitigation and monitoring, synthetic

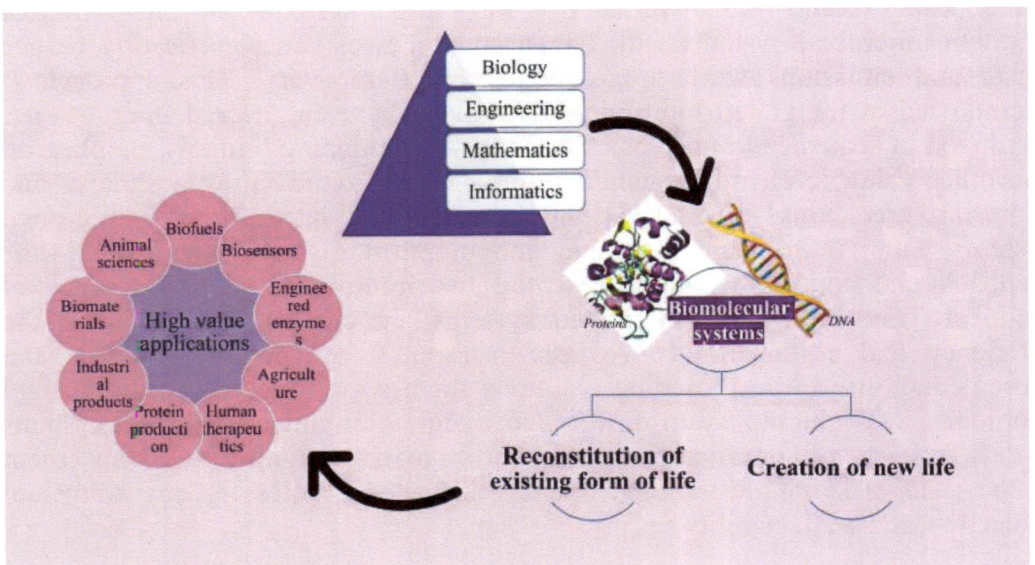

Fig. (2). Overview and scope of synthetic biology.

biology contributes to sustainable resource recovery and waste management. Engineered microbes are being utilized to recover valuable elements such as nitrogen, phosphorus, and rare earth metals from waste streams [29]. Similarly, synthetic biology facilitates the creation of biodegradable plastics and the development of systems for breaking down existing plastic waste [30]. These innovations promote circular economy principles by transforming waste into resources, reducing environmental impacts, and fostering sustainable practices. As synthetic biology continues to evolve, its applications in environmental science hold immense potential to reshape how we interact with and protect the natural world.

SYNTHETIC BIOLOGY FOR POLLUTION REMEDIATION

Synthetic biology has emerged as a transformative tool in pollution remediation, offering innovative approaches to address environmental challenges through the design and engineering of biological systems. By reprogramming microorganisms and enhancing their metabolic pathways, synthetic biology enables the precise degradation, detoxification, and sequestration of harmful pollutants such as heavy metals, hydrocarbons, and pesticides [4]. Engineered microbes can be tailored to function in diverse environments, from contaminated soils to aquatic ecosystems, where traditional remediation methods are less effective [1]. Moreover, synthetic biology enables the creation of microbial consortia that work synergistically to break down complex pollutant mixtures, mimicking natural ecosystems to enhance efficiency. Although challenges such as biosafety, scalability, and ecological risks remain, advancements in genetic engineering and computational modeling are paving the way for safer and more efficient applications, making synthetic biology a cornerstone of sustainable pollution management [31].

Engineered Microorganisms in Pollution Control

Synthetic biology has revolutionized pollution control by creating engineered microorganisms tailored to detect, transform, and remove environmental pollutants with high specificity and efficiency. These microorganisms are designed using tools such as CRISPR-Cas9, metabolic engineering, and synthetic gene circuits, enabling them to perform tasks beyond the capabilities of naturally occurring species [32, 33]. Engineered microorganisms offer innovative solutions to combat environmental degradation, as they can degrade, detoxify, or sequester harmful substances, including heavy metals, organic pollutants, and microplastics. One of the most impactful applications of synthetic biology in pollution control involves the customization of metabolic pathways in microbes [18]. Natural microbial pathways are limited in their ability to process complex pollutants, but synthetic biology enables the addition, removal, or enhancement of genes to overcome these limitations [34]. For instance, *Pseudomonas* and *Escherichia coli* have been engineered to produce specific enzymes for the breakdown of recalcitrant compounds like polychlorinated biphenyls (PCBs) and Polycyclic Aromatic Hydrocarbons (PAHs) [35]. Additionally, novel genetic circuits allow these engineered organisms to act only in the presence of a specific pollutant, enhancing their safety and reducing the risks of unintended ecological effects. A major advantage of engineered microorganisms is their ability to function in diverse environments, from contaminated soils to aquatic ecosystems. This adaptability is achieved by optimizing their genetic makeup to thrive under extreme conditions, such as high salinity, pH variations, or toxic concentrations of pollutants [36]. Furthermore, genetic modifications can enhance biofilm formation in engineered microorganisms, increasing their stability and efficiency in real-world applications [37]. Synthetic biology also allows the development of microbial consortia, where multiple engineered species work together to tackle complex contamination scenarios. Each member of the consortium can be tailored to target specific pollutants or perform complementary roles, such as pollutant degradation, nutrient cycling, or environmental monitoring [38]. These systems mimic natural ecosystems, providing a holistic approach to pollution control while enhancing efficiency through cooperation.

Biodegradation of Toxic Pollutants

Heavy Metals: Heavy metal contamination is a pervasive environmental problem resulting from industrial activities, mining, and agricultural runoff. Unlike organic pollutants, heavy metals cannot be degraded but require removal or transformation into less toxic forms [39]. Synthetic biology has enabled the engineering of microorganisms with enhanced metal uptake, sequestration, or biotransformation capabilities. Engineered bacteria such as *Escherichia coli* and *Shewanella oneidensis* have been modified to produce metallothioneins or phytochelatins, proteins that bind and sequester heavy metals like cadmium, lead, and mercury [40]. In addition to sequestration, synthetic biology enables microbial transformation of metals into less bioavailable or less toxic forms [41]. For instance, mercury can be reduced from its highly toxic methylated form to elemental mercury by engineered Pseudomonas putida. These processes leverage engineered enzymatic pathways to achieve precise control over metal detoxification [42]. The use of bioreactors populated with such engineered strains has shown promise in treating industrial wastewater contaminated with heavy metals, offering a sustainable alternative to chemical treatment methods [43].

Pesticides: The widespread use of pesticides has led to persistent contamination of soil and water, posing risks to ecosystems and human health [44]. Many pesticides, including organophosphates and carbamates, are resistant to natural degradation processes [34]. Synthetic biology offers solutions by engineering microorganisms with enhanced enzymatic pathways for pesticide degradation. For example, E. coli and Bacillus subtilis have been modified to express enzymes like organophosphorus hydrolase, which breaks down organophosphate pesticides into harmless compounds [45, 46]. Additionally, synthetic consortia of microorganisms have been designed to degrade complex pesticide mixtures that require multi-step biochemical pathways [47]. These consortia utilize cross-feeding mechanisms, where intermediate degradation products are further metabolized by different microbial strains, ensuring complete detoxification [48].

Hydrocarbons: Hydrocarbons, including petroleum and its derivatives, are among the most common pollutants resulting from oil spills and industrial discharges [49]. Synthetic biology has enabled the creation of microorganisms capable of degrading hydrocarbons with unprecedented efficiency. Engineered Alcanivorax borkumensis, a naturally occurring hydrocarbon-degrading bacterium, has been modified to express additional enzymes, enabling the breakdown of a wider range of hydrocarbons, including long-chain alkanes and polyaromatic hydrocarbons (PAHs) [50]. Furthermore, synthetic biology tools have been used to enhance the survivability and activity of these microorganisms in harsh environments, such as the deep sea or Arctic regions, where oil spills often occur [51]. Genetic modifications that increase the production of biosurfactants, compounds that facilitate hydrocarbon emulsification, have further improved their efficacy in oil spill remediation [52].

CARBON CAPTURE AND GREENHOUSE GAS MITIGATION

The accelerating rise of atmospheric greenhouse gases, primarily carbon dioxide (CO_2) and methane (CH_4), is a critical driver of climate change [53]. Synthetic biology offers innovative approaches to capture, store, and utilize these gases, transforming them into valuable products while reducing their atmospheric concentrations [23]. One of the most promising strategies involves engineering photosynthetic microorganisms, such as cyanobacteria and algae, to enhance their carbon fixation capabilities. These organisms naturally convert CO_2 into organic compounds during photosynthesis, but synthetic biology can optimize this process by modifying key enzymes like RuBisCO and introducing more efficient carbon capture pathways [54]. Engineered strains of Synechococcus elongatus, for example, have demonstrated increased CO_2 uptake and conversion into biofuels and bioplastics [55]. Methane, a potent greenhouse gas with a global warming potential significantly higher than CO_2, can also be targeted using synthetic biology. Methanotrophic bacteria, which metabolize methane as their primary carbon source, have been engineered to produce industrially relevant compounds, such as biopolymers and organic acids. These engineered systems not only mitigate methane emissions but also contribute to the circular bioeconomy by generating valuable products from waste gases. In addition to direct carbon capture, synthetic biology facilitates the development of systems that integrate carbon utilization with renewable energy technologies. Microbial electrochemical systems, for instance, employ engineered microbes to convert CO_2 into chemicals like ethanol or formate, using electricity generated from renewable sources. This approach effectively links carbon capture with energy storage and utilization, addressing multiple environmental challenges simultaneously [26]. Advances in computational modeling and machine learning are playing a critical role in optimizing synthetic biology systems for large-scale deployment, ensuring that they are economically viable and environmentally sustainable.

BIOSENSORS FOR ENVIRONMENTAL MONITORING

Biosensors for environmental monitoring represent a cutting-edge application of synthetic biology, offering real-time, efficient, and cost-effective solutions for detecting pollutants and tracking environmental changes. These biosensors are typically designed by engineering microorganisms, enzymes, or cellular components to respond to specific environmental stimuli, such as chemical pollutants, heavy metals, or temperature variations. When exposed to target contaminants, the biosensor generates a measurable signal, often in the form of fluorescence, color change, or electrical activity that allows for rapid detection and analysis [28, 56]. This ability to provide on-site, real-time data makes biosensors invaluable for monitoring the health of ecosystems and ensuring the quality of water, air, and soil.

One of the key advantages of biosensors over traditional environmental monitoring techniques is their high sensitivity and specificity. Through genetic engineering, researchers can tailor these sensors to detect even trace amounts of pollutants, enabling early detection of contamination before it reaches harmful levels. Biosensors have been developed to detect heavy metals such as arsenic, mercury, and lead in water sources, which are critical to human health and environmental sustainability [57]. Similarly, biosensors are being designed to detect agricultural runoff, including nitrates and pesticides, which can lead to eutrophication and biodiversity loss in aquatic systems [58]. The integration of biosensors with digital technologies further enhances their applicability in environmental monitoring [59]. Data collected by biosensors can be transmitted to central systems for real-time analysis, allowing for immediate action in response to environmental threats [60]. By enabling real-time, localized monitoring, biosensors have the potential to revolutionize environmental management, offering a more proactive and efficient approach to protecting natural resources and public health.

APPLICATIONS IN ENVIRONMENTAL MANAGEMENT

Synthetic biology has emerged as a transformative tool in environmental management, offering innovative solutions for addressing some of the most pressing environmental challenges. The ability to engineer microorganisms and biosystems to perform specific tasks has opened up new possibilities in the treatment of environmental contaminants, the remediation of polluted sites, and the management of resources in a sustainable manner [61]. From wastewater treatment to soil bioremediation and air quality management, synthetic biology provides versatile tools for mitigating the impact of human activities on the environment [62]. These applications leverage advances in metabolic engineering, genetic circuit design, and microbial consortia to create environmentally friendly processes that improve resource recovery, reduce pollution, and foster ecological balance.

Wastewater Treatment

Wastewater treatment is one of the critical applications of synthetic biology, given the increasing global demand for clean water and the growing challenges posed by industrialization, urbanization, and climate change. Traditional wastewater treatment processes, although effective, are often energy-intensive and produce large amounts of sludge, which can be difficult and costly to manage [63]. One key approach in synthetic biology-based wastewater treatment is the use of engineered bacteria and algae for nutrient removal and organic waste degradation [64]. Genetically modified Escherichia coli and Pseudomonas species have been engineered to degrade complex organic pollutants, such as surfactants, dyes, and pesticides, which are often difficult to remove with conventional methods [65]. Synthetic biology also enables the development of microorganisms capable of converting organic waste into valuable byproducts, such as biofuels, bioplastics, and biogas, providing an additional benefit of resource recovery [24]. Additionally, Microbial Electrochemical Technologies (MET), which harness engineered microbes to generate electricity while removing pollutants, are increasingly being applied in wastewater treatment plants to improve energy efficiency and reduce environmental impact [66]. Furthermore, the integration of synthetic biology with biological nutrient removal processes holds great potential in the treatment of nitrogen and phosphorus, which are major contributors to eutrophication and water quality deterioration. Genetically engineered strains of Nitrosomonas and Nitrobacter, involved in nitrification and denitrification processes, have been optimized to enhance nitrogen removal efficiency [67]. Similarly, synthetic biology-based systems can be used for the efficient removal of phosphorus by engineering bacteria that accumulate phosphorus in excess, thereby preventing its release into water bodies [68]. Overall, synthetic biology-based wastewater treatment not only addresses the limitations of conventional systems but also offers a sustainable approach to managing water resources, providing both environmental and economic benefits.

Soil Bioremediation

Soil bioremediation is a critical environmental management tool for cleaning up contaminated land, particularly in industrial areas and locations affected by agricultural runoff. Soil contamination by heavy metals, pesticides, hydrocarbons, and other hazardous chemicals has significant consequences for ecosystem health and human well-being [69]. Microbial-based soil bioremediation is grounded in the natural ability of soil microorganisms to degrade a wide range of organic pollutants, including petroleum hydrocarbons, Polycyclic Aromatic Hydrocarbons (PAHs), and pesticides [70]. Synthetic biology allows for the enhancement and optimization of these natural processes by creating genetically engineered microorganisms with enhanced capabilities for pollutant degradation. These genetically modified microbes are designed to express enzymes capable of breaking down complex pollutants into less harmful byproducts, such as water and carbon dioxide [25]. Moreover, synthetic biology enables the design of microbial consortia that work synergistically to degrade a broader spectrum of pollutants. In soil environments, these consortia can be engineered to perform complementary functions, such as breaking down hydrocarbons, sequestering heavy metals, and improving soil health by promoting plant growth [71]. Engineered Arthrobacter strains have been used in conjunction with Trichoderma fungi to degrade petroleum contaminants while simultaneously enhancing soil nutrient availability [72]. Additionally, synthetic biology approaches can be used to engineer plants that are more resilient to contaminants, a process known as phytoremediation. Plants engineered to express specific enzymes or proteins can aid in the absorption and detoxification of pollutants, thus enhancing the bioremediation process [73, 74].

Air Quality Management

Air quality management is an essential component of environmental protection, particularly in the context of urbanization, industrial emissions, and climate change. Poor air quality, resulting from the release of pollutants such as Particulate Matter (PM), Volatile Organic Compounds (VOCs), Nitrogen Oxides (NOx), and sulfur dioxide (SO2), has serious health consequences, including respiratory diseases, cardiovascular issues, and premature death [75]. One of the most promising applications of synthetic biology in air quality management is the development of engineered microorganisms capable of degrading airborne pollutants. Genetically modified strains of Pseudomonas and Bacillus have been engineered to degrade VOCs, such as benzene, toluene, and xylene, which are common air pollutants emitted from industrial processes and vehicle exhaust [75, 76]. These engineered microbes can be integrated into biofilters, where they actively metabolize pollutants as air passes through, providing an eco-friendly alternative to conventional air purification methods. Similarly, synthetic biology can be used to design plants with enhanced abilities to absorb and neutralize air pollutants through phytoremediation [75]. Plants engineered to express specific enzymes, such as cytochrome P450s, can metabolize and detoxify toxic substances, improving air quality in urban and industrial environments [77]. Moreover, engineered microorganisms that produce a detectable signal in the presence of pollutants like nitrogen dioxide or ozone could be used to monitor air pollution in urban areas or near industrial sites [78]. This technology has the potential to complement existing air quality monitoring systems, enabling more precise and localized monitoring, as well as providing early warnings of harmful pollution levels. Moreover, synthetic biology can contribute to carbon capture and the mitigation of greenhouse gas emissions. Engineered microorganisms, such as algae and bacteria, can be designed to capture carbon dioxide from industrial emissions or the atmosphere and convert it into valuable products, such as biofuels or bioplastics [79]. These bio-based solutions not only help reduce atmospheric CO_2 concentrations but also provide an opportunity for resource recovery, making air quality management more sustainable [80]. Through these innovative approaches, synthetic biology plays a crucial role in addressing the global challenge of air pollution and climate change.

ETHICAL AND RISK CONSIDERATIONS

As synthetic biology for environmental applications advances, ethical and risk considerations must be carefully addressed. One of the main concerns is the potential unintended ecological consequences of releasing Genetically Modified Organisms (GMOs) into natural environments [81]. Engineered organisms, such as microbes designed for pollution remediation, could disrupt ecosystems by outcompeting native species or transferring modified traits to non-target organisms. This poses risks to biodiversity, and careful risk assessments and monitoring are required to minimize such impacts. Biosafety is another significant issue. The creation of synthetic organisms capable of environmental interventions raises concerns about containment and control [82]. If these organisms escape into the wild, they may persist and spread, potentially causing unforeseen environmental harm. Strict containment protocols and monitoring mechanisms must be established to prevent such scenarios. Moreover, issues of equity in access to synthetic biology technologies must be considered to ensure that the benefits are shared globally and not limited by economic or geographic disparities.

FUTURE DIRECTIONS IN SYNTHETIC BIOLOGY FOR ENVIRONMENTAL APPLICATIONS

The future of synthetic biology in environmental applications holds tremendous promise, driven by continuous advances in genetic engineering, computational modeling, and synthetic biology tools. One key area of growth is the development of more efficient and scalable bioremediation technologies. In the coming years, researchers will focus on designing microorganisms with enhanced capabilities to degrade a wider range of pollutants, including complex chemicals such as pharmaceuticals, pesticides, and persistent organic pollutants. These organisms will be optimized for deployment in diverse environments, from contaminated industrial sites to oceans and freshwater bodies, providing a sustainable solution to pollution. Another exciting direction is the enhancement of carbon capture and climate change mitigation strategies through synthetic biology. Advances in photosynthetic organisms like algae, cyanobacteria, and engineered plants will likely improve their efficiency in sequestering carbon dioxide from the atmosphere. The future may see the development of carbon-negative systems that not only capture and store carbon but also convert it into valuable byproducts such as biofuels or biodegradable plastics. Additionally, the engineering of synthetic methanotrophs to consume methane, a potent greenhouse gas, may become a critical tool in reducing global warming potential. Synthetic biology will also play a pivotal role in advancing environmental monitoring and management. The development of advanced biosensors capable of real-time detection of pollutants, toxins, and environmental changes will become more widespread. These biosensors, integrated with digital technologies such as artificial intelligence, will provide continuous data on environmental conditions, enabling faster response times and better decision-making. This could revolutionize environmental management by allowing for early detection and proactive intervention in pollution and ecosystem health. Synthetic biology will also facilitate the development of sustainable solutions for wastewater treatment, where engineered microbes could selectively remove harmful pollutants while recovering valuable nutrients like nitrogen and phosphorus. These advancements will contribute to the creation of a circular economy, where waste products are minimized and resources are continuously reused.

CONCLUSION

In conclusion, synthetic biology presents transformative potential for addressing a wide array of environmental challenges. From pollution remediation and climate change mitigation to advanced monitoring and sustainable resource recovery, the applications of synthetic biology offer innovative solutions that could reshape how we manage and protect the environment. As research and technological advancements continue, the scope for designing highly efficient organisms and systems will expand, providing new tools to tackle pressing environmental issues such as pollution, biodiversity loss, and resource depletion. The growing integration of synthetic biology with fields like biotechnology, materials science, and data analytics will likely accelerate progress, creating a more sustainable and resilient future. However, as these technologies evolve, ethical, safety, and regulatory considerations must remain at the forefront of development. Careful monitoring, risk assessment, and public engagement are essential to ensure that the deployment of synthetic organisms and other innovations does not unintentionally harm ecosystems or human health. With thoughtful regulation, transparent practices, and global collaboration, synthetic biology can be a powerful tool for environmental sustainability, contributing to a cleaner, healthier planet while fostering economic and societal progress.

REFERENCES

[1] Shapira P, Kwon S, Youtie J. Tracking the emergence of synthetic biology. Scientometrics 2017; 112(3): 1439-69.
[http://dx.doi.org/10.1007/s11192-017-2452-5] [PMID: 28804177]

[2] Ausländer S, Ausländer D, Fussenegger M. Synthetic biology—the synthesis of biology. Angew Chem Int Ed 2017; 56(23): 6396-419.

[http://dx.doi.org/10.1002/anie.201609229] [PMID: 27943572]

[3] Qumsani AT. The contribution of microorganisms to sustainable development: towards a green future through synthetic biology and systems biology. J Umm Al-Qura Univ Appl Sci 2024 Jul 26; 1-7.
[http://dx.doi.org/10.1007/s43994-024-00180-8]

[4] Aminian-Dehkordi J, Rahimi S, Golzar-Ahmadi M, *et al.* Synthetic biology tools for environmental protection. Biotechnol Adv 2023; 68: 108239.
[http://dx.doi.org/10.1016/j.biotechadv.2023.108239] [PMID: 37619824]

[5] Bhatnagar A, Masih J, Kumar R. Biosensor: Application in Environmental Management InHarnessing Microbial Potential for Multifarious Applications 2024 Apr 10. Singapore: Springer Nature Singapore 2024; pp. 455-88.

[6] DeLisi C. The role of synthetic biology in climate change mitigation. Biol Direct 2019; 14(1): 24.

[7] HP JS, Latif AD, Saeed DN, Shinde VM, Dillibabu SP, Suganthi D. Innovative Approaches to Sustainable E-Waste Management Through Bio-Based Products. In: Environmental Applications of Carbon-Based Materials. 2024; pp. 380-470.

[8] Kathi S, Singh S, Yadav R, Singh AN, Mahmoud AED. Wastewater and sludge valorisation: a novel approach for treatment and resource recovery to achieve circular economy concept. Front Chem Eng 2023; 5: 1129783.
[http://dx.doi.org/10.3389/fceng.2023.1129783]

[9] Breyer D, Baldo A, Johansen TK, *et al.* Synthetic Biology-Risk assessments and recommendations for future governance guidelines. 2024.

[10] Nikel PI, de Lorenzo V. Metabolic Engineering. In: Nielsen J, Stephanopoulos G, Lee SY, Eds. Metabolic Engineering: Concepts and Applications 2021; 13: 859-90.
[http://dx.doi.org/10.1002/9783527823468.ch22]

[11] English MA, Gayet RV, Collins JJ. Designing biological circuits: synthetic biology within the operon model and beyond. Annu Rev Biochem 2021; 90(1): 221-44.
[http://dx.doi.org/10.1146/annurev-biochem-013118-111914] [PMID: 33784178]

[12] Slusarczyk AL, Lin A, Weiss R. Foundations for the design and implementation of synthetic genetic circuits. Nat Rev Genet 2012; 13(6): 406-20.
[http://dx.doi.org/10.1038/nrg3227] [PMID: 22596318]

[13] Xie M, Fussenegger M. Designing cell function: assembly of synthetic gene circuits for cell biology applications. Nat Rev Mol Cell Biol 2018; 19(8): 507-25.
[http://dx.doi.org/10.1038/s41580-018-0024-z] [PMID: 29858606]

[14] Naseri G, Koffas MAG. Application of combinatorial optimization strategies in synthetic biology. Nat Commun 2020; 11(1): 2446.
[http://dx.doi.org/10.1038/s41467-020-16175-y] [PMID: 32415065]

[15] Gayathiri E, Prakash P, Kumaravel P, *et al.* Computational approaches for modeling and structural design of biological systems: A comprehensive review. Prog Biophys Mol Biol 2023; 185: 17-32.
[http://dx.doi.org/10.1016/j.pbiomolbio.2023.08.002] [PMID: 37821048]

[16] Rollié S, Mangold M, Sundmacher K. Designing biological systems: Systems Engineering meets Synthetic Biology. Chem Eng Sci 2012; 69(1): 1-29.
[http://dx.doi.org/10.1016/j.ces.2011.10.068]

[17] Dasgupta A, Chowdhury N, De RK. Metabolic pathway engineering: Perspectives and applications. Comput Methods Programs Biomed 2020; 192: 105436.
[http://dx.doi.org/10.1016/j.cmpb.2020.105436] [PMID: 32199314]

[18] Dangi AK, Sharma B, Hill RT, Shukla P. Bioremediation through microbes: systems biology and metabolic engineering approach. Crit Rev Biotechnol 2019; 39(1): 79-98.
[http://dx.doi.org/10.1080/07388551.2018.1500997] [PMID: 30198342]

[19] Maqsood Q, Hussain N, Sumrin A, Ali SW, Tariq MR, Mahnoor M. Monitoring and abatement of synthetic pollutants using engineered microbial systems. Discover Life 2024; 54(1): 9.
[http://dx.doi.org/10.1007/s11084-024-09652-7]

[20] Aggarwal N, Pham HL, Ranjan B, *et al.* Microbial engineering strategies to utilize waste feedstock for sustainable bioproduction. Nat Rev Bioeng 2023; 2(2): 155-74.
[http://dx.doi.org/10.1038/s44222-023-00129-2]

[21] Jagadevan S, Banerjee A, Banerjee C, *et al.* Recent developments in synthetic biology and metabolic engineering in microalgae towards biofuel production. Biotechnol Biofuels 2018; 11(1): 185.
[http://dx.doi.org/10.1186/s13068-018-1181-1] [PMID: 29988523]

[22] Joshi A, Verma KK, D Rajput V, Minkina T, Arora J. Recent advances in metabolic engineering of microorganisms for advancing lignocellulose-derived biofuels. Bioengineered 2022; 13(4): 8135-63.
[http://dx.doi.org/10.1080/21655979.2022.2051856] [PMID: 35297313]

[23] Tiwari T, Kaur GA, Singh PK, Balayan S, Mishra A, Tiwari A. Emerging bio-capture strategies for greenhouse gas reduction: Navigating challenges towards carbon neutrality. Sci Total Environ 2024; 929: 172433.
[http://dx.doi.org/10.1016/j.scitotenv.2024.172433] [PMID: 38626824]

[24] Ashokkumar V, Flora G, Venkatkarthick R, *et al.* Advanced technologies on the sustainable approaches for conversion of organic waste to valuable bioproducts: Emerging circular bioeconomy perspective. Fuel 2022; 324: 124313.
[http://dx.doi.org/10.1016/j.fuel.2022.124313]

[25] Rafeeq H, Afsheen N, Rafique S, *et al.* Genetically engineered microorganisms for environmental remediation. Chemosphere 2023; 310: 136751.
[http://dx.doi.org/10.1016/j.chemosphere.2022.136751] [PMID: 36209847]

[26] Zahed MA, Movahed E, Khodayari A, Zanganeh S, Badamaki M. Biotechnology for carbon capture and fixation: Critical review and future directions. J Environ Manage 2021; 293: 112830.
[http://dx.doi.org/10.1016/j.jenvman.2021.112830] [PMID: 34051533]

[27] Strong PJ, Xie S, Clarke WP. Methane as a resource: can the methanotrophs add value? Environ Sci Technol 2015; 49(7): 4001-18.
[http://dx.doi.org/10.1021/es504242n] [PMID: 25723373]

[28] Bilal M, Iqbal HMN. Microbial-derived biosensors for monitoring environmental contaminants: Recent advances and future outlook. Process Saf Environ Prot 2019; 124: 8-17.
[http://dx.doi.org/10.1016/j.psep.2019.01.032]

[29] Dev S, Sachan A, Dehghani F, Ghosh T, Briggs BR, Aggarwal S. Mechanisms of biological recovery of rare-earth elements from industrial and electronic wastes: A review. Chem Eng J 2020; 397: 124596.
[http://dx.doi.org/10.1016/j.cej.2020.124596]

[30] Ali SS, Elsamahy T, Al-Tohamy R, *et al.* Plastic wastes biodegradation: Mechanisms, challenges and future prospects. Sci Total Environ 2021; 780: 146590.
[http://dx.doi.org/10.1016/j.scitotenv.2021.146590] [PMID: 34030345]

[31] Fatima G, Magomedova A, Parvez S. Biotechnology and sustainable development. Shineeks Publishers 2024. Apr 14

[32] Tang TC, An B, Huang Y, *et al.* Materials design by synthetic biology. Nat Rev Mater 2020; 6(4): 332-50.
[http://dx.doi.org/10.1038/s41578-020-00265-w]

[33] Czajka J, Wang Q, Wang Y, Tang YJ. Synthetic biology for manufacturing chemicals: constraints drive the use of non-conventional microbial platforms. Appl Microbiol Biotechnol 2017; 101(20): 7427-34.
[http://dx.doi.org/10.1007/s00253-017-8489-9] [PMID: 28884354]

[34] Bhatt P, Gangola S, Bhandari G, *et al.* New insights into the degradation of synthetic pollutants in contaminated environments. Chemosphere 2021; 268: 128827.
[http://dx.doi.org/10.1016/j.chemosphere.2020.128827] [PMID: 33162154]

[35] Kumari S, Das S. Bacterial enzymatic degradation of recalcitrant organic pollutants: catabolic pathways and genetic regulations. Environ Sci Pollut Res Int 2023; 30(33): 79676-705.
[http://dx.doi.org/10.1007/s11356-023-28130-7] [PMID: 37330441]

[36] Wani AK, Akhtar N, Sher F, Navarrete AA, Américo-Pinheiro JHP. Microbial adaptation to different environmental conditions: molecular perspective of evolved genetic and cellular systems. Arch Microbiol 2022; 204(2): 144.
[http://dx.doi.org/10.1007/s00203-022-02757-5] [PMID: 35044532]

[37] Li Z, Wang X, Wang J, *et al.* Bacterial biofilms as platforms engineered for diverse applications. Biotechnol Adv 2022; 57: 107932.
[http://dx.doi.org/10.1016/j.biotechadv.2022.107932] [PMID: 35235846]

[38] Che S, Men Y. Synthetic microbial consortia for biosynthesis and biodegradation: promises and challenges. J Ind Microbiol Biotechnol 2019; 46(9-10): 1343-58.
[http://dx.doi.org/10.1007/s10295-019-02211-4] [PMID: 31278525]

[39] Ajiboye TO, Oyewo OA, Onwudiwe DC. Simultaneous removal of organics and heavy metals from industrial wastewater: A review. Chemosphere 2021; 262: 128379.
[http://dx.doi.org/10.1016/j.chemosphere.2020.128379] [PMID: 33182079]

[40] Mishra S, Patel A, Bhatt P, Chen S, Srivastava PK. Perspective Evaluation of Synthetic Biology Approaches for Effective Mitigation of Heavy Metal Pollution. Rev Environ Contam Toxicol 2024; 262(1): 21.
[http://dx.doi.org/10.1007/s44169-024-00072-2]

[41] Jones EM, Marken JP, Silver PA. Synthetic microbiology in sustainability applications. Nat Rev Microbiol 2024; 22(6): 345-59.
[http://dx.doi.org/10.1038/s41579-023-01007-9] [PMID: 38253793]

[42] Kumari S, Amit , Jamwal R, Mishra N, Singh DK. Recent developments in environmental mercury bioremediation and its toxicity: A review. Environ Nanotechnol Monit Manag 2020; 13: 100283.
[http://dx.doi.org/10.1016/j.enmm.2020.100283]

[43] Chandran P, Suresh S, Balasubramain B, *et al.* Biological treatment solutions using bioreactors for environmental contaminants from industrial waste water. J Umm Al-Qura Univ Appl Sci 2023 Jul 21; 1-23.

[44] Zhou W, Li M, Achal V. A comprehensive review on environmental and human health impacts of chemical pesticide usage. Emerg Contam 2024; 11(1): 100410.
[http://dx.doi.org/10.1016/j.emcon.2024.100410]

[45] Anjum N, Ridwan Q, Rashid S, Akhter F, Hanief M. Microbial degradation of organophosphorus pesticides. InBioremediation and Phytoremediation Technologies in Sustainable Soil ManagementApple Academic Press 159-85. 2022 Jun 30
[http://dx.doi.org/10.1201/9781003281207-7]

[46] Tanveer S, Ilyas N, Akhtar N, *et al.* Unlocking the interaction of organophosphorus pesticide residues with ecosystem: Toxicity and bioremediation. Environ Res 2024; 249: 118291.
[http://dx.doi.org/10.1016/j.envres.2024.118291] [PMID: 38301757]

[47] Adamu KS, Bichi YH, Nasiru AY, *et al.* Synthetic microbial consortia in bioremediation and biodegradation. Int J Res Innov Appl Sci 2023; VIII(VII): 232-41.
[http://dx.doi.org/10.51584/IJRIAS.2023.8727]

[48] Cao Z, Yan W, Ding M, Yuan Y. Construction of microbial consortia for microbial degradation of complex compounds. Front Bioeng Biotechnol 2022; 10: 1051233.
[http://dx.doi.org/10.3389/fbioe.2022.1051233] [PMID: 36561050]

[49] Mohammadi L, Rahdar A, Bazrafshan E, Dahmardeh H, Susan MABH, Kyzas GZ. Petroleum hydrocarbon removal from wastewaters: a review. Processes (Basel) 2020; 8(4): 447.
[http://dx.doi.org/10.3390/pr8040447]

[50] Pete AJ. Bioremediation of petroleum-based contaminants by alkane-degrading bacterium Alcanivorax borkumensis. Louisiana State University and Agricultural & Mechanical College. 2022.

[51] Andrey F, Lenar A, Irina P, Inna S. Removal of oil spills in temperate and cold climates of Russia experience in the creation and use of biopreparations based on effective microbial consortia. In: Biodegradation, Pollutants and Bioremediation Principles 2021 Apr 19 137-59.
[http://dx.doi.org/10.1201/9780429293931-7]

[52] Patel S, Homaei A, Patil S, Daverey A. Microbial biosurfactants for oil spill remediation: pitfalls and potentials. Appl Microbiol Biotechnol 2019; 103(1): 27-37.
[http://dx.doi.org/10.1007/s00253-018-9434-2] [PMID: 30343430]

[53] Ramirez-Corredores MM, Goldwasser MR, Falabella de Sousa Aguiar E. Carbon dioxide and climate change. In: Decarbonization as a Route Towards Sustainable Circularity 2023 Jan 2 1-14. Cham: Springer International Publishing
[http://dx.doi.org/10.1007/978-3-031-19999-8_1]

[54] Kubis A, Bar-Even A. Synthetic biology approaches for improving photosynthesis. J Exp Bot 2019; 70(5): 1425-33.
[http://dx.doi.org/10.1093/jxb/erz029] [PMID: 30715460]

[55] Roh H, Lee JS, Choi HI, *et al.* Improved CO_2-derived polyhydroxybutyrate (PHB) production by engineering fast-growing cyanobacterium Synechococcus elongatus UTEX 2973 for potential utilization of flue gas. Bioresour Technol 2021; 327: 124789.
[http://dx.doi.org/10.1016/j.biortech.2021.124789] [PMID: 33556769]

[56] Ma Z, Meliana C, Munawaroh HSH, *et al.* Recent advances in the analytical strategies of microbial biosensor for detection of pollutants. Chemosphere 2022; 306: 135515.
[http://dx.doi.org/10.1016/j.chemosphere.2022.135515] [PMID: 35772520]

[57] Salek Maghsoudi A, Hassani S, Mirnia K, Abdollahi M. Recent advances in nanotechnology-based biosensors development for detection of arsenic, lead, mercury, and cadmium. Int J Nanomedicine 2021; 16: 803-32.
[http://dx.doi.org/10.2147/IJN.S294417] [PMID: 33568907]

[58] Ramakrishnan S, Sathvara P, Tripathi S, Jayaraman A. Water pollutants, sensor types, and their advantages and challenges. InSensors for Environmental Monitoring, Identification, and Assessment. IGI Global. 2024; pp. 78-101.

[59] Wu J, Liu H, Chen W, Ma B, Ju H. Device integration of electrochemical biosensors. Nat Rev Bioeng 2023; 1(5): 346-60.
[http://dx.doi.org/10.1038/s44222-023-00032-w] [PMID: 37168735]

[60] Williams A, Aguilar MR, Pattiya Arachchillage KGG, *et al.* Biosensors for public health and environmental monitoring: the case for sustainable biosensing. ACS Sustain Chem& Eng 2024; 12(28): 10296-312.
[http://dx.doi.org/10.1021/acssuschemeng.3c06112] [PMID: 39027730]

[61] Bhavya G, Belorkar SA, Mythili R, *et al.* Remediation of emerging environmental pollutants: A review based on advances in the uses of eco-friendly biofabricated nanomaterials. Chemosphere 2021; 275: 129975.
[http://dx.doi.org/10.1016/j.chemosphere.2021.129975] [PMID: 33631403]

[62] Kuppan N, Padman M, Mahadeva M, Srinivasan S, Devarajan R. A comprehensive review of sustainable bioremediation techniques: eco friendly solutions for waste and pollution management. Waste Manag Bull 2024; 2(3): 154-71.
[http://dx.doi.org/10.1016/j.wmb.2024.07.005]

[63] Raheem A, Sikarwar VS, He J, *et al.* Opportunities and challenges in sustainable treatment and resource reuse of sewage sludge: A review. Chem Eng J 2018; 337: 616-41.
[http://dx.doi.org/10.1016/j.cej.2017.12.149]

[64] Webster LJ, Villa-Gomez D, Brown R, Clarke W, Schenk PM. A synthetic biology approach for the treatment of pollutants with microalgae. Front Bioeng Biotechnol 2024; 12: 1379301.
[http://dx.doi.org/10.3389/fbioe.2024.1379301] [PMID: 38646010]

[65] Sharma S, Pathania S, Bhagta S, *et al.* Microbial remediation of polluted environment by using recombinant E. coli: a review. Biotech Enviro 2024; 1(1): 8.
[http://dx.doi.org/10.1186/s44314-024-00008-z]

[66] Parambath JB, Ahmad AA, Mohamed AA. Application of Microbial Electrochemical Technologies as Biosensors for the Detection of Inorganic Water Pollutants. In: Selvasembian R, Mal J, Das S, Verma DK, Anastopoulos I, Eds. Emerging Trends in Microbial Electrochemical Technologies for Sustainable Mitigation of Water Resources Contamination: Microbial Electrochemical Technologies in Wastewater Treatment. Cham: Springer Nature Switzerland 2024; pp. 83-107.
[http://dx.doi.org/10.1007/978-3-031-74636-9_4]

[67] Xi H, Zhou X, Arslan M, *et al.* Heterotrophic nitrification and aerobic denitrification process: Promising but a long way to go in the wastewater treatment. Sci Total Environ 2022; 805: 150212.
[http://dx.doi.org/10.1016/j.scitotenv.2021.150212] [PMID: 34536867]

[68] Pathom-aree W, Sattayawat P, Inwongwan S, *et al.* Microalgae growth-promoting bacteria for cultivation strategies: Recent updates and progress. Microbiol Res 2024; 286: 127813.
[http://dx.doi.org/10.1016/j.micres.2024.127813] [PMID: 38917638]

[69] Barinova GM, Gaeva DV, Krasnov EV. Hazardous chemicals and air, water, and soil pollution and contamination. Good Health and Well-Being. 2020; pp. 255-66.
[http://dx.doi.org/10.1007/978-3-319-95681-7_48]

[70] Premnath N, Mohanrasu K, Guru Raj Rao R, *et al.* A crucial review on polycyclic aromatic Hydrocarbons - Environmental occurrence and strategies for microbial degradation. Chemosphere 2021; 280: 130608.
[http://dx.doi.org/10.1016/j.chemosphere.2021.130608] [PMID: 33962296]

[71] Xiang L, Harindintwali JD, Wang F, *et al.* Integrating biochar, bacteria, and plants for sustainable remediation of soils contaminated with organic pollutants. Environ Sci Technol 2022; 56(23): 16546-66.
[http://dx.doi.org/10.1021/acs.est.2c02976] [PMID: 36301703]

[72] Pozdnyakova N, Muratova A, Bondarenkova A, Turkovskaya O. Degradation of a Model mixture of PAHs by bacterial–fungal co-cultures. Front Biosci (Elite Ed) 2023; 15(4): 26.
[http://dx.doi.org/10.31083/j.fbe1504026] [PMID: 38163938]

[73] Kurade MB, Ha YH, Xiong JQ, Govindwar SP, Jang M, Jeon BH. Phytoremediation as a green biotechnology tool for emerging environmental pollution: A step forward towards sustainable rehabilitation of the environment. Chem Eng J 2021; 415: 129040.
[http://dx.doi.org/10.1016/j.cej.2021.129040]

[74] World Health Organization WHO global air quality guidelines: particulate matter (PM2. 5 and PM10), ozone, nitrogen dioxide, sulfur dioxide and carbon monoxide. World Health Organization 2021.Sep 7.

[75] Maurya A, Sharma D, Partap M, Kumar R, Bhargava B. Microbially-assisted phytoremediation toward air pollutants: Current trends and future directions. Environ Technol Innov 2023; 31: 103140.
[http://dx.doi.org/10.1016/j.eti.2023.103140]

[76] Isha AS, Ali S, Khalid A, Naseer IA, Raza H, Chang Y-C. Bioremediation of Smog: Current Trends and Future Perspectives. Processes (Basel) 2024; 12(10): 2266.
[http://dx.doi.org/10.3390/pr12102266]

[77] Kumar S. Engineering cytochrome P450 biocatalysts for biotechnology, medicine and bioremediation.

Expert Opin Drug Metab Toxicol 2010; 6(2): 115-31.
[http://dx.doi.org/10.1517/17425250903431040] [PMID: 20064075]

[78] Seesaard T, Kamjornkittikoon K, Wongchoosuk C. A comprehensive review on advancements in sensors for air pollution applications. Sci Total Environ 2024; 951: 175696.
[http://dx.doi.org/10.1016/j.scitotenv.2024.175696] [PMID: 39197792]

[79] Kumar M, Sundaram S, Gnansounou E, Larroche C, Thakur IS. Carbon dioxide capture, storage and production of biofuel and biomaterials by bacteria: A review. Bioresour Technol 2018; 247: 1059-68.
[http://dx.doi.org/10.1016/j.biortech.2017.09.050] [PMID: 28951132]

[80] Ondrasek G, Meriño-Gergichevich C, Manterola-Barroso C, *et al.* Bio-based resources: systemic & circular solutions for (agro)environmental services. RSC Advances 2024; 14(32): 23466-82.
[http://dx.doi.org/10.1039/D4RA03506B] [PMID: 39055268]

[81] Tsatsakis AM, Nawaz MA, Kouretas D, *et al.* Environmental impacts of genetically modified plants: A review. Environ Res 2017; 156: 818-33.
[http://dx.doi.org/10.1016/j.envres.2017.03.011] [PMID: 28347490]

[82] Soleimani Sasani M. The Importance of Biosecurity in Emerging Biotechnologies and Synthetic Biology. Avicenna J Med Biotechnol 2024; 16(4): 223-32.
[http://dx.doi.org/10.18502/ajmb.v16i4.16738] [PMID: 39606684]

CHAPTER 3

Mechanisms and Advanced Processes in Wastewater Treatment Technologies

R Dhinesh[1,*] and S S Kirthiga[1]

[1] *Department of Aquatic Environment Management, Faculty of Fisheries Science, Kerala University of Fisheries and Ocean Studies, Kochi, Kerala, India*

Abstract: Effective wastewater treatment is essential for protecting the environment and public health, as it transforms contaminated water into a cleaner, safer form that can be safely returned to the environment or reused. This chapter provides a comprehensive overview of the various processes involved in wastewater treatment, covering all the physical, chemical, and biological mechanisms and methodologies. The initial preliminary and primary treatment methods focus on removing large solids and debris through screening, sedimentation, and grit removal, and then the secondary treatment processes, which utilize biological methods to degrade organic matter and pollutants. Key techniques discussed include activated sludge systems, trickling filters, and sequencing batch reactors, along with advancements in biological nutrient removal, such as nitrification/denitrification and enhanced phosphorus removal. The chapter further explores tertiary treatment technologies designed to achieve higher levels of purification. These include advanced filtration methods, disinfection techniques such as ultraviolet (UV) irradiation and ozonation, and chemical treatments like coagulation and advanced oxidation processes. The sludge management, covering the generation, treatment, and disposal or reuse of sludge produced during wastewater treatment, is highly insisted upon. Furthermore, the advanced and emerging technologies, including membrane bioreactors, electrochemical treatments, and resource recovery methods like energy and nutrient recovery, were also discussed in detail. Besides, the chapter also provides a detailed analysis of the system integration, optimization strategies, regulatory considerations, and the future directions of wastewater treatment processes, highlighting the importance of innovation and sustainability in advancing effective wastewater management solutions.

Keywords: Biological, Chemical, Eco-friendly, Methods, Physical, Sustainable, Treatment, Wastewater.

---

* **Corresponding author R Dhinesh:** Department of Aquatic Environment Management, Faculty of Fisheries Science, Kochi, Kerala, India; E-mail: dhinesh7143@gmail.com

Rajneesh Kumar, Ram Sharan Singh & Maulin P. Shah. (Eds.)
All rights reserved-© 2026 Bentham Science Publishers

INTRODUCTION

Wastewater treatment plays a critical role in safeguarding both public health and the environment. As populations grow and industrialization continues to expand, the demand for effective wastewater management solutions becomes ever more pressing. Untreated or poorly treated wastewater poses a significant threat to both human health and the integrity of natural ecosystems. Inadequate treatment can lead to the contamination of water bodies, making them unsafe for drinking, agriculture, and recreational use, while also contributing to environmental degradation. The introduction of various pollutants, including heavy metals, organic contaminants, pathogens, and nutrients, into water systems not only compromises aquatic biodiversity but also exacerbates issues such as eutrophication and the spread of diseases [1, 2]. Historically, wastewater management systems focused on the removal of organic matter and suspended solids [3]. However, as scientific understanding has evolved, it has become clear that wastewater is a complex matrix containing a wide range of emerging contaminants. These include pharmaceuticals, personal care products, pesticides, endocrine-disrupting compounds, microplastics, and industrial chemicals, many of which are resistant to traditional treatment methods [4]. As a result, there has been a growing emphasis on advancing wastewater treatment technologies to address these challenges and meet increasingly stringent regulatory standards. Innovations in treatment processes, particularly in the realms of advanced oxidation, membrane filtration, and biological systems, have the potential to improve removal efficiencies for a broad spectrum of pollutants [5]. The scope of wastewater treatment extends far beyond municipal systems, encompassing industrial wastewater, agricultural runoff, and even stormwater. Each type of wastewater presents unique challenges, depending on its composition and the specific contaminants it carries. Municipal wastewater typically contains organic materials, nutrients, and pathogens, while industrial effluents may include a variety of toxic substances, ranging from heavy metals to solvents and acids [6]. Agricultural runoff, often laden with nutrients such as nitrogen and phosphorus from fertilizers, as well as pesticides, has become a major contributor to the degradation of freshwater bodies [7]. Stormwater, though often less concentrated in pollutants, can carry a wide array of contaminants due to its exposure to urban surfaces and pollutants deposited on roads, rooftops, and industrial areas [8].

In recent years, there has been a concerted effort to integrate emerging technologies and methodologies into wastewater treatment processes. This includes the application of nanotechnology, bioengineering, and membrane bioreactor systems, all of which hold the promise of significantly enhancing treatment capabilities. Nanomaterials have been shown to improve the removal of heavy metals, organic pollutants, and pathogens through adsorption, photocatalysis, and other mechanisms [9]. Similarly, bioelectrochemical systems, which utilize microbial fuel cells to treat wastewater while simultaneously generating electricity, represent an exciting frontier in sustainable treatment technologies [10]. The incorporation of artificial intelligence and machine learning is also revolutionizing the way wastewater treatment plants are operated, allowing for real-time optimization and predictive maintenance that can improve overall efficiency and reduce energy consumption [11]. Despite these advancements, challenges remain in ensuring the scalability and economic viability of these technologies. Furthermore, the complexity of treating wastewater with emerging contaminants, particularly those that are difficult to degrade biologically or chemically, means that additional research is needed to develop more effective treatment processes. This chapter aims to provide an overview of the principles, advancements, and challenges associated with wastewater treatment technologies, while highlighting the importance of integrating new approaches to promote a circular economy and sustainable water management.

WASTEWATER TREATMENT PROCESSES AND THEIR IMPORTANCE

Wastewater treatment is a vital process that ensures the safe management of liquid waste generated from domestic, industrial, and agricultural activities. It plays a critical role in protecting public health, preserving environmental integrity, and facilitating water reuse in the face of growing global water scarcity. The importance of wastewater treatment extends beyond environmental protection to include human health and economic benefits. Untreated or poorly treated wastewater can lead to severe water pollution, endangering aquatic ecosystems and causing the spread of waterborne diseases [12]. Proper wastewater management prevents contamination of freshwater resources, ensuring clean water availability for drinking, irrigation, and industrial processes. Moreover, wastewater treatment contributes to sustainable development by facilitating resource recovery. Nutrients such as nitrogen and phosphorus can be reclaimed as fertilizers, while biogas generated during anaerobic digestion can serve as a renewable energy source [13]. In the context of a circular economy, wastewater treatment supports the efficient use of water resources, reduces environmental footprints, and fosters resilience against climate change and water scarcity.

The wastewater treatment process typically involves three main stages: primary, secondary, and tertiary treatments, each targeting specific contaminants to achieve comprehensive pollutant removal [14], and the complete process involved in the wastewater treatment is represented in Fig. (**1**). These stages work together to convert wastewater into a safer effluent, suitable for discharge into natural ecosystems or for various reuse applications. In the primary stage, physical processes such as screening, grit removal, and sedimentation are employed to remove large debris, sand, and settleable solids. This stage significantly reduces the pollutant load, preventing damage and clogging of downstream equipment. However, it does not address dissolved organic matter or fine particulates, necessitating further treatment in subsequent stages [15]. The secondary treatment focuses on the biological degradation of dissolved organic matter and nutrients. Utilizing microorganisms, this stage breaks down organic pollutants, significantly reducing Biochemical Oxygen Demand (BOD) and Chemical Oxygen Demand (COD) in the effluent. Technologies such as activated sludge systems, biofilters, and anaerobic digestion are widely applied [16]. This stage is essential for maintaining the ecological balance of receiving water bodies, as it prevents oxygen depletion and reduces the risk of eutrophication. In the tertiary stage, also known as advanced treatment, the focus shifts to removing finer and more persistent contaminants, such as nutrients (nitrogen and phosphorus), pathogens, heavy metals, and emerging pollutants like pharmaceuticals, pesticides, and microplastics. Advanced techniques, including membrane filtration, ultraviolet disinfection, and advanced oxidation processes, are employed to ensure the effluent meets stringent discharge or reuse standards [17]. This stage is particularly important for water-stressed regions or industries requiring high-quality water for specific purposes.

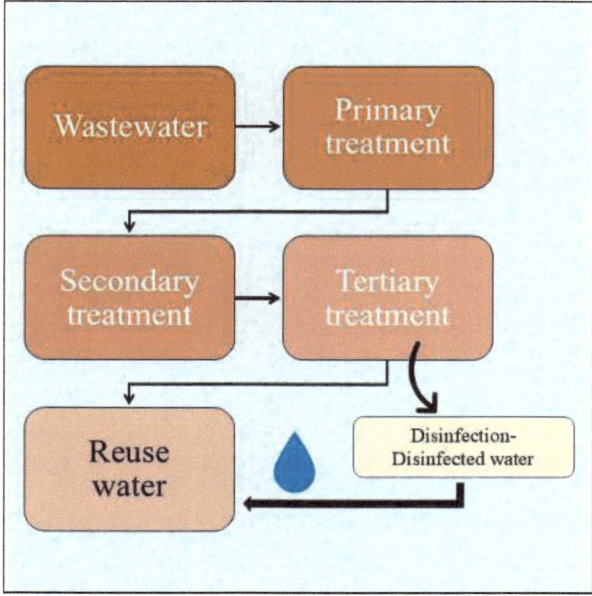

Fig. (1). Flowchart representing different stages of wastewater treatment.

PRELIMINARY AND PRIMARY TREATMENT

Wastewater treatment involves a series of processes aimed at removing contaminants to make water suitable for reuse or safe disposal. The initial stages, referred to as preliminary and primary treatments, are crucial for reducing the load on secondary and tertiary treatment processes, ensuring their effectiveness, and minimizing operational costs [18]. The mechanisms and processes involved in the primary wastewater treatment are represented in Fig. (**2**).

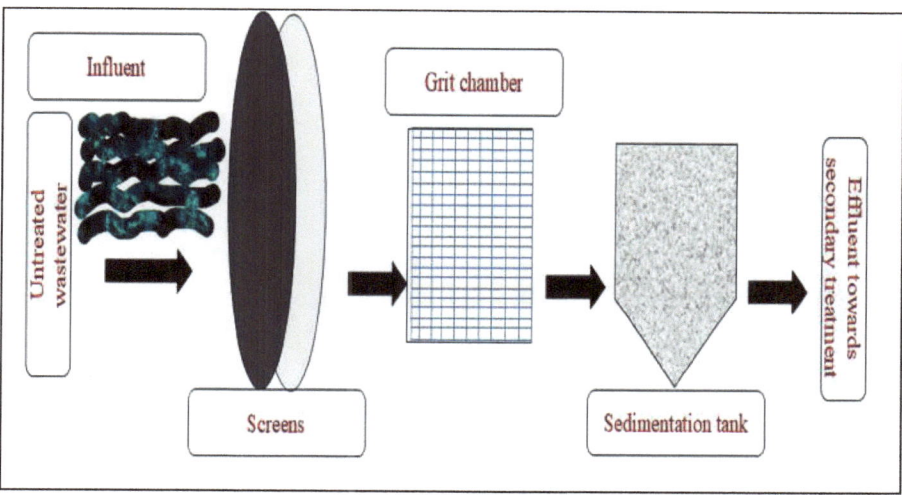

Fig. (2). Schematic representation of the primary wastewater treatment process.

Screening and Removal of Large Solids

Screening is the first step in wastewater treatment, and its primary purpose is to remove large debris and solids that could damage equipment or obstruct pipes in subsequent treatment processes. Screening involves passing wastewater through a series of screens or meshes to trap large particles such as plastics, sticks, leaves, and other coarse debris [19]. Screens are typically classified into two categories: coarse and fine. Coarse screens remove larger debris with openings ranging from 6 mm to 25 mm, while fine screens, with openings between 0.25 mm and 6 mm, capture smaller particles [20]. In some systems, the wastewater passes through several stages of screening, progressively capturing finer particles. The efficiency of screening depends on the mesh size, flow velocity, and the type of screen used. Mechanical systems, including rotary, drum, and band screens, are commonly used in large treatment plants due to their ability to handle large volumes of wastewater and their self-cleaning capabilities [21]. Fine screens are especially important in removing small solids, such as hair, fibers, and smaller debris, which could otherwise clog pipes or interfere with later treatment stages [6]. While screening removes large debris effectively, it has limitations in terms of removing smaller solids, organic matter, or dissolved contaminants. These small particles may pass through the screen, contributing to the pollutant load in subsequent treatment stages. Therefore, screening is generally followed by additional treatment processes such as sedimentation or filtration to achieve more thorough removal.

Sedimentation and Grit Removal Mechanisms

Sedimentation, or settling, is the next primary treatment process and is used to remove suspended solids that are heavier than water. During sedimentation, wastewater is allowed to sit in a large tank, causing solid particles to settle to the bottom due to gravity. This process is particularly effective for removing particulate matter such as sand, silt, and larger suspended solids [22]. The efficiency of sedimentation depends on factors like the size and density of the particles, the detention time (the time wastewater remains in the tank), and the design of the sedimentation tank [23]. To enhance sedimentation, wastewater treatment plants often incorporate mechanisms such as lamella settlers or inclined plate settlers, which increase the surface area for settling, in which these systems are particularly effective in removing fine suspended particles [24]. Settling tanks are typically designed with a hydraulic profile that allows for the slow flow of water, promoting the downward movement of particles and their accumulation at the bottom of the tank for easy removal [25]. Grit removal is a specific form of sedimentation designed to remove heavier particles such as sand, gravel, and other inorganic materials that do not decompose, in which these particles, commonly known as grit, can cause significant wear and tear on equipment, clog pipes, and reduce the efficiency of subsequent treatment stages [26]. Grit removal is achieved through a combination of sedimentation and mechanical agitation. Grit chambers are typically designed with a slow flow velocity, allowing heavier particles to settle at the bottom while the lighter organic matter remains in suspension [27]. The grit removal process is highly important in wastewater treatment, as the presence of grit can significantly hinder the performance of biological treatment systems. In addition to physical removal, grit management involves regular cleaning of grit chambers to ensure efficient removal and disposal. Modern grit removal systems incorporate advanced technologies such as vortex-based systems and aerated grit chambers, which improve efficiency by utilizing centrifugal forces and aeration to separate grit particles from the wastewater [28]. While sedimentation and grit removal effectively reduce the volume of suspended solids in the wastewater, these processes are not capable of removing dissolved contaminants or organic matter. Therefore, further treatment steps are required to achieve comprehensive pollutant removal.

Limitations and Efficiency of Primary Treatment

Primary treatment provides significant initial reductions in the suspended solids and organic load in wastewater. However, this stage has several limitations that affect its overall efficiency and scope. One of the main limitations is that primary treatment is primarily focused on removing larger suspended solids, and it does not address the removal of dissolved pollutants, nutrients (such as nitrogen and phosphorus), or pathogens [26]. Another limitation is the relatively low efficiency in removing fine particles, colloids, and soluble pollutants that are present in wastewater. These smaller particles, including microplastics, pharmaceuticals, and endocrine-disrupting compounds, are not effectively captured during the sedimentation process. Furthermore, the effectiveness of grit removal can be affected by factors such as particle size distribution and wastewater flow conditions [29]. One of the main concerns with primary treatment is the generation of sludge. The removal of suspended solids results in the accumulation of primary sludge, which requires further treatment and disposal. This sludge often contains organic matter, nutrients, and pathogens, which need to be addressed in subsequent treatment stages to prevent environmental contamination and health risks. The management of sludge can be a challenging and costly aspect of primary treatment [30]. However, there is still a need for innovations in primary treatment that can address the growing concerns associated with emerging contaminants and the complexities of modern wastewater composition.

SECONDARY TREATMENT: BIOLOGICAL METHODS

Secondary treatment processes, particularly biological methods, are essential for further purifying wastewater by targeting dissolved and suspended organic matter, nutrients, and other pollutants, and the overall process involved is represented in Fig. (**3**). Biological treatment relies on microorganisms to break down contaminants through metabolic processes, and these methods are generally more cost-effective and environmentally sustainable compared to chemical treatment techniques [31].

Principles of Biological Wastewater Treatment

Biological wastewater treatment relies on the use of microorganisms, particularly bacteria, fungi, and algae, to degrade organic matter, nutrients, and other contaminants present in wastewater. The primary principle of biological treatment is biodegradation, where microorganisms metabolize pollutants for growth and energy, and these processes can occur aerobically, anaerobically, or under anoxic conditions, depending on the nature of the microorganisms and the available oxygen [32]. In aerobic biological treatment, oxygen is supplied to the system, allowing aerobic bacteria to decompose organic matter into carbon dioxide and

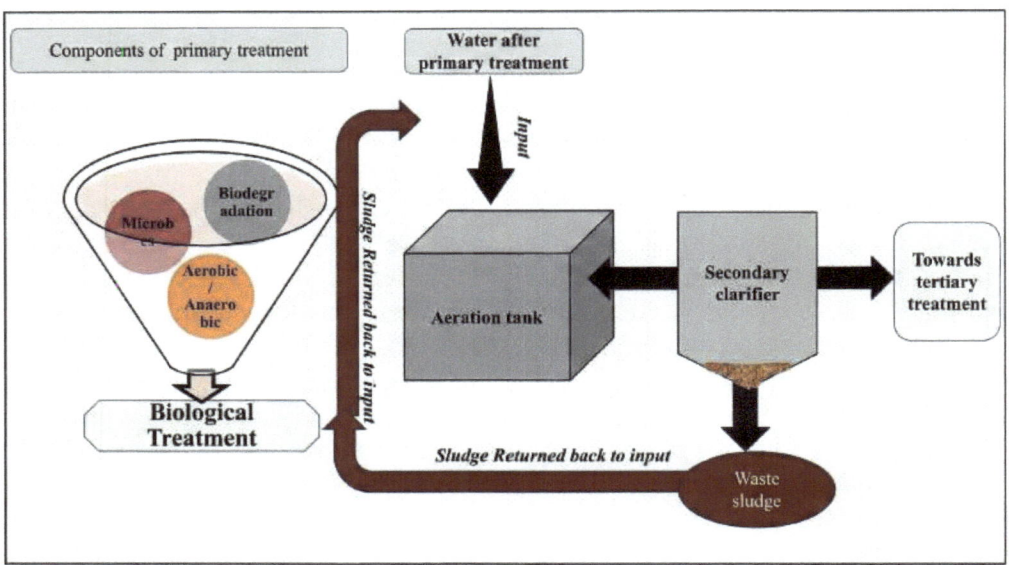

Fig. (3). Overall process involved in the secondary stage of wastewater treatment.

water, in which these bacteria utilize the organic matter as a food source, breaking it down through oxidative reactions [33]. Anaerobic conditions, on the other hand, promote the activity of anaerobic bacteria, which decompose organic matter without oxygen, producing methane and carbon dioxide as byproducts. The anoxic conditions, often found in denitrification processes, support denitrifying bacteria that convert nitrate to nitrogen gas, reducing nitrogen pollution in wastewater [34]. The efficiency of biological wastewater treatment systems is governed by several factors, including the concentration of microorganisms, the temperature, pH, and nutrient availability [33]. Optimal conditions must be maintained to ensure that microbial processes can effectively break down contaminants. Biological treatment systems are often employed in stages, with secondary treatment following primary treatment to remove dissolved organic material and further reduce the pollutant load. This reliance on biological activity makes these systems highly effective for treating domestic and industrial effluents, reducing the environmental impact of wastewater discharge [35].

Activated Sludge Systems: Design and Operation

Activated sludge is one of the most widely used biological treatment methods, particularly for treating municipal and industrial wastewater. This process involves the aeration of wastewater to promote the growth of microorganisms, which break down organic pollutants in the presence of oxygen [36]. The activated sludge system consists of a bioreactor, typically an aeration tank, where microorganisms (activated sludge) are mixed with the wastewater to promote microbial growth and biodegradation. The mixture is then allowed to settle in a secondary clarifier, where the activated sludge is separated from the treated effluent [37]. The design of an activated sludge system involves careful consideration of several factors. The Hydraulic Retention Time (HRT), or the time the wastewater spends in the system, is critical for ensuring adequate contact between microorganisms and contaminants [38]. The Sludge Retention Time (SRT), or the time the microorganisms remain in the system, is also essential for maintaining an active population of microbes capable of degrading organic matter [39]. These parameters, along with aeration requirements, influence the overall performance of the system in terms of contaminant removal efficiency. Activated sludge systems are designed to operate under varying conditions, depending on the specific wastewater characteristics. The use of fine bubbles in aeration tanks improves oxygen transfer efficiency, allowing for optimal microbial activity [40]. A key operational challenge in activated sludge systems is maintaining a balance between the microorganisms' growth rate and the system's ability to remove organic matter. This balance is influenced by factors such as nutrient availability (nitrogen, phosphorus), temperature, and dissolved oxygen concentrations. Inadequate aeration or overloading of the system can lead to sludge bulking, where excess biomass accumulates, reducing the treatment efficiency [41]. To improve the performance of activated sludge systems, variations of the basic process, such as extended aeration and contact stabilization, have been developed. These systems allow for more extended contact times and are particularly useful for treating wastewater with high organic content [36]. Additionally, advancements in process control technologies, such as real-time monitoring of parameters like dissolved oxygen and sludge concentration, have further enhanced the efficiency and adaptability of activated sludge systems [42].

Trickling Filters and Sequencing Batch Reactors

Trickling filters and Sequencing Batch Reactors (SBRs) are two alternative biological treatment methods that provide distinct advantages and are suited to different types of wastewaters. Trickling filters are fixed-film biological reactors in which wastewater flows over a bed of microbial-supporting media, typically made of plastic or stone. The media provide a surface for microorganisms to grow and form a biofilm that decomposes organic matter as the wastewater passes through [43]. As the wastewater flows over the media, microorganisms attached to the surface consume organic pollutants, and the treated effluent flows out of the system. Trickling filters can operate in both aerobic and anoxic conditions, depending on the design and aeration configuration, which are often used as a secondary treatment method following primary sedimentation [44]. Trickling filters are relatively simple to operate and maintain, with lower energy requirements than activated sludge systems, making them more cost-effective for smaller municipalities and rural areas [45]. However, their performance can be influenced by factors such as media clogging, biofilm sloughing, and variations in wastewater quality. To overcome these limitations, modern trickling filters are often combined with other treatment technologies, such as aeration tanks or biological nutrient removal processes, to enhance overall treatment efficiency [46]. Sequencing Batch Reactors (SBRs) are a type of batch reactor that operate in a series of distinct phases. Unlike continuous flow systems like activated sludge or trickling filters, SBRs treat wastewater in discrete batches. Each batch undergoes a sequence of processes, including fill, react, settle, and decant phases [47]. During the "react" phase, aeration occurs to promote the growth of microorganisms and the degradation of organic matter, and the wastewater is then allowed to settle, separating the treated effluent from the biomass. The decant phase removes the clarified effluent, and the cycle begins again. SBRs offer several advantages, including flexibility, ease of operation, and the ability to treat varying volumes of wastewater. They are particularly well-suited for applications with fluctuating inflows or small communities with lower wastewater volumes [48].

Biological Nutrient Removal (BNR)

Biological Nutrient Removal (BNR) is a critical component of secondary treatment processes aimed at reducing the levels of nitrogen and phosphorus, which are the two key nutrients that contribute to eutrophication and water quality degradation in receiving waters [49]. Nitrification and denitrification are two essential processes for nitrogen removal, while phosphorus removal often requires specialized techniques. Both nitrification-denitrification and enhanced phosphorus removal techniques are critical for meeting stringent water quality standards, especially in areas where nutrient pollution is a concern.

Nitrification and Denitrification: Nitrification is an aerobic process in which ammonia (NH_3) is oxidized to nitrite (NO_2) and then to nitrate (NO_3) by nitrifying bacteria, primarily Nitrosomonas and Nitrobacter. This process requires the presence of oxygen, and the efficiency of nitrification depends on factors such as pH, temperature, and dissolved oxygen levels [50]. Nitrification is typically followed by denitrification, a process that occurs under anoxic conditions, where denitrifying bacteria such as *Pseudomonas* reduce nitrate to nitrogen gas (N_2), which is then released into the atmosphere. This process requires an organic carbon source, which is typically provided by the biodegradable organic material in the wastewater [51].

Enhanced Phosphorus Removal: Phosphorus removal can be achieved through biological processes that promote the uptake of phosphorus by microorganisms. Enhanced Biological Phosphorus Removal (EBPR) is commonly used in BNR systems, where certain bacteria, particularly *Accumulibacter spp.*, accumulate phosphorus under anaerobic conditions and then release it under aerobic conditions. This cycle, known as the anaerobic-aerobic cycle, facilitates the removal of phosphorus as the bacteria take up excessive amounts of phosphate and store it in intracellular polyphosphate granules [52]. To optimize phosphorus removal, this process is typically combined with the nitrification-denitrification cycle, allowing for the simultaneous removal of nitrogen and phosphorus [53].

TERTIARY TREATMENT TECHNOLOGIES

Tertiary treatment technologies are advanced treatment methods used to further purify wastewater following secondary treatment processes. The purpose of tertiary treatment is to remove residual contaminants that remain after biological treatment, including microorganisms, nutrients, fine suspended solids, and chemicals that cannot be removed efficiently by previous treatment stages [5]. These technologies are crucial for ensuring the highest possible water quality before discharge into the environment or for reuse. Tertiary treatment methods are diverse and include advanced filtration techniques, disinfection methods, and chemical treatments, each designed to address specific contaminants and achieve stringent water quality standards.

Advanced Filtration Techniques

Advanced filtration techniques are often employed in tertiary treatment to remove fine particles, suspended solids, and microorganisms that are not effectively captured in primary or secondary treatments. These filtration methods include multimedia filtration, microfiltration, ultrafiltration, and reverse osmosis, each offering different levels of filtration precision depending on the required water quality [54]. Multimedia filtration is one of the most common methods used in tertiary treatment, utilizing layers of different filter media such as sand, gravel, and anthracite coal. These materials have varying pore sizes, allowing for the gradual removal of larger particles as the water passes through each layer. This method is effective for removing suspended solids and some microorganisms, but may not be sufficient for removing dissolved contaminants or very fine particles [55]. Microfiltration (MF) and ultrafiltration (UF) are membrane-based filtration technologies that use semi-permeable membranes to separate contaminants from water. Microfiltration membranes have pore sizes typically ranging from 0.1 to 10 microns, while ultrafiltration membranes are capable of filtering particles as small as 0.01 microns [56]. Both MF and UF are highly effective at removing suspended solids, bacteria, and larger viruses, though UF can also remove colloidal matter and some larger organic molecules [57]. The most advanced filtration method in tertiary treatment is Reverse Osmosis (RO), which utilizes a semi-permeable membrane to remove dissolved salts, heavy metals, and organic contaminants [58]. RO has the ability to produce high-quality effluent by rejecting a wide range of dissolved pollutants, including salts, pharmaceuticals, and pesticides. This method requires a significant amount of energy for membrane operation and is generally used when extremely high-purity water is required, such as for industrial reuse or potable water production [59]. While advanced filtration technologies are effective in removing suspended solids, microorganisms, and certain dissolved pollutants, they often require frequent maintenance, including membrane cleaning and replacement.

Disinfection Methods

Disinfection is a crucial component of tertiary treatment, designed to eliminate pathogenic microorganisms from wastewater before it is discharged into receiving bodies of water or reused. Various disinfection technologies are used, including ultraviolet (UV) irradiation and ozonation, each offering distinct advantages in terms of microbial inactivation, efficiency, and byproduct formation [60].

Ultraviolet (UV) Irradiation

Ultraviolet (UV) irradiation is a widely used disinfection method in tertiary wastewater treatment due to its effectiveness in inactivating a broad range of microorganisms, including bacteria, viruses, and protozoa. UV radiation works by damaging the DNA or RNA of microorganisms, preventing their replication and rendering them incapable of causing infections. UV disinfection systems typically consist of UV lamps or LEDs that emit light in the UV-C spectrum, which is most effective at germicidal action [61]. UV irradiation is particularly advantageous because it does not require the addition of chemicals, thus avoiding the formation of harmful disinfection byproducts, which is a common concern with chlorine-based disinfection methods. Furthermore, UV treatment is rapid and does not alter the taste or odor of water, making it ideal for treating water that is intended for potable use or industrial applications. However, UV disinfection is highly dependent on factors such as the intensity of UV light, the duration of exposure, and the turbidity of the water. High levels of suspended solids or dissolved organic matter can reduce the effectiveness of UV radiation by absorbing or scattering the light, which may necessitate additional filtration steps before UV treatment [62]. Despite its advantages, UV disinfection does not provide a residual disinfecting effect, meaning that treated water could potentially become re-contaminated after treatment if not properly managed. In addition, UV systems require regular maintenance, such as cleaning the lamps and ensuring optimal lamp performance, to maintain their efficiency [63].

Ozonation Processes

Ozonation is another powerful disinfection method used in tertiary treatment, where ozone (O_3), a strong oxidizing agent, is applied to wastewater to kill microorganisms and degrade organic contaminants. Ozone is generated by passing oxygen through an electrical discharge, producing O_3, which is then injected into the wastewater [64]. Ozone is highly effective at inactivating bacteria, viruses, and protozoa by oxidizing their cellular structures and disrupting their metabolic functions. Additionally, ozone can break down a variety of organic pollutants, such as pesticides, pharmaceuticals, and industrial chemicals, making it a versatile treatment option [65]. The major advantage of ozonation over other disinfection methods is its strong oxidative power, which allows it to degrade complex organic compounds and remove color and odor from the water. Ozone also decomposes into oxygen, leaving no residual disinfectant byproducts in the treated water [66]. However, ozonation requires specialized equipment to generate ozone, and the process is energy-intensive, limiting its widespread application in smaller-scale facilities. Ozonation has been shown to be particularly effective when combined with other treatment processes, such as coagulation or advanced oxidation, to enhance pollutant removal [67]. The cost and energy requirements of ozonation are often a concern, particularly for large-scale applications, but advances in ozone generation technology and the development of hybrid treatment systems are helping to make ozonation more efficient and economically viable [68].

Chemical Treatments

Chemical treatments, including coagulation and Advanced Oxidation Processes (AOPs), are essential tertiary treatment technologies used to address specific contaminants that may not be efficiently removed by biological or physical processes [69]. These techniques often target dissolved organic matter, heavy metals, and other pollutants that pose challenges to traditional treatment methods [14]. Coagulation is a chemical process used to destabilize suspended particles and colloids in wastewater, facilitating their aggregation and subsequent removal through sedimentation or filtration [70]. In coagulation, chemicals called coagulants, such as alum (aluminum sulfate), ferric chloride, or polyaluminum chloride, are added to the wastewater to neutralize the charge of suspended particles, causing them to clump together into larger aggregates called flocs. These flocs can then be removed through sedimentation or filtration [71]. Coagulation is often used in conjunction with other treatment processes, such as flocculation, where the flocs are gently mixed to promote further aggregation, followed by filtration or settling [72]. Advanced Oxidation Processes (AOPs) are a group of highly efficient chemical treatments that utilize powerful oxidants, such as hydroxyl radicals (•OH), to degrade organic pollutants and microorganisms. AOPs can be achieved through various methods, including ozonation, Fenton's reagent (a mixture of hydrogen peroxide and iron salts), and ultraviolet (UV) radiation combined with hydrogen peroxide (UV/H_2O_2). These processes generate hydroxyl radicals, which are highly reactive and capable of breaking down complex organic molecules into simpler, non-toxic compounds [73, 74]. AOPs are particularly effective in treating wastewater containing persistent organic pollutants, pharmaceuticals, and other contaminants that are resistant to biological degradation. While AOPs are highly effective in pollutant removal, their main limitation is the high energy and chemical costs associated with the generation of hydroxyl radicals [73]. These processes are generally employed for the treatment of industrial effluents or in situations where conventional treatment methods cannot achieve the desired water quality standards [14]. However, ongoing research into improving the efficiency of AOPs, such as the development of novel catalysts and hybrid systems, holds promise for reducing costs and expanding their applicability in wastewater treatment.

SLUDGE MANAGEMENT AND RESOURCE UTILIZATION

Sludge is a byproduct of wastewater treatment and consists of organic and inorganic materials, microorganisms, and water. Sludge management is a critical component of wastewater treatment systems, ensuring the safe handling, treatment, and disposal of the semi-solid residual material generated during treatment processes. It aims to minimize environmental risks, mitigate health hazards, and ensure compliance with regulatory frameworks [75, 76]. The primary methods employed include thickening, dewatering, stabilization, and drying, which collectively reduce the volume of sludge and improve its physical and chemical characteristics for safe disposal or further use. Stabilization techniques such as anaerobic digestion and composting play a significant role in reducing pathogens, controlling odours, and converting sludge into a more stable and nutrient-rich material [77]. However, sludge management faces challenges such as high moisture content, the presence of hazardous contaminants like heavy metals and persistent organic pollutants, and the rising volume of sludge due to expanding urbanization and industrial activities [75]. These challenges underscore the need for innovative, cost-effective, and sustainable strategies to manage sludge efficiently.

At the same time, sludge is increasingly being recognized as a valuable resource, offering significant potential for resource recovery and sustainable utilization. Rich in organic matter, energy, and essential nutrients like nitrogen and phosphorus, sludge can serve as a feedstock for renewable energy production, agricultural applications, and material recovery [78]. Advanced technologies such as anaerobic digestion not only stabilize sludge but also produce biogas, a renewable energy source that can be used for heating, electricity generation, or as a vehicle fuel [79]. Thermal processes like pyrolysis and gasification further expand the scope of sludge utilization, converting it into biochar, syngas, or bio-oil for use in soil improvement, carbon sequestration, or clean energy applications [80]. However, to unlock the full potential of sludge, treatment processes must effectively neutralize harmful contaminants while ensuring economic feasibility, scalability, and alignment with environmental and safety regulations. This holistic approach transforms sludge management from a waste disposal challenge into an opportunity for advancing sustainability and circular economy objectives.

ADVANCED AND EMERGING WASTEWATER TREATMENT TECHNOLOGIES

The global demand for clean water is escalating due to population growth, urbanization, and industrialization. Traditional wastewater treatment methods, while effective, often fall short in addressing the increasingly complex and diverse nature of wastewater contaminants [69]. This has led to the development and integration of advanced and emerging wastewater treatment technologies, which offer enhanced efficiency, resource recovery, and environmental sustainability. These technologies aim to meet stricter regulatory standards, improve treatment outcomes, and contribute to the circular economy by recovering valuable resources from wastewater.

Membrane Bioreactors

Membrane bioreactors (MBRs) have emerged as a revolutionary technology in wastewater treatment, combining biological treatment with membrane filtration. MBRs utilize semi-permeable membranes to separate treated water from suspended solids, microorganisms, and other contaminants, enabling MBRs to achieve superior effluent quality, often meeting or exceeding stringent regulatory standards for reuse applications in agriculture, industry, and municipal supply. One of the primary advantages of MBRs is their ability to operate at higher biomass concentrations compared to conventional activated sludge systems. This results in a smaller footprint, making MBRs particularly suitable for urban areas where space is limited [81]. Additionally, MBRs produce high-quality effluent that is low in turbidity, pathogens, and total suspended solids, making them ideal for water reclamation and reuse, and the process also eliminates the need for secondary clarifiers, reducing the complexity of plant design [76]. Despite their advantages, MBRs face significant challenges. Membrane fouling, caused by the accumulation of particles, organic matter, and biofilms on the membrane surface, remains a critical issue. Fouling reduces membrane permeability, increases operational costs, and necessitates frequent cleaning and maintenance [82]. Advanced cleaning techniques, such as backwashing and chemical cleaning, have been developed to mitigate fouling, but they can compromise membrane longevity [83]. Another challenge is the high energy consumption associated with MBR operations, particularly for aeration and membrane filtration. Research is ongoing to develop energy-efficient membranes and hybrid systems that integrate MBRs with other technologies to improve sustainability and cost-effectiveness [84].

Electrochemical Treatment Processes

Electrochemical treatment processes are gaining attention for their ability to treat a wide range of contaminants in wastewater, including persistent organic pollutants, heavy metals, and pathogens. These processes utilize electrical energy to drive chemical reactions that degrade or transform pollutants into less harmful forms [85]. Key electrochemical methods include electrocoagulation, electrooxidation, and electroflotation. Electrocoagulation involves the generation of coagulants *in situ* by dissolving sacrificial metal electrodes, such as aluminum or iron, into the wastewater. The coagulants destabilize and aggregate suspended particles and dissolved contaminants, facilitating their removal through sedimentation or flotation [86]. Electrocoagulation is particularly effective in removing heavy metals, oil and grease, and color from industrial wastewater. It offers advantages such as minimal chemical usage, compact equipment, and ease of automation [87]. However, it requires regular electrode replacement and generates sludge that must be managed appropriately. Electrooxidation employs an electric current to generate reactive species, such as hydroxyl radicals, at the electrode surface. These radicals exhibit strong oxidizing properties, enabling the degradation of recalcitrant organic compounds, including pharmaceuticals and pesticides [88]. Advanced electrode materials, such as boron-doped diamond and mixed metal oxides, have improved the efficiency and selectivity of electrooxidation processes [89]. However, challenges remain in scaling up the technology and addressing energy consumption. Electro flotation uses electrolytically generated gas bubbles to lift contaminants to the water surface, where they can be skimmed off, and this process is particularly effective in removing oil, grease, and suspended solids. Electro flotation is often combined with other electrochemical methods to enhance overall treatment efficiency. Research is focused on optimizing electrode design and operating conditions to maximize performance while minimizing energy use [90].

CONCLUSION

Wastewater treatment technologies play a crucial role in ensuring the sustainability and health of both environmental and human systems. From traditional methods like activated sludge processes and chemical treatments to advanced technologies such as membrane filtration, constructed wetlands, and bioreactors, each treatment method offers specific advantages and limitations. The choice of technology depends on various factors, including the type of contaminants, environmental regulations, and available resources. Modern innovations, such as resource recovery from wastewater, highlight the potential of turning waste into valuable resources, like energy, nutrients, and water. This has become increasingly significant in addressing global challenges like water scarcity, nutrient pollution, and the energy demands of wastewater treatment systems. Looking forward, there is a growing need for more efficient, cost-effective, and sustainable wastewater treatment solutions that can be integrated into decentralized and large-scale applications. Emerging technologies such as electrochemical treatment, advanced oxidation processes, and nanotechnology-based approaches are showing promising results in enhancing treatment efficiency, reducing energy consumption, and improving the removal of complex pollutants. The continuous development and adoption of these technologies, combined with an emphasis on resource recovery, are key to addressing future water quality challenges and achieving global sustainability goals. Therefore, wastewater treatment must evolve into a holistic approach, emphasizing not only pollution control but also resource recovery, energy efficiency, and minimal environmental impact.

REFERENCES

[1] Sonone SS, Jadhav S, Sankhla MS, Kumar R. Water contamination by heavy metals and their toxic effect on aquaculture and human health through food Chain. Lett. Appl. NanoBioScience 2020; 10(2): 2148-66.

[2] Singh V. Water Pollution. In: Vir Singh, ed Textbook of Environment and Ecology 2024 Mar 23. 253-66.
[http://dx.doi.org/10.1007/978-981-99-8846-4_17]

[3] Lofrano G, Brown J. Wastewater management through the ages: A history of mankind. Sci Total Environ 2010; 408(22): 5254-64.
[http://dx.doi.org/10.1016/j.scitotenv.2010.07.062] [PMID: 20817263]

[4] Kuchangi SN, Mruthunjayappa MH, Sanna Kotrappanavar N. An overview of water pollutants in present scenario. In: Pandey JK, Manna S, Patel RK, eds. 3D Printing Technology for Water Treatment Applications. Singapore: Springer; 2023. p. 83-105.

[5] Fernandes J, Ramísio PJ, Puga H. A Comprehensive Review on Various Phases of Wastewater Technologies: Trends and Future Perspectives. Eng 2024; 5(4): 2633-61.
[http://dx.doi.org/10.3390/eng5040138]

[6] Sathya K, Nagarajan K, Carlin Geor Malar G, Rajalakshmi S, Raja Lakshmi P. A comprehensive review on comparison among effluent treatment methods and modern methods of treatment of

[7] Singh NK, Sanghvi G, Yadav M, Padhiyar H, Christian J, Singh V. Fate of pesticides in agricultural runoff treatment systems: Occurrence, impacts and technological progress. Environ Res 2023; 237(Pt 2): 117100.
[http://dx.doi.org/10.1016/j.envres.2023.117100] [PMID: 37689336]

[8] Goonetilleke A, Lampard JL. Stormwater quality, pollutant sources, processes, and treatment options. InApproaches to water sensitive urban design 2019 Jan 1 49-74. Woodhead Publishing
[http://dx.doi.org/10.1016/B978-0-12-812843-5.00003-4]

[9] Kokkinos P, Mantzavinos D, Venieri D. Current trends in the application of nanomaterials for the removal of emerging micropollutants and pathogens from water. Molecules 2020; 25(9): 2016.
[http://dx.doi.org/10.3390/molecules25092016] [PMID: 32357416]

[10] Kumar SS, Kumar V, Malyan SK, et al. Microbial fuel cells (MFCs) for bioelectrochemical treatment of different wastewater streams. Fuel 2019; 254: 115526.
[http://dx.doi.org/10.1016/j.fuel.2019.05.109]

[11] Alprol AE, Mansour AT, Ibrahim MEED, Ashour M. Artificial intelligence technologies revolutionizing wastewater treatment: Current trends and future prospective. Water 2024; 16(2): 314.
[http://dx.doi.org/10.3390/w16020314]

[12] Tariq A, Mushtaq A. Untreated wastewater reasons and causes: A review of most affected areas and cities. Int J Chem Biochem Sci 2023; 23(1): 121-43.

[13] Diaz-Elsayed N, Rezaei N, Guo T, Mohebbi S, Zhang Q. Wastewater-based resource recovery technologies across scale: A review. Resour Conserv Recycling 2019; 145: 94-112.
[http://dx.doi.org/10.1016/j.resconrec.2018.12.035]

[14] Saravanan A, Senthil Kumar P, Jeevanantham S, et al. Effective water/wastewater treatment methodologies for toxic pollutants removal: Processes and applications towards sustainable development. Chemosphere 2021; 280: 130595.
[http://dx.doi.org/10.1016/j.chemosphere.2021.130595] [PMID: 33940449]

[15] Purohit A. A review on treatment processes for municipal waste water. Int J Adv Res Eng Sci Technol. 2017; 4(2): 184-90.

[16] Krzeminski P, Tomei MC, Karaolia P, et al. Performance of secondary wastewater treatment methods for the removal of contaminants of emerging concern implicated in crop uptake and antibiotic resistance spread: A review. Sci Total Environ 2019; 648: 1052-81.
[http://dx.doi.org/10.1016/j.scitotenv.2018.08.130] [PMID: 30340253]

[17] Rathi BS, Kumar PS. Application of adsorption process for effective removal of emerging contaminants from water and wastewater. Environ Pollut 2021; 280: 116995.
[http://dx.doi.org/10.1016/j.envpol.2021.116995] [PMID: 33789220]

[18] Ramalho RS. Introduction to wastewater treatment processes. 2nd ed. New York: Academic Press; 2012.

[19] Ding S, Li X, Qiao X, Liu Y, Wang H, Ma C. Identification and screening of priority pollutants in printing and dyeing industry wastewater and the importance of these pollutants in environmental management in China. Environ Pollut 2024; 362: 124938.
[http://dx.doi.org/10.1016/j.envpol.2024.124938] [PMID: 39265766]

[20] Rezai B, Allahkarami E. Wastewater treatment processes—techniques, technologies, challenges faced, and alternative solutions. In: Karri RR, Ravindran G, Dehghani MH, eds. Soft computing techniques in solid waste and wastewater management. Amsterdam: Elsevier; 2021. p. 35-53.

[21] Riffat R, Husnain T. Fundamentals of Wastewater Treatment and Engineering. 2nd ed. London: CRC Press; 2022.
[http://dx.doi.org/10.1201/9781003134374]

[22] Kosar S, Isik O, Cicekalan B, *et al.* Impact of primary sedimentation on granulation and treatment performance of municipal wastewater by aerobic granular sludge process. J Environ Manage 2022; 315: 115191.
[http://dx.doi.org/10.1016/j.jenvman.2022.115191] [PMID: 35526399]

[23] Lasaki BA, Maurer P, Schönberger H. Uncovering the reasons behind high-performing primary sedimentation tanks for municipal wastewater treatment: An in-depth analysis of key factors. J Environ Chem Eng 2024; 12(2): 112460.
[http://dx.doi.org/10.1016/j.jece.2024.112460]

[24] Reyes C, Apaz F, Niño Y, Barraza B, Arratia C, Ihle CF. A review on steeply inclined settlers for water clarification. Miner Eng 2022; 184: 107639.
[http://dx.doi.org/10.1016/j.mineng.2022.107639]

[25] Gao H, Stenstrom MK. Development and applications in computational fluid dynamics modeling for secondary settling tanks over the last three decades: A review. Water Environ Res 2020; 92(6): 796-820.
[http://dx.doi.org/10.1002/wer.1279] [PMID: 31782964]

[26] Kolosovska T. A comprehensive evaluation of wastewater grit characteristics and its application in the design of a facility-specific grit removal solution. Rowan University. 2021.

[27] Plana Puig Q. Characterization and modelling of grit chambers based on particle settling velocity distributions [Phd dissertation]. Québec (QC): Université Laval; 2011.

[28] Mekuria MD. Operational performance of Bahir Dar Textile Wastewater Treatment plant, Ethiopia [master's thesis]. Bahir Dar: Bahir Dar University; 2018.

[29] Obi LU, Olisaka FN, Onyia FC, Innocent IH, Onyemaechi P. Physical and biological removal of the mass load of emergent pollutants from waste treatment facilities. In: Emergent Pollutants in Freshwater Plankton Communities CRC Press 121-47. 2024 Oct 11
[http://dx.doi.org/10.1201/9781003362975-10]

[30] Rorat A, Courtois P, Vandenbulcke F, Lemiere S. Sanitary and environmental aspects of sewage sludge management. In: Industrial and Municipal Sludge. Butterworth-Heinemann 2019 Jan 1.
[http://dx.doi.org/10.1016/B978-0-12-815907-1.00008-8]

[31] Nsenga Kumwimba M, Meng F. Roles of ammonia-oxidizing bacteria in improving metabolism and cometabolism of trace organic chemicals in biological wastewater treatment processes: A review. Sci Total Environ 2019; 659: 419-41.
[http://dx.doi.org/10.1016/j.scitotenv.2018.12.236] [PMID: 31096373]

[32] Doukani K, Boukirat D, Boumezrag A, Bouhenni H, Bounouira Y. Fundamentals of biodegradation process. InHandbook of biodegradable materials. Cham: Springer International Publishing 2022 Sep 17; pp. 1-27.
[http://dx.doi.org/10.1007/978-3-030-83783-9_73-1]

[33] Aziz A, Basheer F, Sengar A, Irfanullah , Khan SU, Farooqi IH. Biological wastewater treatment (anaerobic-aerobic) technologies for safe discharge of treated slaughterhouse and meat processing wastewater. Sci Total Environ 2019; 686: 681-708.
[http://dx.doi.org/10.1016/j.scitotenv.2019.05.295] [PMID: 31195278]

[34] Lohani SP, Havukainen J. Anaerobic digestion: factors affecting anaerobic digestion process. In: Chandel AK, Silva BAM, editors. Waste Bioremediation. Boca Raton (FL): CRC Press; 2017. p. 343-59.

[35] Ahmed SF, Mofijur M, Nuzhat S, *et al.* Recent developments in physical, biological, chemical, and hybrid treatment techniques for removing emerging contaminants from wastewater. J Hazard Mater 2021; 416: 125912.
[http://dx.doi.org/10.1016/j.jhazmat.2021.125912] [PMID: 34492846]

[36] Skouteris G, Rodriguez-Garcia G, Reinecke SF, Hampel U. The use of pure oxygen for aeration in

aerobic wastewater treatment: A review of its potential and limitations. Bioresour Technol 2020; 312: 123595
[http://dx.doi.org/10.1016/j.biortech.2020.123595] [PMID: 32506043]

[37] Ouyang J, Li C, Wei L, *et al*. Activated sludge and other aerobic suspended culture processes. Water Environ Res 2020; 92(10): 1717-25.
[http://dx.doi.org/10.1002/wer.1427] [PMID: 32762078]

[38] Ejhed H, Fång J, Hansen K, *et al*. The effect of hydraulic retention time in onsite wastewater treatment and removal of pharmaceuticals, hormones and phenolic utility substances. Sci Total Environ 2018; 618: 250-61.
[http://dx.doi.org/10.1016/j.scitotenv.2017.11.011] [PMID: 29128774]

[39] Zhang L, Guo K, Wang L, Xu R, Lu D, Zhou Y. Effect of sludge retention time on microbial succession and assembly in thermal hydrolysis pretreated sludge digesters: Deterministic *versus* stochastic processes. Water Res 2022; 209: 117900.
[http://dx.doi.org/10.1016/j.watres.2021.117900] [PMID: 34902758]

[40] Drewnowski J, Remiszewska-Skwarek A, Duda S, Łagód G. Aeration process in bioreactors as the main energy consumer in a wastewater treatment plant. Review of solutions and methods of process optimization. Processes (Basel) 2019; 7(5): 311.
[http://dx.doi.org/10.3390/pr7050311]

[41] Waqas S, Harun NY, Sambudi NS, *et al*. Effect of operating parameters on the performance of integrated fixed-film activated sludge for wastewater treatment. Singapore: World Scientific; 2020.
[http://dx.doi.org/10.3390/membranes13080704] [PMID: 37623765]

[42] Zhang W, Tooker NB, Mueller AV. Enabling wastewater treatment process automation: leveraging innovations in real-time sensing, data analysis, and online controls. Environ Sci Water Res Technol 2020; 6(11): 2973-92.
[http://dx.doi.org/10.1039/D0EW00394H]

[43] Gray NF. Fixed-film reactors in wastewater treatment. Singapore: World Scientific 2020.
[http://dx.doi.org/10.1142/q0271]

[44] Bressani-Ribeiro T, Almeida PGS, Volcke EIP, Chernicharo CAL. Trickling filters following anaerobic sewage treatment: state of the art and perspectives. Environ Sci Water Res Technol 2018; 4(11): 1721-38.
[http://dx.doi.org/10.1039/C8EW00330K]

[45] Deng Y. Improvements to the performance of trickling filters by inclusion of alternative surface-active media (Doctoral dissertation, Loughborough University) 2018.

[46] Hussain A, Kumari R, Sachan SG, Sachan A. Biological wastewater treatment technology: Advancement and drawbacks. In: Microbial Ecology of Wastewater Treatment Plants. Amsterdam: Elsevier; 2021. p. 175-92.

[47] Askari SS, Giri BS, Basheer F, Izhar T, Ahmad SA, Mumtaz N. Enhancing sequencing batch reactors for efficient wastewater treatment across diverse applications: A comprehensive review. Environ Res 2024; 260: 119656.
[http://dx.doi.org/10.1016/j.envres.2024.119656] [PMID: 39034021]

[48] Lawal IM, Soja UB, Hussaini A, *et al*. Sequential batch reactors for aerobic and anaerobic dye removal: A mini-review. Case Stud Chem Environ Eng 2023; 8: 100547.
[http://dx.doi.org/10.1016/j.cscee.2023.100547]

[49] Parde D, Ghosh R, Rajpurohit P, Bhaduri S, Behera M. Nutrient retrieval techniques in wastewater treatment. In: Biological and Hybrid Wastewater Treatment Technology: Recent Developments in India 2024 Aug 3; 159-95.
[http://dx.doi.org/10.1007/978-3-031-63046-0_7]

[50] Abd Rahman WN, Hamdan R. Nitrification-denitrification Process in a Combined AVFF and UAHFF

System for Ammonia Removal. Recent Trends Civ Eng Built Environ 2021; 2(1): 438-46.

[51] Robles-Porchas GR, Gollas-Galván T, Martínez-Porchas M, Martínez-Cordova LR, Miranda-Baeza A, Vargas-Albores F. The nitrification process for nitrogen removal in biofloc system aquaculture. Rev Aquacult 2020; 12(4): 2228-49.
[http://dx.doi.org/10.1111/raq.12431]

[52] Zhang H, Zhang SS, Zhu L, Li YP, Chen L. Phosphorus recovery in the alternating aerobic/anaerobic biofilm system: Performance and mechanism. Sci Total Environ 2022; 810: 152297.
[http://dx.doi.org/10.1016/j.scitotenv.2021.152297] [PMID: 34896486]

[53] Shukla S, Rajta A, Setia H, Bhatia R. Simultaneous nitrification–denitrification by phosphate accumulating microorganisms. World J Microbiol Biotechnol 2020; 36(10): 151.
[http://dx.doi.org/10.1007/s11274-020-02926-y] [PMID: 32924078]

[54] Lember E, Kuusik A. Guidelines For The Selection Of Tertiary Wastewater Treatment Technology 2020. Available from: https://bestbalticproject.eu/wp-content/uploads/2020/11/BEST_report_TUT_27_11.pdf

[55] Cescon A, Jiang JQ. Filtration process and alternative filter media material in water treatment. Water 2020; 12(12): 3377.
[http://dx.doi.org/10.3390/w12123377]

[56] Urošević T, Trivunac K. Achievements in low-pressure membrane processes microfiltration (MF) and ultrafiltration (UF) for wastewater and water treatment. In: Drioli E, Giorno L, eds. Current trends and future developments on (bio-) membranes. Amsterdam: Elsevier; 2020. p. 67-107.

[57] Bardhan A, Akhtar A, Subbiah S. Microfiltration and ultrafiltration membrane technologies. In: Nayak SK, Dutta K, Gohil JM, eds. Advancement in Polymer-Based Membranes for Water Remediation. 2022. p. 3-42.
[http://dx.doi.org/10.1016/B978-0-323-88514-0.00001-2]

[58] Faroon MA, Al Saad ZA, Albadran FA, Ahmed LA. Review on technology-based on reverse osmosis. Anbar Journal of Engineering Sciences 2023; 14(1)
[http://dx.doi.org/10.37649/aengs.2023.139414.1047]

[59] Hailemariam RH, Woo YC, Damtie MM, Kim BC, Park KD, Choi JS. Reverse osmosis membrane fabrication and modification technologies and future trends: A review. Adv Colloid Interface Sci 2020; 276: 102100.
[http://dx.doi.org/10.1016/j.cis.2019.102100] [PMID: 31935555]

[60] Bodzek M, Konieczny K, Rajca M. Membranes in water and wastewater disinfection. Arch Environ Prot 2019; 45(1): 3-18.

[61] Li X, Cai M, Wang L, Niu F, Yang D, Zhang G. Evaluation survey of microbial disinfection methods in UV-LED water treatment systems. Sci Total Environ 2019; 659: 1415-27.
[http://dx.doi.org/10.1016/j.scitotenv.2018.12.344] [PMID: 31096352]

[62] González Y, Gómez G, Moeller-Chávez GE, Vidal G. UV Disinfection Systems for wastewater treatment: Emphasis on reactivation of microorganisms. Sustainability (Basel) 2023; 15(14): 11262.
[http://dx.doi.org/10.3390/su151411262]

[63] Kebbi Y, Muhammad AI, Sant'Ana AS, do Prado-Silva L, Liu D, Ding T. Recent advances on the application of UV-LED technology for microbial inactivation: Progress and mechanism. Compr Rev Food Sci Food Saf 2020; 19(6): 3501-27.
[http://dx.doi.org/10.1111/1541-4337.12645] [PMID: 33337035]

[64] Meher P, Deshmukh N, Mashalkar A, Kumar D. Ozone (O3) generation and its applications: A review. In AIP Conference Proceedings AIP Publishing. 2023; 2764.(1)

[65] Epelle EI, Macfarlane A, Cusack M, *et al.* Ozone application in different industries: A review of recent developments. Chem Eng J 2023; 454: 140188.
[http://dx.doi.org/10.1016/j.cej.2022.140188] [PMID: 36373160]

[66] Tripathi S, Hussain T. Water and wastewater treatment through ozone-based technologies. In: Development in wastewater treatment research and processes Elsevier 2022 Jan 1; 139-72.
[http://dx.doi.org/10.1016/B978-0-323-85583-9.00015-6]

[67] Rekhate CV, Srivastava JK. Recent advances in ozone-based advanced oxidation processes for treatment of wastewater- A review. Chem Eng J Adv 2020; 3: 100031.
[http://dx.doi.org/10.1016/j.ceja.2020.100031]

[68] Das PP, Dhara S, Samanta NS, Purkait MK. Advancements on ozonation process for wastewater treatment: A comprehensive review. Chem Eng Process 2024; 202: 109852.
[http://dx.doi.org/10.1016/j.cep.2024.109852]

[69] Rout PR, Zhang TC, Bhunia P, Surampalli RY. Treatment technologies for emerging contaminants in wastewater treatment plants: A review. Sci Total Environ 2021; 753: 141990.
[http://dx.doi.org/10.1016/j.scitotenv.2020.141990] [PMID: 32889321]

[70] Wang LK, Wang MH, Shammas NK, Hahn HH. Physicochemical treatment consisting of chemical coagulation, precipitation, sedimentation, and flotation. In: Integrated natural resources research 2021; 1265-397.
[http://dx.doi.org/10.1007/978-3-030-61002-9_6]

[71] Pillai SB, Thombre NV. Coagulation, flocculation, and precipitation in water and used water purification. In: Handbook of water and used water purification 2024 Mar 27; 3-27.
[http://dx.doi.org/10.1007/978-3-319-78000-9_63]

[72] Iwuozor KO. Prospects and challenges of using coagulation-flocculation method in the treatment of effluents. Adv J Chemistry-Section A 2019; 2(2): 105-27.
[http://dx.doi.org/10.29088/SAMI/AJCA.2019.2.105127]

[73] Pandis PK, Kalogirou C, Kanellou E, *et al.* Key points of advanced oxidation processes (AOPs) for wastewater, organic pollutants and pharmaceutical waste treatment: A mini review. ChemEngineering 2022; 6(1): 8.
[http://dx.doi.org/10.3390/chemengineering6010008]

[74] Jennifer ZA, Sonia AD, Francesca EB, Nadia FM, Suanny MR. Advanced oxidation processes by UV/H2O2 for the removal of anionic surfactants in a decentralized wastewater treatment plant in Ecuador. Water Sci Technol 2024; 90(8): 2340-51.
[http://dx.doi.org/10.2166/wst.2024.311]

[75] Kobe J, Shafiq MD, Alkarimiah R, Yaser AZ, Shukor H, Mohd Zaini Makhtar M. Overview of Sludge in Waste Treatment Plant. InMicrobial Fuel Cell (MFC) Applications for Sludge Valorization 2023 Jun 9; 1-22.
[http://dx.doi.org/10.1007/978-981-99-1083-0_1]

[76] Qrenawi LI, Rabah FKJ. Sludge management in water treatment plants: literature review. Int J Environ Waste Manag 2021; 27(1): 93-125.
[http://dx.doi.org/10.1504/IJEWM.2021.111909]

[77] Dehal A, Rathika K, Yadav BR, Kumar AR. The Wholistic Approach for Sewage Sludge Management. In: Management of Wastewater and Sludge 2023 May 12; 203-32. CRC Press
[http://dx.doi.org/10.1201/9781003202431-12]

[78] Raheem A, Sikarwar VS, He J, *et al.* Opportunities and challenges in sustainable treatment and resource reuse of sewage sludge: A review. Chem Eng J 2018; 337: 616-41.
[http://dx.doi.org/10.1016/j.cej.2017.12.149]

[79] Kabeyi MJB, Olanrewaju OA. Biogas production and applications in the sustainable energy transition. J Energy 2022; 2022(1): 1-43.
[http://dx.doi.org/10.1155/2022/8750221]

[80] Gururani P, Bhatnagar P, Bisht B, *et al.* Recent advances and viability in sustainable thermochemical conversion of sludge to bio-fuel production. Fuel 2022; 316: 123351.

[http://dx.doi.org/10.1016/j.fuel.2022.123351]

[81] Bis M, Montusiewicz A, Piotrowicz A, Łagód G. Modeling of wastewater treatment processes in membrane bioreactors compared to conventional activated sludge systems. Processes (Basel) 2019; 7(5): 285.
[http://dx.doi.org/10.3390/pr7050285]

[82] Du X, Shi Y, Jegatheesan V, Haq IU. A review on the mechanism, impacts and control methods of membrane fouling in MBR system. Membranes (Basel) 2020; 10(2): 24.
[http://dx.doi.org/10.3390/membranes10020024] [PMID: 32033001]

[83] Rajendran DS, Devi EG, Subikshaa VS, et al. Recent advances in various cleaning strategies to control membrane fouling: a comprehensive review. Clean Technol Environ Policy 2024; 27: 649-64.

[84] Osman AI, Chen Z, Elgarahy AM, et al. Membrane technology for energy saving: principles, techniques, applications, challenges, and prospects. Adv Energy Sustain Res 2024; 5(5): 2400011.
[http://dx.doi.org/10.1002/aesr.202400011]

[85] Fitch A, Balderas-Hernandez P, Ibanez JG. Electrochemical technologies combined with physical, biological, and chemical processes for the treatment of pollutants and wastes: A review. J Environ Chem Eng 2022; 10(3): 107810.
[http://dx.doi.org/10.1016/j.jece.2022.107810]

[86] Yasri N, Hu J, Kibria MG, Roberts EP. Electrocoagulation separation processes. In: Multidisciplinary Advances in Efficient Separation Processes. American Chemical Society. 2020; pp. 167-203.
[http://dx.doi.org/10.1021/bk-2020-1348.ch006]

[87] Shokri A, Fard MS. A critical review in electrocoagulation technology applied for oil removal in industrial wastewater. Chemosphere 2022; 288(Pt 2): 132355.
[http://dx.doi.org/10.1016/j.chemosphere.2021.132355] [PMID: 34582927]

[88] Nair G, Soni B, Shah M. A comprehensive review on electro-oxidation and its types for wastewater treatment. Groundw Sustain Dev 2023; 23: 100980.
[http://dx.doi.org/10.1016/j.gsd.2023.100980]

[89] He Y, Lin H, Guo Z, Zhang W, Li H, Huang W. Recent developments and advances in boron-doped diamond electrodes for electrochemical oxidation of organic pollutants. Separ Purif Tech 2019; 212: 802-21.
[http://dx.doi.org/10.1016/j.seppur.2018.11.056]

[90] Can OT, Gengec E, Kobya M, Demirbas E, Khataee A. Electroflotation Process: Principles and Applications. In: Shah MP, ed. Emerging Technologies in Wastewater Treatment. Boca Raton (FL): CRC Press; 2023. p. 115-32.

CHAPTER 4

Bioaugmentation and Biostimulation

Azimul Hasan[1,*] and **Arun Kumar K.**[1]

[1] *School of Biosciences, Engineering and Technology, VIT Bhopal University, Sehore, Madhya Pradesh, India*

Abstract: Bioaugmentation and biostimulation are procedures utilized for bioremediation to upgrade the quality of soil and water by utilizing microorganisms, nutrient composition, and development variables. In bioaugmentation, the utilisation of particular diverse sorts of microorganisms for the degradation of poisons, and in the case of Biostimulation, supplement composition can be included at the location for the degradation of poisons. Through the expansion of one or more restricting supplements or biosurfactants to the framework, biostimulation accelerates the speed of degradation. As an illustration, bentazone, mecoprop, and dichlorprop biodegradation were invigorated in anaerobic aquifer fabric following the expansion of oxygen. The study of assorted organisms from different areas is influenced by several factors, including the source of the organisms (indigenous populations), the specific strains used, the size of the inoculum, the culture media, and the genetic engineering of the organisms. Bioaugmentation can be impacted by different components such as temperature, pH, composition, and concentration of the toxins and microbial inoculum, which is required for the greatest action of the treatment. Local microorganisms can be fortified by giving supplements and growth-promoting substances for quickening the breakdown of contaminants like heavy metals, natural and inorganic particles, PFAS, chlorinated compounds, *etc*. Metagenomics and metabolomics, two cutting-edge atomic procedures, have shed vital light on the complex connections between local and obtrusive microbial populations. The utilisation of heritarily adjusted life forms, or GMOs, has brought about an increase in the productivity of toxin resistance and resistance to stress. Next-generation sequencing and metagenomics can select appropriate microbial communities. These have been demonstrated to be advantageous in managing contaminants like man-made plastics and natural compounds. Besides, bioaugmentation and biostimulation are becoming increasingly important for the recovery of sullied ranges and feasible natural administration. Integration of phytoremediation and bioaugmentation can be accomplished to a degree by joining rhizodegradation and phytoaccumulation.

Keywords: Bioaugmentation, Biostimulation, Genetically modified organism, Integration with phytoremediation, Next-gen sequencing, Technical approaches.

---

[*] **Corresponding author Azimul Hasan:** School of Biosciences, Engineering and Technology, VIT Bhopal University, Sehore, Madhya Pradesh, India; E-mail: azimulhasan23phd10040@vitbhopal.ac.in

INTRODUCTION

Biostimulation refers to the process of introducing additional nutrients to existing microorganisms on-site to enhance their ability to break down the chemical contaminants present in the affected media, *i.e.*, soil or water bodies, while bioaugmentation is the technology centred on the introduction and proliferation of microbial consortia in the contaminated site for the remediation of organic and inorganic contaminants, heavy metals, and so forth. In recent research, biotechnology is improving each year, such as metabolomic, genomic studies, CRISPR, and Genetically Modified Organisms (GMOs) [1]. Bioaugmentation is the addition of enhanced specific microbial strains or consortia to a polluted environment for improving a bioremediation process intended to enhance pollutant biodegradation. It is applied when natural weakening takes much time, or native microorganisms fail to digest particular contaminants at an acceptable rate. The bioaugmentation of polluted soil, water, and sludges with natural agents like pesticides, petroleum hydrocarbons, and chemical compounds has been well accepted for treatment [2, 3]. The progression relies on exogenous microbes-often genetically engineered or occurring naturally-whose specialized enzymatic systems can degrade recalcitrant pollutants [4]. It is a practice very much like biostimulation- enhancement of the native microbial community's activities with supplementation of nutrients. Both focus on the enhancement of bioremediation, though bioaugmentation involves the direct introduction of pollutant-degrading organisms. Biostimulation enhances the native microbial populations [5].

Principle of Bioaugmentation

The key goal of bioaugmentation is to improve the rate or efficiency of degrading pollutants by dosing with microorganisms that possess the necessary catabolic pathways. It can be autochthonous, *i.e.*, indigenous to the specific environment, or allochthonous, *i.e.*, from another source. This is based on the principle that these microbes, with their unique enzymatic potential, should significantly enhance the rate of decomposition of complex pollutants in environments where the natural microbes are incompetent [6, 7]. Thus, in bioaugmentation, success may be related to the habitat and strain selection, and the ability of newly introduced microorganisms to outcompete other microorganisms. On the other hand, if conducted correctly, bioaugmentation may improve the genetic composition of the microbial population as well as its enzymatic activity within the contaminated site.

Microorganisms used in Bioaugmentation

Species used for bioaugmentation (Tables **1** and **2**) are usually selected depending on their capacity to degrade specific pollutants. Based on the origins and genera,

Table 1. Specific microorganisms used in the degradation of different contaminants.

Sr No.	Bacterial Strain	Contaminant	Ref.
1	Ralstonia eutropha JMP134	2,4-D, 2-Chloromaleycetic acid	[24]
2	E. Coli D11	2- chloromaleyacetate	
3	Comamonas testosterone BR60	Chlorine and methyl substituted benzoate	
4	Pseudomonas sp. H1	Cadmium	
5	B. thuringiensis B3 and B. cereus B6	Diesel oil and Crude oil	[25]
6	Deinococcus radiodurans	Ionic mercury and toluene	[26]
7	Citrobacter spp.	Copper, nickel, cadmium, cobalt	[27]
8	Citrobacter spp. and Chlorella vulgaris	Cd, Pb, Ag, Hg, Vr, Cu	[28, 29]
9	Sphingomonas xenophaga	Bromoamine	[30]
10	Cryptococcus humicolus,	Cyanide	[31]
11	Pseudomonas PCT01 (JF721324) and PTS02 (JF721325), Paracoccus Denitrifican	combination of carbazole, naphthalene, quinoline, pyridine, and phenol	[32]
12	Arthrobacter spp. W1	Naphthalene (NAP), carbazole (CA), dibenzofuran (DBF), and dibenzothiophene combined with phenol (PH)	[33]
13	Stenotrophomonas spp., Ochrobactrum spp.	Chromium	[34]

Table 2. Different types of fungi involved in the degradation of contaminants.

Sr no.	Fungi	Contaminant	Ref.
1	Aspergillus niger	Cd, Th, Ur	[35 - 39]
2	Pleurotus ostreatus	Cd, Cu, Zn	
3	Rhizopus arrhizus	Cd, Ca, Hg, Pb, P, Au	
4	Stereum hirsutum	Cd, Co, Cu, Ni	
5	Phormidium valderium	Cd, Pb	
5	Ganoderma applantus	Cu, Pb, Hg	
6	Cladosporium cladosporioide, Rhizopus stolonifer and Penicillium purpurogenum	atrazine	[40]

common species include *Pseudomonas, Burkholderia, Rhodococcus, and Bacillus* [8]. Several studies have been conducted on the catabolism of these bacteria with different contaminants, including heavy metals, hydrocarbons, and chlorinated solvents. For example, research has shown *Burkholderia spp.* to strongly degrade nitrophenolic pesticides more effectively at slightly alkaline pH conditions and temperatures of approximately 30 A°C. Also, while Pseudomonas species have already been employed in bioaugmentation in oil-polluted areas apart from their petroleum-derived hydrocarbon elastomers, such species can be incorporated into this kind of bioremediation. Choosing the right strain of microorganisms is critical; in some cases, it is required to obtain bacteria from environments with the same pollutant to guarantee the activeness of the bacterium in the intended site [9].

Factors Influencing Bioaugmentation

Several environmental and biological factors influence the accomplishment of bioaugmentation. The introduced microorganisms must be capable of surviving and competing in the new environment. Key factors that affect this process include:

Environmental Parameters

Some other factors that affect the survival and activity of introduced microorganisms in the environment include pH, temperature, moisture content, and nutrients, among others. For example, it has been inferred that high organic matter content in soil has the potential to cause the breakdown of pollutants such as 2,4-dichlorophenoxyacetic acid (2,4-D) because the compound can easily be broken down biologically. Again, the concentration of water in the medium affects the speed of microbial action; a very low concentration of water would inhibit the strain of bacteria's metabolic activity and also hinder the permeation of the substrate [9].

Competition with Indigenous Microorganisms

The introduced microbes must compete with the native microbial community for nutrients and ecological niches. The ability to outcompete indigenous species is crucial for the success of bioaugmentation. In some cases, introducing pre-adapted or genetically engineered strains that are highly efficient at degrading specific pollutants can overcome this challenge [10].

Predation and Abiotic Factors

Predation by protozoa and the existence of poisonous compounds can reduce the survival rates of the introduced microorganisms. Protective measures, such as encapsulating microbes in protective matrices like alginate beads, can enhance their survival and prolong their activity [11].

Recent Advances in Bioaugmentation

Recent emphasis in improvement efforts for bioaugmentation has been oriented towards efficient overall process performance, regulation of environmental conditions, and enhancement of the selection and survival of microbial strains. Advances in microbial ecology and molecular biology have led to the design of genetically manipulated microorganisms with enhanced capacities for degradative activity [12].

Genetic Engineering of Microbial Strains

Genetic alterations have been employed to enhance the metabolic pathways of microbes and, therefore, enable the microbes to break down a wider variety of contaminants. For example, strains of *Pseudomonas putida* [10, 13] have been engineered to degrade both organophosphate and organochlorine pesticides, enabling simultaneous remediation of multiple contaminants. Additionally, research has shown that inducing mutations in specific enzymes, such as lactonases, can increase their affinity for certain pesticides, further enhancing the biodegradation method [14, 15].

Encapsulation and Immobilization Techniques

To improve the survival and persistence of bioaugmented microbes, researchers have developed encapsulation techniques that protect microorganisms from environmental stressors. Encapsulation in materials such as polyvinyl alcohol or calcium alginate beads has been used to extend the lifespan of biodegrading microbes, allowing for more sustained bioremediation efforts [16, 17].

Co-cultures and Synergistic Microbial Communities

The other method that shows capacity is the application of mixed microbial consortia capable of degrading complex pollutants synergistically. For example, it has been suggested that the use of bacterial and fungal co-cultures removes such pollutants as diuron and 2,6-dichlorobenzamide much more effectively than the use of bacterial cultures [18 - 21].

Metagenomics and Environmental Genomics

The application of metagenomics has provided new insights into the microbial communities involved in bioaugmentation. By analyzing the genetic diversity of microbial populations, researchers can identify key catabolic genes and microbial species that contribute to the degradation of pollutants. This information can be used to optimize bioaugmentation strategies and select the most effective microbial consortia [22, 23].

BIOSTIMULATION IN ENVIRONMENTAL REMEDIATION: RECENT ADVANCES AND APPROACHES

Biostimulation is the most widely practised bioremediation technique that accelerates the rate of biodegradation of the pollutants present in polluted environments by the addition of nutrients or substrates, favourably stimulating the natural microbial activities. This technology is good for cleaning up all forms of contamination, more so hydrocarbon contamination of water and land. The article highlights recent technical advancements in the field of bioremediation by presenting the principles of biostimulation and the latest advances, challenges, and industrial applications in environmental rehabilitation [43 - 45].

Principles of Biostimulation

The biostimulation bioremediation technique revolves around augmenting the indigenous microbial populations present at a contaminated site by the addition of limiting nutrients like nitrogen, phosphorus, or oxygen, and electron acceptors or donors. Such stimulation enhances the metabolism of the microbes, thus increasing the rate at which contaminants, particularly heavy metals, pesticides, and hydrocarbons, are degraded [43]. In summary, bioaugmentation is the provision of material to be added to existing decaying micro-populations that will enhance their capability rather than supplementing with exogenous microorganisms. Primary objectives of biostimulation include the remediation of hydrocarbon contamination, wherein the indigenous microbial populations are capable of acquiring carbon sources from petroleum products but usually lack other essential nutrients. Fertilizers or biosolids, which are often nutrient-rich, have been considered improvements in some published research. For example, there is a 96% enhancement observed in the enhancement of biodegradation of petroleum hydrocarbon contaminated soils through the use of nitrogen and phosphorus fertilizer-enriched biosolids [46 - 48].

Recent Advances in Biostimulation

There have been recent developments in molecular biology, bioinformatics, and environmental genomics that enhance the efficiency and applicability of biostimulation. These advances in today's molecular biology and advanced environmental genomics, along with bioinformatics, allow a deeper understanding of microbial communities that, in turn, facilitate better-targeted strategies for remediation [49, 50].

Molecular Techniques and Genomics

One of the most important benefits of the most recent high-throughput molecular technologies, including metagenomics, metaproteomics, and stable isotope probing, is that they make possible the tracing of the composition of a microbial community in real-time mode during the process of biostimulation. Besides, there exists the possibility of following those special taxa of microorganisms responsible for the degradation of pollution under various environmental conditions. For instance, it has isolated particular bacteria capable of breaking down hydrocarbons and has been able to trace the movement of carbon and nitrogen across microbial communities. In this study, researchers were able to identify which of the functionally encoded genes can break down pollutants; this work thus opens new routes for optimizing the levels of nutrient supply according to the maximal output that can be attained from the genetic makeup of the microorganisms [51 - 53].

Systems Biology and Predictive Modeling

Nowadays, systems biology approaches drive the development and improvement of biostimulation strategy designs. These approaches combine molecular, geochemical, and hydrological studies and enable predictive models to simulate microbial activity under different conditions and rates of pollutant degradation. Predictive models enable improvements in designing plans for and executing biostimulation strategies based on guidelines like optimal nutrient combinations, application rates, and environmental conditions for effective bioremediation [54 - 57].

APPLICATIONS OF BIOSTIMULATION AT CONTAMINATED SITES

Examples of polluted habitats where biostimulation has been applied include soil, groundwater, and marine ecosystems. Biostimulation has successfully remediated a number of the most important pollutants, including heavy metals, insecticides, and hydrocarbons.

Hydrocarbon Contamination

Remediation of hydrocarbon-contaminated environments using biostimulation techniques: Biostimulation is arguably the most studied application of biotechnology. Nitrogen and phosphorus-rich fertilizers or organic additions like compost provide a significant enhancement of the degradation of petroleum hydrocarbons both in soils and waters. In biostimulation, temperature appears to be an important factor, and even more so in the subantarctic temperature forests. Since biostimulation elevates the activity of local dormitory microorganisms believed to be responsible for the hydrocarbon degradation, the rate of hydrocarbon degradation in soil with increased nutrients is enhanced with increased temperature. It might therefore be applicable in the clean-up of polluted environments.

Degradation of Pesticides and Industrial Chemicals

Biostimulation has also been proven to promote the degradation of several pesticides and industrial chemicals. For example, bentazone and mecoprop biodegradation of aquifer material was promoted by the supplementation of oxygen; thus, it proves how biostimulation can be used in the remediation of pesticide-polluted water [58 - 60].

Remediation of Heavy Metal

Aside from the applications discussed earlier, biostimulation has also been used in minimizing the adverse impact of soils that are polluted with heavy metals, albeit not as extensively. The introduction of organic materials or electron donors, for instance, enhanced some microbial activity such as sulfate reduction, making it easy to precipitate and immobilize some heavy metals like arsenic, chromium, and so on [61 - 63].

CHALLENGES AND LIMITATIONS OF BIOSTIMULATION

Although these advantages have their place, biostimulation has various disadvantages that greatly limit its applications.

Environmental Factors

Environmental parameters such as pH, temperature, moisture content, and the presence of other microbes significantly impact biostimulation. In some situations, nutrients are present in excess due to the overproduction of algae (Table **3**) that results in eutrophication, thus depleting oxygen levels in the water and affecting the aquatic ecosystem [64].

Site Specificity

Biostimulation is very site-specific, and its effectiveness can be site-specific. For instance, the wave action in a marine setting could wash away or dilute added nutrients, making it challenging to maintain the conditions necessary for effective biostimulation.

Table 3. Different *algal spp.* used in the degradation of pollutants.

Sr no.	Microalgae/ cyanobacteria	Contaminant	Ref.
1	*Spirulina*	Crude oil	[41, 42]
2	*Chlorella*	Crude oil	
3	*spirogyra*	Petroleum products	
4	*Scenedesmus*	Crude oil	
5	*Oscillatoria*	Hydrocarbon-rich compound	
6	*Chlorococcum*	Aromatic compounds	

Contaminant Characteristics

The efficacy of biostimulation will be subjected to the physical and chemical characteristics of the contaminants. Hydrophobic contaminants like PAHs are not available for degradation by microbes because they are strongly bound to particles in the soil. Such contaminants may need further physical or chemical treatment to enhance their bioavailability before being aptly treated by biostimulation [9].

RECENT INNOVATIONS AND ADVANCEMENTS IN BIOSTIMULATION

Some of these novelties have tried to fill some of the shortcomings encountered during the application of traditional biostimulation technologies. They consequently led to much more efficient and effective remediation techniques.

Integrated Bioaugmentation and Biostimulation

Probably, the most promising approach is the integrated one, combining bioaugmentation with biostimulation. Their concomitant application of specific pollutant-degrading microorganisms along with nutrient addition has resulted in a higher degradation rate and the removal of a greater number of contaminants. The same method has also been quite successful in the remediation of recalcitrant pollutants such as PAHs. In these cases, natural microbial populations are usually too scanty to allow complete degradation [43, 65, 66].

While the usage of biostimulation and bioaugmentation creates a possibility of enhanced breakdown of pollutants by bioremediation due to the advantages of both approaches, there are also instances where biostimulation, *i.e.*, addition of nutrients or other substances to encourage the activity of existing microorganisms, is in play through the addition of specific microbial strains in some processes. Studies were devoted to that issue, and some evidences were produced. For example, one of the studies tried to prove whether anaerobic microflora, which was previously nourished on biochar, might stimulate the disintegration of the PAH phenanthrene (PHE) within soiled substrates. The current experiment demonstrates the potential effectiveness of the treatment of the soil with the PHE phenanthrene pollutant, which involves combining treatment with biochar amendment, and biostimulation and bioaugmentation techniques, as adding biochar, which is the treatment for the PHE degradation, carries out its treatment [67 - 69].

Another case study showed that the additive effect of both methods could be used for the cleaning of hydrocarbon-contaminated lake sediments. The study used three different types of batch incubations, which included controls that did not receive any treatment, biostimulation and bioaugmentation, and biostimulation alone, and all were incubated for 32 days. Data indicated that more favourable microorganisms were observed at higher populations in biostimulated and bioaugmented conditions and with increased efficiencies of biodegradation. Advantages of the Integrated Biostimulation and Bioaugmentation combination increase microbial resilience and redundancy, thus lowering the chance that treatment will fail due to microbial competition or environmental changes. This integrated approach would work if properly implemented to appraise a site, choose the right microbial consortia, control nutrients, and monitor continuously. All factors considered, this set of strategies provides for a practical, sustainable, and effective clean-up of contaminated areas, potentially providing a more robust response to pollution issues [70 - 73].

Genetic Engineering and Synthetic Biology

Advances in genetic engineering and synthetic biology are providing means of exploring new avenues towards improvement in biostimulation and the degradation of microorganisms. For instance, genetically engineered bacteria and fungi with high metabolic routes for pollutant degradation have been designed for specific contamination targeting. Nutrient-addition-stimulated genetically engineered microorganisms, therefore, provide a much more customized and efficient system for bioremediation [74, 75]. Development of both systems biology and genetic engineering will be required for developing appropriate biostimulation and bioaugmentation techniques for the cleanup of the environment. The current work has further shown that bioaugmentation has evolved into an indigenous site-specific approach involving advanced microbial technology to treat multiple categories of environmental contaminants. This ranges to the utilisation of consortia of microbes developed based on systems biology principles and genetically engineered organisms.

Advances in genetic bioaugmentation have recently shown that genetically engineered microbes are more efficient at degradation due to their enhanced stability and survival against environmental stresses. For example, plasmids with catabolic genes have been shown to help in degrading complex pollutants such as triclocarban into less harmful ones when introduced into strains such as *Pseudomonas putida* [76]. This is how genetically modified strains could be used to enhance routes for biodegradation. Systems biology researches microbial communities, which represent an innovative area that highlights how microorganisms interact and can survive in contaminated environments. The hydrocarbon-degradation rates in contaminated sediments show how the introduction of microbial consortia, which often comprises more than one dominating taxon such as Firmicutes or Proteobacteria, can greatly shift the biodegradation dynamics [77, 78]. Potential synergistic relationships that may be exploited to enhance remediation efficacies could be manifested through intensive characterization of these microbiological interactions by systems biology approaches. Furthermore, systems biology can aid in the selection of appropriate microbiological inoculants for bioaugmentation. Through metagenomics, scientists can recover and characterize the active consortia obtained from different ecological niches such as agricultural byproducts and polluted sites [79, 80].

Synergistic Fungi and Bacteria

Another novel application in biostimulation is the harmonious practice of fungi and bacteria. The former could degrade complex organic molecules, and the latter showed metabolic versatility [81]. Together, therefore, this pair of microbes is more robust for degrading a wide range of pollutants. This integrated approach was highly effective for oil-contaminated soils and industrial chemicals using fungal (Table **2**) bioaugmentation and nitrogen biostimulation with improved degradation rates and maintained microbial diversity [9, 82].

Nutrient and Substrate Amendment

Organic substrates that are added to the medium to support microbial growth include molasses, vegetable oils, and dung. Inorganic nutrients are also added to balance the C: N ratio, which is crucial for microbial metabolism, such as nitrogen and phosphorus.

Use of Nanoparticles

Metallic nanoparticles (*e.g.*, iron, silver) improve the degradation of pollutants. Nanocatalysts like nano-titanium dioxide (TiO_2) and carbon-based nanomaterials can boost microbial activity by accelerating electron transfer [83]. Out of all the advanced nanomaterials, nanoparticles (NPs) seem to be very effective tools for biostimulation and bioaugmentation to enhance the existing strategies of bioremediation. The high surface area volume ratio and reactivity of some nanoparticles enhance the effect of bioremediation when combined with other techniques. This is what is called nano-bioremediation, and it seeks to eliminate the problem of pollutants in the environment economically and efficiently [84].

In biostimulation, nanoparticles can be used by providing the necessary electron donors or other growth-limiting components to accelerate the activity of the local microbial population. For instance, Zero-Valent Iron (ZVI) nanoparticles have been found as an effective catalyst for hastening the process of reductive dechlorination, one of the main mechanisms of dissolving refractory pollutants, in particular, for instance, examples include polychlorinated biphenyls, and also polycyclic aromatic hydrocarbons. Nanotubes and other nanoparticles increase the activities of microorganisms; in this case, oil-degrading bacteria like Pseudomonas are stimulated more vigorously. Nanoparticles also enable an elevated nutrient supply that, in turn, supports the localized microbial activities in the degradation of pollutants more efficiently [85]. Bioaugmentation is the process of introducing strains of microorganisms into a contaminated area, where they are presumably protected, therefore helping to survive detrimental environmental conditions for longer periods. Some microorganisms have grown better due to the addition of carbon nanotubes and other nanomaterials in order to increase their pollutant hydrocarbon and heavy metal degrading abilities. Additionally, nanoparticles may potentially enable long-term degrading microbic activity by the continuous supply of electrons necessary to facilitate such an event. However, despite these facts, nanoparticles were shown to greatly enhance the efficacy of biostimulation and bioaugmentation, though careful evaluation of the ecotoxicological impact is essential. The interaction between nano-forms and local microbial communities must be considered along with their potential toxicity to non-target species for safer application of nano-bioremediation techniques (Table 4) [85].

Biochar Amendment

Biochar is an amendment that is used to enhance the physical properties of soils and the microbial activity of the associated soils because of its carbonaceous composition. It provides more surface area for microbes to attach, enhances contaminant sorption, and improves their biodegradation. The goal of this work by Chen *et al.*, 2024 deals with the transformations of As with respect to methylation and volatilization in arsenic-contaminated paddy soils under different water management regimes: bioaugmentation with a genetically engineered *Pseudomonas putida* KT2440 (GE P. putida) and biostimulation with rice straw and Biochar treatments [86, 87]. Such alterations of dissolved oxygen in flooded

Table 4. Differences between biostimulation and bioaugmentation.

Characteristic	Biostimulation	Bioaugmentation	References
Definition	Enhancing indigenous microbial activity by adjusting environmental factors like soil temperature, pH, moisture, and nutrients to improve contaminant degradation.	To degrade the pollutants, exogenous microbial strains, particularly concentrated populations, are added.	[34, 90 - 97]
Microbial Source	Utilizes the existing native microbial populations naturally present in the contaminated environment.	Relies on added, specialized microbial cultures introduced to the site.	
Contaminant Degradation	Stimulates the activity of native organisms, improving their ability to degrade contaminants, provided the environmental conditions are optimized.	Introduces organisms specifically capable of degrading particular contaminants, enhancing the biodegradation process where native microbes are insufficient.	
Cost	Typically lower, as it primarily involves the adjustment of environmental conditions and nutrient addition.	Higher, due to the need for cultivating and introducing specialized microbial populations.	
Effectiveness	Effective when indigenous microbial populations can degrade the contaminants with proper stimulation.	More effective in cases where native microbes are inadequate for contaminant degradation, targeting specific contaminants with specialized microorganisms.	
Environmental Fit	Environmentally friendly and has minimal impact since it works with native microbes that are already adapted to the site.	May face adaptation issues as introduced microbes might struggle to survive or thrive in the new environment.	
Risk of Failure	Lower, as native microbes are already adapted to the site's specific conditions.	Higher, as introduced microbes may fail to thrive if environmental conditions are not conducive to their survival and activity.	
Application	suitable for many polluted areas where, with a little stimulation, indigenous microorganisms may degrade contaminants.	More targeted for specific contaminants where native microbes are insufficient, such as in highly polluted or complex environments.	
Rate of Degradation	The degradation rate may be slower since it relies on the efficiency and adaptability of the existing microbial populations.	The degradation rate is often faster if the introduced microbes are well-suited to degrade the contaminants.	
Time Frame	Generally slower, as it depends on the growth and action of the innate microbial community.	Potentially faster, as the introduced microbes are pre-selected and optimized for contaminant degradation.	
Examples	Adding nutrients like oxygen or nitrogen to stimulate microbial growth.	Introducing specific bacteria, such as those used to degrade oil spills (*e.g.*, *Pseudomonas* for hydrocarbon degradation).	

soils were confirmed, as the total and methylated Arsenic (As) concentration was higher than in the controlled non-flooded treatments. Nevertheless, the combination of GE *P. putida* and non-flooded soils greatly enhanced As methylation and volatilization, which was shown to be better than the flooded soils treatment. Therefore, this indicates that it is possible to undertake remediation in a very efficient way by the use of bioaugmentation and biostimulation approaches in soils that are not flooded, which will contribute a lot to the solutions to the arsenic problem in rice fields [87]. The following was exhibited as the key benefit of the biochar:

- Increased Microbial Activity Biochar increased the microbial metabolism and increased the surface area of microbes, neutralizing the pH value of the soil. It helped in speeding up the consumption of organic matter, hence the rapid acceleration of biodegradation.
- Improved Bioavailability of Organic Matter: The addition of biochar improved the bioavailability of dissolved organic carbon, as manifested by increased soluble chemical oxygen demand SCOD values in some of the treatment groups. Mechanisms enhanced microbial growth responsible for degrading organic compounds and further degrading PAH.
- Support of Anaerobic activities: Biochar also mediated the induction of anaerobic microbial communities, particularly in acetogenesis and other associated functions that are of extreme importance to the microbial breakdown of PAH. This, therefore, caused the uptake of VFAs, which enhanced the performance of the microbes.
- The synergistic effect was observed between the biochar and bioaugmentation approaches applied for the PAH biodegradation to better remove PHE from polluted soils in this study.

Oxygen and Hydrogen Peroxide Injection

Oxygen-Releasing Compounds (ORCs) like calcium peroxide (CaO_2) are used to provide a sustained oxygen source for aerobic biodegradation. Hydrogen peroxide (H_2O_2) can be used to enhance microbial growth by behaving as an electron acceptor in anaerobic environments.

Electro-bioremediation (Bioelectrochemical Systems)

This involves applying an electrical current to contaminated environments, stimulating the metabolic activities of microbes through Microbial Fuel Cells (MFCs) or bioelectrochemical reactors. The process enhances the degradation of pollutants like hydrocarbons and other pollutants [88, 89].

CONCLUSION

The recent advances in biotechnology have meaningfully improved both bioaugmentation and biostimulation, which impart highly promising prospects for environmental remediation. Bioaugmentation added specific microbial strains to degrade pollutants; genetic engineering allowed the development of microbes modified for enhanced degradative abilities and permitted the degradation of complex contaminants, such as organophosphates and organochlorines, achieved by adding genetically engineered *Pseudomonas putida*. More so, materials like calcium alginate have also been incorporated into the methods of encapsulation to facilitate the introduced microbes to survive longer, improving the sustainability of the remediation in question. Biostimulation is a process that has been related to the addition of nutrients to enhance the microbial activities of naturally existing microorganisms. With metagenomics, biostimulatory practices have progressed further to not only comprehend but also investigate the dynamics of a community of microbes within the system being treated without interference. Bioaugmentation and biostimulation practices have been synergistically applied, which has been found to be more effective than either one in degrading pollutants. Finally, the enrichment of microorganisms in both methodologies has been achieved through the addition of biochar and nanoparticles in the two approaches. The new developments also serve as a foundation for further advancement through systems biology and synthetic biology, which can offer more efficient and targeted approaches in bioremediation towards a more sustainable and effective cleanup of polluted environments. However, careful management with ecological considerations should not be overlooked lest unintended negative consequences be unleashed.

REFERENCES

[1] Mrozik A, Piotrowska-Seget ZJMr. Bioaugmentation as a strategy for cleaning up of soils contaminated with aromatic compounds Microbiol Res 2010; 165(5): 363-75.
[http://dx.doi.org/10.1016/j.micres.2009.08.001]

[2] Dai Y, Li J, Wang S, *et al*. Unveiling the synergistic mechanism of autochthonous fungal bioaugmentation and ammonium nitrogen biostimulation for enhanced phenanthrene degradation in oil-contaminated soils. J Hazard Mater 2024; 465: 133293.
[http://dx.doi.org/10.1016/j.jhazmat.2023.133293] [PMID: 38141301]

[3] Lim MW, Lau EV, Poh PE. A comprehensive guide of remediation technologies for oil contaminated soil - Present works and future directions. Mar Pollut Bull 2016; 109(1): 14-45.
[http://dx.doi.org/10.1016/j.marpolbul.2016.04.023] [PMID: 27267117]

[4] Helbling DEJCOiB. Bioremediation of pesticide-contaminated water resources: the challenge of low concentrations Curr Opin Biotechnol 2015; 33: 142-8.
[http://dx.doi.org/10.1016/j.copbio.2015.02.012]

[5] Liu L, Helbling DE, Kohler HPE, Smets BF. A model framework to describe growth-linked biodegradation of trace-level pollutants in the presence of coincidental carbon substrates and microbes. Environ Sci Technol 2014; 48(22): 13358-66.

[http://dx.doi.org/10.1021/es503491w] [PMID: 25321868]

[6] Ueno A, Ito Y, Yumoto I, Okuyama H. Isolation and characterization of bacteria from soil contaminated with diesel oil and the possible use of these in autochthonous bioaugmentation. World J Microbiol Biotechnol 2007; 23(12): 1739-45.
[http://dx.doi.org/10.1007/s11274-007-9423-6] [PMID: 27517830]

[7] Nikolopoulou M, Eickenbusch P, Pasadakis N, Venieri D, Kalogerakis N. Microcosm evaluation of autochthonous bioaugmentation to combat marine oil spills. N Biotechnol 2013; 30(6): 734-42.
[http://dx.doi.org/10.1016/j.nbt.2013.06.005] [PMID: 23835403]

[8] Shen W, Zhu N, Cui J, et al. Ecotoxicity monitoring and bioindicator screening of oil-contaminated soil during bioremediation. Ecotoxicol Environ Saf 2016; 124: 120-8.
[http://dx.doi.org/10.1016/j.ecoenv.2015.10.005] [PMID: 26491984]

[9] Tribedi P, Goswami M, Chakraborty P, et al. Bioaugmentation and biostimulation: a potential strategy for environmental remediation. J Microbiol Exp 2018; 6(5): 223-31.

[10] Chettri D, Verma AK, Verma AKJB. Bioaugmentation: an approach to biological treatment of pollutants. Biodegradation 2024; 35(2): 117-35.
[http://dx.doi.org/10.1007/s10532-023-10050-5]

[11] El Fantroussi S, Agathos SNJCoim. Is bioaugmentation a feasible strategy for pollutant removal and site remediation? Curr Opin Microbiol 8(3): 268-75.
[http://dx.doi.org/10.1016/j.mib.2005.04.011]

[12] Thompson IP, Van Der Gast CJ, Ciric L, Singer AC. Bioaugmentation for bioremediation: the challenge of strain selection. Environ Microbiol. 2005; 7(7): 909-15.
[http://dx.doi.org/10.1111/j.1462-2920.2005.00804.x]

[13] Rafeeq H, Afsheen N, Rafique S, et al. Genetically engineered microorganisms for environmental remediation. Chemosphere. 2023; 310: 136751.
[http://dx.doi.org/10.1016/j.chemosphere.2022.136751]

[14] Gentry T, Rensing C, Pepper I. New Approaches for Bioaugmentation as a Remediation Technology. Crit Rev Environ Sci Technol 2004; 34(5): 447–94.
[http://dx.doi.org/10.1080/10643380490452362]

[15] Michalska J, Piński A, Żur J, Mrozik A. Analysis of the bioaugmentation potential of Pseudomonas putida OR45a and Pseudomonas putida KB3 in the sequencing batch reactors fed with the phenolic landfill leachate. Water 2020; 12(3): 906.
[http://dx.doi.org/10.3390/w12030906]

[16] Nwankwegu AS, Onwosi CO. Microbial cell immobilization: a renaissance to bioaugmentation inadequacies. A review. Environ Technol Rev. 2017; 6(1): 186–98.
[http://dx.doi.org/10.1080/21622515.2017.1356877]

[17] Siripattanakul S, Khan E. Fundamentals and applications of entrapped cell bioaugmentation for contaminant removal. In: Shah V, ed. Emerging environmental technologies, Volume II. Dordrecht: Springer; 2010. p. 129–43.
[http://dx.doi.org/10.1007/978-90-481-3352-9_7]

[18] Valdez-Vazquez I, Castillo-Rubio LG, Pérez-Rangel M, Sepúlveda-Gálvez A, Vargas A. Enhanced hydrogen production from lignocellulosic substrates via bioaugmentation with Clostridium strains. Ind Crops Prod. 2019; 137: 105-11.
[http://dx.doi.org/10.1016/j.indcrop.2019.05.023]

[19] Espinosa-Ortiz EJ, Rene ER, Gerlach RJCrib. Potential use of fungal-bacterial co-cultures for the removal of organic pollutants Critical Rev Biotech 2022; 42(3): 361–383.
[http://dx.doi.org/10.1080/07388551.2021.1940831]

[20] Mauroo E. Bio-augmentation Of Activated Carbon For Enhanced Removal Of 2, 6-dichlorobenzamide (BAM) [master's thesis]. Ghent (Belgium): Ghent University; 2017.

[21]   Ellegaard-Jensen L, Knudsen BE, Johansen A, Albers CN, Aamand J, Rosendahl S. Fungal–bacterial consortia increase diuron degradation in water-unsaturated systems. Sci Total Environ 2014; 466-467: 699-705.
[http://dx.doi.org/10.1016/j.scitotenv.2013.07.095]

[22]   Bharagava RN, Purchase D, Saxena G, Mulla SI. Applications of metagenomics in microbial bioremediation of pollutants: from genomics to environmental cleanup. In: Microbial diversity in the genomic era. Elsevier 2019; pp. 459-77.
[http://dx.doi.org/10.1016/B978-0-12-814849-5.00026-5]

[23]   Kumar V, Thakur IS, Singh AK, Shah MP. Application of metagenomics in remediation of contaminated sites and environmental restoration. In: Emerging technologies in environmental bioremediation. Elsevier 2020; pp. 197-232.
[http://dx.doi.org/10.1016/B978-0-12-819860-5.00008-0]

[24]   Pepper IL, Gentry TJ, Newby DT, Roane TM, Josephson KL. The role of cell bioaugmentation and gene bioaugmentation in the remediation of co-contaminated soils. Environ Health Perspect. 2002; 110(Suppl 6): 943-6.
[http://dx.doi.org/10.1289/ehp.02110s6943]

[25]   Raju MN, Leo R, Herminia SS, MorAn REB, Venkateswarlu K, Laura S. Biodegradation of Diesel, Crude Oil and Spent Lubricating Oil by Soil Isolates of Bacillus spp. Bull Environ Contam Toxicol 2017; 98(5): 698-705.
[http://dx.doi.org/10.1007/s00128-017-2039-0] [PMID: 28210752]

[26]   Brim H, McFarlan SC, Fredrickson JK, et al. Engineering Deinococcus radiodurans for metal remediation in radioactive mixed waste environments. Nat Biotechnol 2000; 18(1): 85-90.
[http://dx.doi.org/10.1038/71986] [PMID: 10625398]

[27]   Kanekar PP, Kanekar SP. Metallophilic, Metal-Resistant, and Metal-Tolerant Microorganisms. In: Kanekar PP, Kanekar SP, Eds. Diversity and Biotechnology of Extremophilic Microorganisms from India. Singapore: Springer Nature Singapore 2022; pp. 187-213.
[http://dx.doi.org/10.1007/978-981-19-1573-4_6]

[28]   Ghosh S, Selvakumar G, Ajilda AK, Webster TJ. Microbial biosorbents for heavy metal removal. In: Shah MP, Rodriguez Couto S, Kumar V, eds. New Trends in Removal of Heavy Metals from Industrial Wastewater. Boca Raton (FL): CRC Press; 2021. p. 213-62.
[http://dx.doi.org/10.1016/B978-0-12-822965-1.00010-6]

[29]   Cañizares-Villanueva RO, Martínez Roldán AJ, Perales-Vela HV, Vázquez-Hernández M, Melchy-Antonio O. Bioremediation of Copper and Other Heavy Metals using Microbial Biomass. In: Das S, Dash HR, eds. Handbook of Metal-Microbe Interactions and Bioremediation. 1st ed. Boca Raton (FL): CRC Press; 2017. p. 18.
[http://dx.doi.org/10.1201/9781315153353-41]

[30]   Qu Y, Zhou J, Wang J, Song Z, Xing L, Fu X. Bioaugmentation of bromoamine acid degradation with Sphingomonas xenophaga QYY and DNA fingerprint analysis of augmented systems. Biodegradation 2006; 17(1): 83-91.
[http://dx.doi.org/10.1007/s10532-005-3544-0] [PMID: 16453174]

[31]   Park D, Lee DS, Kim YM, Park JM. Bioaugmentation of cyanide-degrading microorganisms in a full-scale cokes wastewater treatment facility. Bioresour Technol 2008; 99(6): 2092-6.
[http://dx.doi.org/10.1016/j.biortech.2007.03.027] [PMID: 17513106]

[32]   Zhu X, Liu R, Liu C, Chen L. Bioaugmentation with isolated strains for the removal of toxic and refractory organics from coking wastewater in a membrane bioreactor. Biodegradation 2015; 26(6): 465-74.
[http://dx.doi.org/10.1007/s10532-015-9748-z] [PMID: 26510738]

[33]   Quan X, Shi H, Liu H, Lv P, Qian Y. Enhancement of 2,4-dichlorophenol degradation in conventional activated sludge systems bioaugmented with mixed special culture. Water Res 2004; 38(1): 245-53.

[http://dx.doi.org/10.1016/j.watres.2003.09.003] [PMID: 14630123]

[34] Yan X, Yan Z, Zhu X, *et al.* Comparing Different Strategies for Cr(VI) Bioremediation. Bioaugmentation, Biostimulation, and Bioenhancemen. Sustainability 2023; 15(16): 12522.
[http://dx.doi.org/10.3390/su151612522]

[35] Malik NA, Kumar J, Wani MS, Tantray YR, Ahmad T. Role of Mushrooms in the Bioremediation of Soil. In: Dar GH, Bhat RA, Mehmood MA, Hakeem KR, eds. Microbiota and Biofertilizers, Vol 2. Cham: Springer; 2021. p. 77–102.
[http://dx.doi.org/10.1007/978-3-030-61010-4_4]

[36] Vaksmaa A, Guerrero-Cruz S, Ghosh P, Zeghal E, Hernando-Morales V, Niemann H. Role of fungi in bioremediation of emerging pollutants. Front Mar Sci. 2023; 10: 107090.
[http://dx.doi.org/10.3389/fmars.2023.1070905]

[37] Verma S, Kuila A. Bioremediation of heavy metals by microbial process. Environ Technol Innov 2019; 14: 100369.
[http://dx.doi.org/10.1016/j.eti.2019.100369]

[38] Sarikurkcu C, Yildiz D, Akata I, Tepe B. Evaluation of the metal concentrations of wild mushroom species with their health risk assessments. Environ Sci Pollut Res Int 2021; 28(17): 21437-54.
[http://dx.doi.org/10.1007/s11356-020-11685-0] [PMID: 33415633]

[39] Balaji V, Arulazhagan P, Ebenezer PJJEB. Enzymatic bioremediation of polyaromatic hydrocarbons by fungal consortia enriched from petroleum contaminated soil and oil seeds. J Environ Biol 2014; 35(3): 521-9.

[40] Gonçalves MS, Sampaio SC, Sene L, Suszek FL, Coelho SRM, Bravo CEC. Isolation of filamentous fungi present in swine wastewater that are resistant and with the ability to remove atrazine. Afr J Biotechnol 2012: 11(50): 11074-11077.

[41] Dell' Anno F, Rastelli E, Sansone C, Brunet C, Ianora A, Dell' Anno A. Bacteria, Fungi and Microalgae for the Bioremediation of Marine Sediments Contaminated by Petroleum Hydrocarbons in the Omics Era. Microorganisms 2021; 9(8): 1695.
[http://dx.doi.org/10.3390/microorganisms9081695] [PMID: 34442774]

[42] Farrington J, Takada H. Persistent Organic Pollutants (POPs), Polycyclic Aromatic Hydrocarbons (PAHs), and Plastics: Examples of the Status, Trend, and Cycling of Organic Chemicals of Environmental Concern in the Ocean. Oceanography (Wash DC) 2014; 27(1): 196-213.
[http://dx.doi.org/10.5670/oceanog.2014.23]

[43] Adams GO, Fufeyin PT, Okoro SE, Ehinomen I. Bioremediation, biostimulation and bioaugmention: a review. Int J Environ Bioremed Biodegrad 2015; 3(1): 28-39.
https://www.semanticscholar.org/paper/Bioremediation%2C-Biostimulation-and-Bioaugmention%3A-A-Adams-Fufeyin/1d31fc82c4b82c3b83639b1568c07dc6f241baa9

[44] Tyagi M, da Fonseca MMR, de Carvalho CCJB. Bioaugmentation and biostimulation strategies to improve the effectiveness of bioremediation processes 2011; 35(1): 28-39.
[http://dx.doi.org/10.1007/s10532-010-9394-4]

[45] Kumari P, Sankhyan NK, Walia A. Bioremediation, biostimulation and bioaugmentation: a new perspective for pollution free environment. In: Naik YD, Nayak JK, Panda D, eds. Advances in agricultural biotechnology. New Delhi: Daya Publishing House; 2023. p. 123–138.

[46] Ite AE, Ibok UJ. Role of Plants and Microbes in Bioremediation of Petroleum Hydrocarbons Contaminated Soils. Int J Environ Bioremed Biodegrad 2019; 7(1): 1-19. Available from: https://pubs.sciepub.com/ijebb/7/1/1/

[47] Sayed K, Baloo L, Sharma NK. Bioremediation of Total Petroleum Hydrocarbons (TPH) by Bioaugmentation and Biostimulation in Water with Floating Oil Spill Containment Booms as Bioreactor Basin. Int J Environ Res Public Health 2021; 18(5): 2226.
[http://dx.doi.org/10.3390/ijerph18052226]

[48] Wu M, Dick WA, Li W, *et al*. Bioaugmentation and biostimulation of hydrocarbon degradation and the microbial community in a petroleum-contaminated soil. Int Biodeterior Biodegradation 2016; 107: 158-64.
[http://dx.doi.org/10.1016/j.ibiod.2015.11.019]

[49] Bartucca ML, Cerri M, Del Buono D, Forni C. Use of biostimulants as a new approach for the improvement of phytoremediation performance. Plants 2022; 11(15): 1946.
[http://dx.doi.org/10.3390/plants11151946] [PMID: 35893650]

[50] Garg S, Nain P, Kumar A, *et al*. Next generation plant biostimulants genome sequencing strategies for sustainable agriculture development Front Microbiol 2024; 15: 1439561.
[http://dx.doi.org/10.3389/fmicb.2024.1439561]

[51] Wawrik B. Stable isotope probing the N cycle: current applications and future directions. In: Marco D, Ed. Metagenomics of the microbial nitrogen cycle: theory, methods and applications 2014; 87-110.
https://www.caister.com/hsp/abstracts/n2/05.html

[52] Xie J, He Z, Liu X, *et al*. GeoChip-based analysis of the functional gene diversity and metabolic potential of microbial communities in acid mine drainage. Appl Environ Microbiol 2011; 77(3): 991-9.
[http://dx.doi.org/10.1128/AEM.01798-10]

[53] Murrell JC. DNA stable isotope probing, whole genome amplification and metagenomics. In: Young LY, Zylstra GJ, Eds. Conference-EC-US Task Force Joint US-EU Workshop on Metabolomics and Environmental Biotechnology. Palma de Mallorca, Spain. Washington (DC): 2008.Washington (DC) 2008. https://digital.library.unt.edu/ark:/67531/metadc831572/m1/12/

[54] Maluleka K, Roopchund R, Seedat N, Sillanpää M. Developing a predictive machine learning model and a kinetic model for the bioremediation of terrestrial diesel spills. Results Eng 2024; 23: 102378.
[http://dx.doi.org/10.1016/j.rineng.2024.102378]

[55] Hazen TC. Cometabolic bioremediation. In: Timmis KN, ed. Handbook of hydrocarbon and lipid microbiology. Berlin: Springer; 2010. p. 2503–14.
[http://dx.doi.org/10.1007/978-3-540-77587-4_185]

[56] Romantschuk M, Lahti-Leikas K, Kontro M, *et al*. Bioremediation of contaminated soil and groundwater by *in situ* biostimulation Frontiers in Microbiology 2023; 14: 1258148.
[http://dx.doi.org/10.3389/fmicb.2023.1258148]

[57] Xu G, Zhao S, Liu J, He J. Bioremediation of organohalide pollutants: progress, microbial ecology, and emerging computational tools. Curr Opin Environ Sci Health. 2023; 32: 100452.
[http://dx.doi.org/10.1016/j.coesh.2023.100452]

[58] Raj A, Dubey A, Malla MA, Kumar A. Pesticide pestilence: Global scenario and recent advances in detection and degradation methods. J Environ Manage. 2023; 338: 117680.
[http://dx.doi.org/10.1016/j.jenvman.2023.117680]

[59] Aliyu GO, Anyanwu CU, Nnamchi CI, Onwosi CO. Evaluation of the effectiveness of bioaugmentation and biostimulation in atrazine removal in a polluted matrix using degradation kinetics. Soil Sediment Contam 2023; 32(1): 105–24.
[http://dx.doi.org/10.1080/15320383.2022.2059444]

[60] Bhuyan B, Kotoky R, Pandey P. Impacts of rhizoremediation and biostimulation on soil microbial community, for enhanced degradation of petroleum hydrocarbons in crude oil-contaminated agricultural soils. Environ Sci Pollut Res 2023; 30: 94649-668.
[http://dx.doi.org/10.1007/s11356-023-29033-3]

[61] Yan X, Yan Z, Zhu X, Zhou Y, Ma G, Li S. Comparing Different Strategies for Cr(VI) Bioremediation: Bioaugmentation, Biostimulation, and Bioenhancement. Sustainability 2023; 15(16): 12522.
[http://dx.doi.org/10.3390/su151612522]

[62] Mao S, Ma S, Zhao Q, *et al*. Carbohydrate based biostimulation regulates the structure, function and

remediation of Cr(VI) pollution by SRBs flora. Environ Res 2024; 263(2): 120088.
[http://dx.doi.org/10.1016/j.envres.2024.120088]

[63] Kumar A, Song HW, Mishra S, *et al.* Application of microbial-induced carbonate precipitation (MICP) techniques to remove heavy metal in the natural environment: A critical review. Chemosphere 2023; 318: 137894.
[http://dx.doi.org/10.1016/j.chemosphere.2023.137894] [PMID: 36657570]

[64] Azubuike CC, Chikere CB, Okpokwasili GC. Bioremediation techniques–classification based on site of application: principles, advantages, limitations and prospects. World J Microbiol Biotechnol 2016; 32(180): 180.
[http://dx.doi.org/10.1007/s11274-016-2137-x]

[65] Jayaprakash K, Govarthanan M, Mythili R, Selvankumar T, Chang Y-C. Bioaugmentation and Biostimulation Remediation Technologies for Heavy Metal Lead Contaminant. In: Chang Y-C, ed. Microbial Biodegradation of Xenobiotic Compounds. 1st ed. Boca Raton (FL): CRC Press; 2019. p 13.
[http://dx.doi.org/10.1201/b22151-2]

[66] Lladó S, Covino S, Solanas AM, Viñas M, Petruccioli M, D'annibale A. Comparative assessment of bioremediation approaches to highly recalcitrant PAH degradation in a real industrial polluted soil. J Hazard Mater 2013; 248-249: 407-14.
[http://dx.doi.org/10.1016/j.jhazmat.2013.01.020]

[67] Gu H, Yan J, Liu Y, *et al.* Autochthonous bioaugmentation accelerates phenanthrene degradation in acclimated soil. Environ Res. 2023; 224: 115543.
[http://dx.doi.org/10.1016/j.envres.2023.115543]

[68] Xue H, Shi Y, Qiao J, Li X, Liu R. Enhancing Anaerobic Biodegradation of Phenanthrene in Polluted Soil by Bioaugmentation and Biostimulation: Focus on the Distribution of Phenanthrene and Microbial Community Analysis. Sustainability. 2024; 16(1): 366.
[http://dx.doi.org/10.3390/su16010366]

[69] Zhang Y, Liu S, Niu L, *et al.* Sustained and efficient remediation of biochar immobilized with Sphingobium abikonense on phenanthrene-copper co-contaminated soil and microbial preferences of the bacteria colonized in biochar. Biochar. 2023; 5(43): 43.
[http://dx.doi.org/10.1007/s42773-023-00241-x]

[70] Harrisson O. Developing and Assessing Efficient Hydrocarbon Bioremediation Strategies for Cold-Temperature Soils and Sediments: Two Case Studies. Northern Canada. Canada: McGill University 2023.

[71] Alaidaroos BA. Advancing Eco-Sustainable Bioremediation for Hydrocarbon Contaminants: Challenges and Solutions. Processes 2023; 11(10): 3036.
[http://dx.doi.org/10.3390/pr11103036]

[72] Kalneniece K, Gudra D D, Lielauss L, *et al.* Batch-mode stimulation of hydrocarbons biodegradation in freshwater sediments from historically contaminated Alūksne lake. J Contam Hydrol. 2023; 253: 104103.
[http://dx.doi.org/10.1016/j.jconhyd.2022.104103]

[73] Sanjana M, Prajna R, Urvi SK, Kavitha RV. Bioremediation–the recent drift towards a sustainable environment. Enviro. Sci. Adv. 2024; 3(8): 1097-110.
[http://dx.doi.org/10.1039/D3VA00358B]

[74] Ahmad A, Mustafa G, Rana A, Zia AR. Improvements in Bioremediation Agents and Their Modified Strains in Mediating Environmental Pollution. Curr Microbiol. 2023; 80: 208.
[http://dx.doi.org/10.1007/s00284-023-03316-x]

[75] Kasanke CP. Microbial ecology and biodegradation potential of a sulfolane-contaminated, subarctic aquifer [Phd dissertation]. Fairbanks (AK): University of Alaska Fairbanks; 2019.

[76] Puranik S, Sruthy KS, Manoj M, Vikram KV, Karijadar P, Singh SK. Microbial Inoculants and Their

Potential Application in Bioremediation: Emphasis on Agrochemicals. In: Kumar A, Shukla L, Singh J, Ferreira LFR, eds. Microbial Bioremediation and Biodegradation: Integrated Approaches for Environmental Management. Cham: Springer; 2023. p. 118-45.
[http://dx.doi.org/10.1002/9781119851158.ch8]

[77] Malik N, Lakhawat SS, Kumar V, Sharma V, Bhatti JS, Sharma PK. Recent advances in the omics-based assessment of microbial consortia in the plastisphere environment: Deciphering the dynamic role of hidden players. Process Saf Environ Prot 2023; 176: 207-25.
[http://dx.doi.org/10.1016/j.psep.2023.06.013]

[78] Meng L, Li W. B/E-enhanced bioremediation of diesel-polluted seawater and microbial community response [Internet]. Rochester (NY): Social Science Research Network; 2023 Oct 17 [cited 2025 Dec 19]. SSRN: 4604278. Available from: https://ssrn.com/abstract=4604278

[79] Muter O. Current Trends in Bioaugmentation Tools for Bioremediation: A Critical Review of Advances and Knowledge Gaps. Microorganisms 2023; 11(3): 710.
[http://dx.doi.org/10.3390/microorganisms11030710]

[80] Olawoyin DC, Odeyemi A, Osemwegie OO, Obayomi V, Akinsanola BA. Bacterial Genetic Bioaugmentation in the Bioremediation of Soil Pollutants: A Mini Review. In: 2024 International Conference on Science, Engineering and Business for Driving Sustainable Development Goals (SEB4SDG). Omu-Aran, Nigeria: IEEE; 2024. p. 1-7.
[http://dx.doi.org/10.1109/SEB4SDG60871.2024.10630000]

[81] Rouphael Y, Colla G. Synergistic biostimulatory action: Designing the next generation of plant biostimulants for sustainable agriculture Front Plant Sci 2018; 9: 1655.
[http://dx.doi.org/10.3389/fpls.2018.01655]

[82] Qiao J, Zhang C, Luo S, Chen W. Bioremediation of highly contaminated oilfield soil: Bioaugmentation for enhancing aromatic compounds removal. Front Environ Sci Eng. 2014; 8(2): 293–304.
[http://dx.doi.org/10.1007/s11783-013-0561-9]

[83] Bodzek M, Konieczny K, Kwiecińska-Mydlak A. Recent advances in water and wastewater disinfection by nano-photocatalysis Desalination and Water Treatment 2023; 305: 2-16.
[http://dx.doi.org/10.5004/dwt.2023.29390]

[84] Kumari B, Singh D. A review on multifaceted application of nanoparticles in the field of bioremediation of petroleum hydrocarbons Ecological Engineering 2016; 97: 98-105.
[http://dx.doi.org/10.1016/j.ecoleng.2016.08.006]

[85] Romantschuk M, Lahti-Leikas K, Kontro M, *et al.* Bioremediation of contaminated soil and groundwater by *in situ* biostimulation. Front Microbiol 2023; 14: 1258148.
[http://dx.doi.org/10.3389/fmicb.2023.1258148] [PMID: 38029190]

[86] Amiri FA, Amiri NA, Karimi P, Bioaugmentation of Bio-Slurry Reactor Containing Pyrene Contaminated Soil by Engineered *Pseudomonas putida* KT2440. 2024; 235(6): 355.
[http://dx.doi.org/10.1007/s11270-024-07186-2]

[87] Chen P, Liu Y, Sun GX. Evaluation of water management on arsenic methylation and volatilization in arsenic-contaminated soils strengthened by bioaugmentation and biostimulation. J Environ Sci (China) 2024; 137: 515-26.
[http://dx.doi.org/10.1016/j.jes.2023.02.023] [PMID: 37980035]

[88] Ambaye TG, Vaccari M, Franzetti A, *et al.* Microbial electrochemical bioremediation of petroleum hydrocarbons (PHCs) pollution: Recent advances and outlook. Chem Eng J. 2023; 452(3): 139372.

[89] Shi X, Duan Z, Wang J, Simultaneous removal of multiple heavy metals using single chamber microbial electrolysis cells with biocathode in the micro-aerobic environment Chemosphere 2023; 318: 137982.
[http://dx.doi.org/10.1016/j.chemosphere.2023.137982]

[90] Abdulsalam S, Bugaje IM, Adefila SS,. Comparison of biostimulation and bioaugmentation for remediation of soil contaminated with spent motor oil. Int J Environ Sci Technol 2011; 8(1): 187–94.
[http://dx.doi.org/10.1007/BF03326208]

[91] Yuan Q-B, Shen Y, Huang Y-M, Hu N. A comparative study of aeration, biostimulation and bioaugmentation in contaminated urban river purification. Environ Technol Innov. 2018; 11: 276-85.
[http://dx.doi.org/10.1016/j.eti.2018.06.008]

[92] Leavitt M, Brown KJ. Biostimulation versus bioaugmentation—three case studies. In: Lewis Pub, ed. Bioremediation: field experience. Boca Raton (FL): Lewis Publishers; 1994. p. 72–9.

[93] Gomez MG, Anderson CM, Graddy CMR, DeJong JT, Nelson DC, Ginn TR. Large-Scale Comparison of Bioaugmentation and Biostimulation Approaches for Biocementation of Sands. J Geotech Geoenviron Eng. 2017; 143(5): 04017004.
[http://dx.doi.org/10.1061/(ASCE)GT.1943-5606.0001640]

[94] Fan W, Xiao Y, Cao B, Shi J, Wu H, Shu S. Comparison of bioaugmentation and biostimulation approaches for biocementation in soil column experiments. J Build Eng 2024; 82: 108335.
[http://dx.doi.org/10.1016/j.jobe.2023.108335]

[95] Mohammadi M, Bayat Z, Hassanshahian M, Mousavi M, Shekarchizadeh F. Microbial community response to biostimulation and bioaugmentation in crude oil-polluted sediments of the Persian Gulf. Environ Res. 2024; 249: 118197.
[http://dx.doi.org/10.1016/j.envres.2024.118197]

[96] Liu H, Wang R, Luo M, *et al.* Coupling of biostimulation and bioaugmentation for benzene, toluene, and trichloroethylene removal from co-contaminated soil. Water Air Soil Pollut. 2024 Sep; 235(9): 665.
[http://dx.doi.org/10.1007/s11270-024-07481-y]

[97] Cunningham C, Philp J. Comparison of bioaugmentation and biostimulation in ex situ treatment of diesel contaminated soil. Land Contamination Reclamation 2000; 8(4): 261-9. Available from: https://www.academia.edu/27334811/

CHAPTER 5

Bioremediation Techniques

Shruti Khanna Ahuja[1,*], **Akriti Gupta**[2] and **Preeti Mehta Kakkar**[2,*]

[1] *Centre for Medical Biotechnology, Amity Institute of Biotechnology, Amity University, Noida, Uttar Pradesh, India*

[2] *Department of Biotechnology, Amity Institute of Biotechnology, Amity University, Noida, Uttar Pradesh, India*

Abstract: Bioremediation techniques are essential for addressing environmental contamination, particularly from oil spills and industrial pollutants. These methods utilize microorganisms to degrade harmful substances, offering a sustainable and cost-effective solution. This book chapter provides a detailed description of bioremediation, its advantages and disadvantages, and the different techniques employed. The techniques can be categorized into *in situ* and *ex situ* approaches, each with distinct applications and benefits. The chapter also discusses the classification of bioremediation techniques, how it is useful for us, and its characteristics. The applications of these techniques for tackling pollution from petroleum hydrocarbons, heavy metals, pesticides, and various other organic and inorganic contaminants are also discussed in this chapter. The last section of the chapter also discusses the advantages, limitations, challenges, future outlook, and environmental impact of these bioremediation techniques.

Keywords: Bioremediation techniques, Environmental impact, Heavy metals, Industrial pollutants, Petroleum hydrocarbons.

INTRODUCTION

The escalation of intensive agriculture, industrial development, and the application of chemicals, nuclear resources, and petrochemicals has led to significant pollution of ecosystems by heavy metals, microplastics, pharmaceutical drugs, organic compounds, petroleum and its derivatives, crude oil, pesticides, chemical fertilizers, xenobiotics, *etc*. These contaminations have found their way into food and water, causing serious health issues. Accumulation of heavy metal ions has caused cancer, coronary diseases, mental retardation, and chronic illness of the kidneys and brain. The carcinogenicity of hydrocarbons has

---

[*] **Corresponding authors Shruti Khanna Ahuja and Preeti Mehta Kakkar:** Centre for Medical Biotechnology, Amity Institute of Biotechnology, Amity University, Noida, Uttar Pradesh, India E-mail: skahuja@amity.edu, Department of Biotechnology, Amity Institute of Biotechnology, Amity University, Noida, Uttar Pradesh, India; E-mail: pmkakkar@amity.edu

Rajneesh Kumar, Ram Sharan Singh & Maulin P. Shah. (Eds.)
All rights reserved-© 2026 Bentham Science Publishers

extensively affected aquatic life through polluted water bodies and, in turn, humans [1]. Increased concentrations of xenobiotic compounds in soil and water can cause ecotoxicological effects in soil, eutrophication in water bodies, oxidative stress, plant genotoxicity, developmental disorders in aquatic life, physiological changes in terrestrial life, and decreased immune response in humans [2].

In the past years, researchers have been working on developing unique strategies to combat the continuous rise in pollution levels across the globe. The objective is to rehabilitate contaminated ecosystems through environmentally conscious, economically viable, time-efficient, sustainable, and efficacious methodologies. Traditional methods involving incineration, excavation, use of chemicals, coagulation, membrane filtration, *etc*, have been used for decades. These techniques do not entirely eradicate pollutants, disrupt the environment, have a high cost of implementation, and, in some cases, form recalcitrant derivatives [3, 4].

Bioremediation embodies an advanced and environmentally sustainable approach that employs biological microorganisms to break down and mitigate pollutants. The basic mechanism of bioremediation is to degrade, mineralize, and detoxify pollutants to decrease the concentration of contaminants in polluted areas and restore the ecosystem [5]. Recently, this has been done using a variety of organisms, such as bacteria, algae, fungi, and plants. Certain species include *Xanthobacter, Arthrobacter, Pseudomonas, Bacillus, Mycobacterium, Pseudomonas aeruginosa, etc* [2]. These microorganisms can be native to the contaminated site or may be added from outside for the process to proceed.

Bioremediation is advancing with technical innovations, enhancing its effectiveness and usefulness in restoring contaminated areas. It poses various challenges in its application, such as the complexity of contaminants at the site, which makes the degradation process difficult; environmental conditions such as temperature, pH, moisture content, and nutrient availability greatly affect the performance of these techniques; the risk of secondary contamination; solubility and bioavailability of contaminants; and microbial dynamics. Despite its challenges, bioremediation utilizes the natural metabolic processes of living organisms to clean up waste. Its effectiveness can be increased by understanding and optimizing the factors affecting bioremediation and selecting appropriate remediation techniques [6].

BIOREMEDIATION TECHNIQUES

Bioremediation procedures can be broadly divided into two categories, as shown in Fig. (**1**), based on where the process is carried out: *ex-situ* bioremediation and *in-situ* bioremediation techniques.

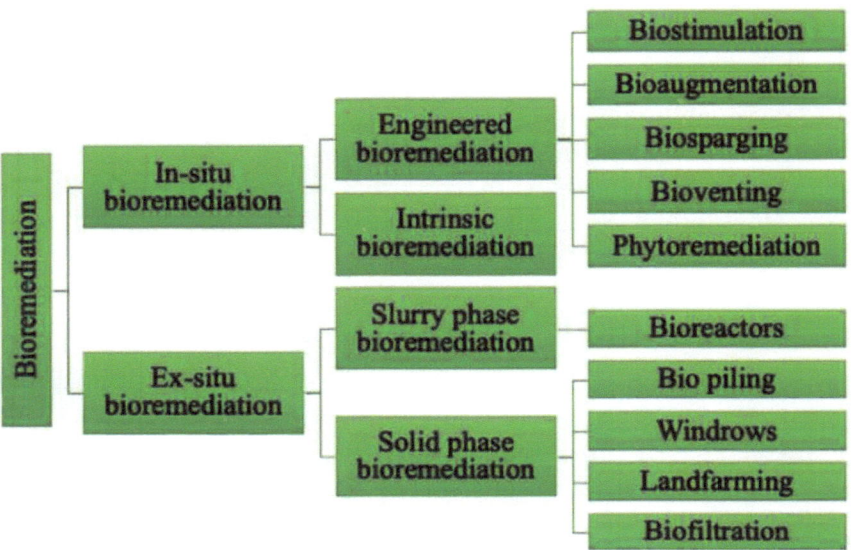

Fig. (1). Diverse bioremediation techniques [3].

In-situ Bioremediation Techniques

These bioremediation techniques involve the treatment of polluted surfaces where they are located. They are utilized when transporting contaminated surfaces, such as soil, from one location to another, which is not feasible. It necessitates no excavation and induces minimal or no surface disturbance, making it more cost-effective than *ex-situ* bioremediation techniques except for designing and installing complex equipment. They are eco-friendly and complete tasks in shorter time frames [3, 4, 7]. It has given promising results in the sustainable management of contaminated sediments [8]. It encompasses various techniques that work by incorporating nutrient supplements and electron acceptors to promote and sustain the activity of native microorganisms to transform pollutants into non-toxic end products [9].

In-situ bioremediation can be distributed into two categories: Intrinsic and Engineered bioremediation.

Intrinsic Bioremediation

Intrinsic bioremediation or natural attenuation is a remediation process that involves the activation of the native microorganisms present in polluted areas. The procedure aims to stimulate the existing aerobic or anaerobic microbial population at the contamination site to degrade the pollutants. It requires no external force, making it less expensive than engineered bioremediation techniques [5, 3]. This is done considering that the indigenous microbes have already adapted to the site environment and have established a microbe-hydrocarbon relationship, making them highly effective for the biodegradation of contaminants [10]. Natural attenuation has been used by many countries, such as the United States, Europe, *etc*, for decades to control and manage the spreading of contamination to other regions [11].

Engineered In-situ bioremediation

This method introduces specific microorganisms into the contamination site to degrade the pollutants. The decomposition ability of indigenous microorganisms is accelerated by improving physicochemical conditions around them by introducing genetically modified microorganisms [5, 3, 12]. Genetically Engineered Microorganisms (GEMs) are created in laboratory settings *via* genetic engineering and used to promote the microbial degradation of chemicals polluting the environment. For example, a genetically modified Pseudomonas aeruginosa, which can absorb Cd (II), and a genetically modified Pseudomonas fluorescens HK44 have been used in the remediation of PAHs from heavy metals contaminated sites [13]. Among other microbes, Xanthobacter autotrophicus was seen to remove DCA (1,2-dichloroethane) from groundwater contamination [14].

Biostimulation

Biostimulation is recognized as an exceptionally effective, cost-efficient, and eco-friendly remediation technique [15]. It employs the application of high amounts of nutrients such as oxygen, phosphorus, and nitrogen, electron donors and acceptors, and optimizing the surrounding pH, temperature, and aeration to enhance naturally occurring microbial communities at the pollution site to facilitate the breakdown of hazardous and toxic compounds into nontoxic and less harmful forms [16, 4]. Stimulations are usually given underground *via* injection wells. This method primarily benefits from including locally occurring microorganisms well adapted to and prevalent in the environment [10]. Some microorganisms that respire chlorinated solvents, tetrachloroethane (PCE), are Firmicutes, Anaeromyxobacter, Desulfitobacterium, Propionibacterium, *etc* [17]. This is primarily successful for the remediation of hydrocarbon and petroleum derivatives, crude oil, and chlorinated solvents contamination. Since petroleum serves as a rich carbon source for the microorganisms to proliferate, a small increase in the concentrations of other rate-limiting nutrients following their addition substantially speeds up the degradation rate [18].

Some challenges that may be faced are that the remediation process will slow down for an old polluted surface area, as very little microbial activity will be available to break down pollutants [17]. Tight clays, grainy material, or fractures in the subsurface area can prevent equal distribution and penetration of nutrients. Biostimulation might also stimulate heterotrophic microbes that may create competition between innate degraders [19].

Bioaugmentation

It involves the inoculation of naturally occurring or genetically modified microbial strains of archaea or bacteria to hazardous waste surfaces, unlike biostimulation, which utilizes nutrients to support native bacteria. This is done to detoxify specific contaminants in polluted soil or groundwater to reduce the levels of unwanted waste [15, 18]. Mixed cultures are preferable over pure bacterial strains due to their flexibility and higher potential to degrade acrylic, phenolic compounds, and other industrial wastes [16, 18]. Bioaugmentation is generally applied to reduce cyanide, chlorinated solvents, nicotine, natural pollutants, xenobiotics, and contaminated sites using *Cryptococcus humicolus, Acinetobacter sp., Achromobacter, Nitrosomonas, Xanthobacter, Rhodococcus spp, Pseudomonas, etc* [18, 19]. Studies suggest that bioaugmentation gave better remediation results than biostimulation on sites with low bioactivity [17].

For a successful bioaugmentation, strain selection is crucial. It must be able to break down most chemicals, maintain its genetic stability and viability, compete with native species, and survive in hostile environments [19]. Failure of bioaugmentation depends upon extreme environmental conditions, such as low temperatures or high pH. Low pollutant concentrations can hinder the proliferation rates of microbes, whereas high concentrations may cause biotoxicity in strains, competition between indigenous microorganisms, high cost of operation for complete irradiation of contaminants over a period of time, and the lack of appropriate regulation and optimization plans [20]. A comparison between biostimulation and bioaugmentation is depicted in Fig. (**2**).

Biosparging

This technique involves the injection of air by installing sparge points below the water table into the saturated environment through boreholes to stimulate the activity of native microbes [12]. Due to the increase in the concentration of oxygen available, the ability of native microbes to break down contaminants is increased. It is typically employed to remove or transfer volatile organic compounds, such as xylene isomers, upward to the unsaturated environment, sometimes necessitating vapor extraction [4]. A more than 70% decrease in BTEX was also seen in BTEX-contaminated water [15].

Bioventing

Bioventing is an environmentally friendly, non-invasive, low-cost bioremediation technique similar to biosparging. This involves a controlled airflow supply to the unsaturated zone to increase indigenous microbial activity and enhance the biodegradation of contaminants [7]. In contrast to biosparging, adding nutrients,

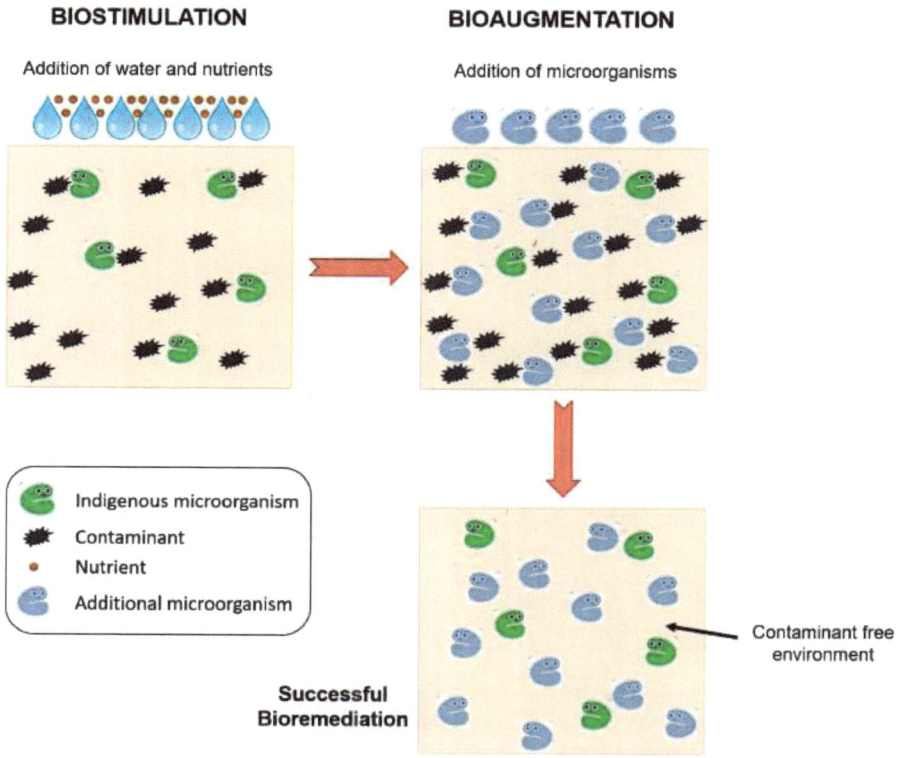

Fig. (2). Comparison between Bioaugmentation and Biostimulation techniques [10].

including nitrogen, phosphorus, and moisture, transforms pollutants into innocuous forms. The parameters affecting this procedure's efficiency are air injection and airflow rates [15]. This technique effectively cleans up motor oil-contaminated sites such as diesel fuel, jet fuel, gasoline, chlorinated, and volatile organic compounds [21].

Phytoremediation

This type of *in-situ* bioremediation implements flora to restore, detoxify, and regulate the effects of toxins. This approach employs plant interactions across physical, biological, chemical, biochemical, and microbiological dimensions to mitigate pollution toxicity [22]. Phytoremediation employs plants primarily because of their resilience in adverse circumstances, ability to absorb pollutants, adaptability to various climatic conditions, and mineralization to reclaim a healthy ecosystem for living organisms [23]. It has shown remarkable results in mitigating air pollutants, heavy metal sewage, and organic pollution of soil caused by formaldehyde, hexane, toluene, mercury, copper, chromium, iron, cobalt, BTEX, *etc*. Some plant species used are *Scindapsus aureus*, *N. obliterata*, *Pistia stratiotes*, *Azolla pinnata*, *etc* [24]. Depending upon the type of target contamination, various ways are utilized, including extraction, filtration, stabilization, volatilization, and degradation, which are illustrated in Fig. (**3**).

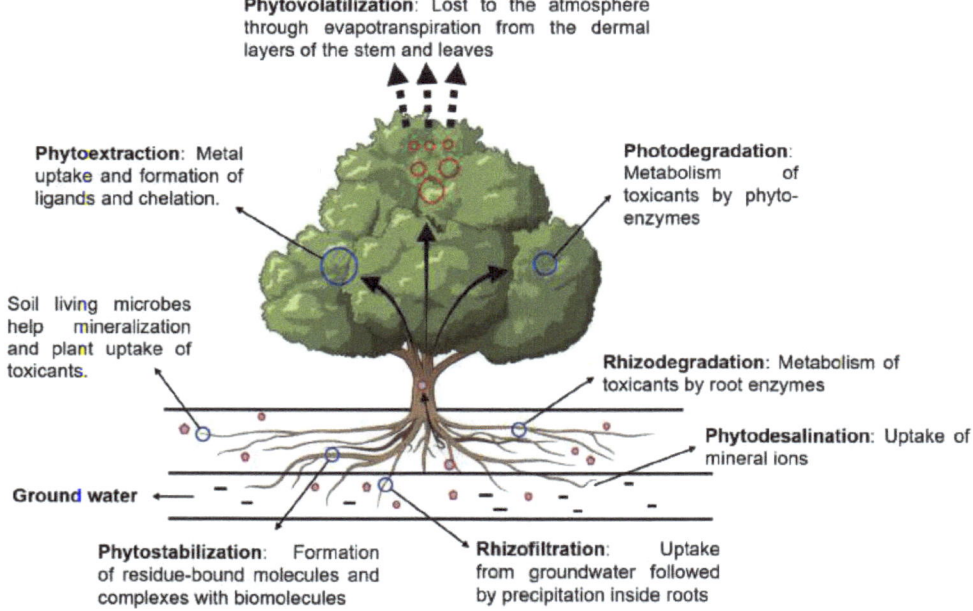

Fig. (3). Phytoremediation processes and their associated functions [29].

Phytoextraction: The extraction or removal of contaminants such as heavy metals using hyperaccumulator plant species at the contamination site. This process is also known as phytoaccumulation due to the transfer of toxins through the root system to the shoot system of hyperaccumulators [19]. Some examples of hyperaccumulators used are *Solanum nigrum L., Xanthium strumarium L., Helianthus annuus L.,etc* [25].

Phytofiltration: Also known as rhizofiltration, it is a cheap and eco-friendly technique to filter out harmful substances or impurities at a specific contaminated site *via* the root systems of hyperaccumulator plant species. These plants selectively absorb and precipitate pollutants, which are then stored in the plant biomass. After they absorb enough contaminants, they are disposed of, removing high amounts of pollutants from the environment [26].

Phytostabilization: Like rhizofiltration, this method employs toxin-tolerant plants to avoid mass erosion, airborne transmission, and leaching of pollutants to nearby areas [27]. It generally requires covering the site with plants to immobilize pollutants below ground. It is a time-consuming procedure with regular monitoring and maintenance, which poses a risk of environmental pollution as the pollutants are not removed but retained at the contamination site [28]. A notable advantage over phytofiltration is that it removes the need for plant biomass disposal [16].

Phytovolatilization: It is a process where plants remove organic pollutants from the soil by absorbing and increasing the volatility of compounds and releasing them into the atmosphere through volatilization. Compounds can volatilize from any plant part, like the stem, leaves, or root-soil interaction. Many compounds move upwards in the shoot system of the plant and are released *via* transpiration [29].

Phytodegradation: Phytodegradation involves harnessing plants' enzymatic activity and their ability to destroy xenobiotics inside plant tissues. This is done by various interactions between soil, microbes, water, and plants, where plants provide surface area for bacteria to colonize and accelerate the breakdown process. Another name for phytodegradation is rhizodegradation, as the degradation process is seen mainly in rhizospheres. Its effectiveness was seen in various pollution sites, including wastewater treatment, groundwater, soil, sludges, industrial waste sites, agricultural land, *etc*, by the degradation of chlorinated hydrocarbons, organic pollutants such as pesticides, industrial effluents, and petroleum hydrocarbons [29, 30].

Phytoremediation has shown great potential for efficiently cleaning the environment using eco-friendly techniques to eradicate contaminants and pollution without needing large-scale machinery or setups. It is economical, cost-effective, and enhances the environment's aesthetics with well-chosen plants [30]. Plant biomass harvested from phytoextraction and phytofiltration can provide an abundant supply for bio-oil, biosorbent, and bioenergy production. Despite its many advantages, it is a lengthy process that majorly depends upon the selected plant species, climatic conditions, soil properties, waste-disposal concerns, concerns for food-chain contamination if neglected, flooding, aging, and pollutant characteristics, which makes it difficult to manage in the long run [31].

Ex-situ Bioremediation

These approaches encompass excavating toxins from the contaminated site and transporting them to a different location for treatment. The *ex-situ* bioremediation technique depends upon a range of factors, including the extent and depth of pollution, cost of operation, type of contamination, location, and geographic characteristics, and performance criteria [7]. These techniques are expensive compared to *in-situ* bioremediation but provide greater control over their progression and can be utilized for an extensive range of contaminants [32, 15]. It is mainly used to treat domestic, agricultural, sewage, sludge, municipal waste, industrial waste, *etc*, to produce compost to enhance soil fertility. While performing these methods, some adjustments are made if required in terms of pH, nutrient availability, moisture, aeration, temperature, *etc*, essential for the growth of microorganisms [33].

Ex-situ bioremediation can be divided into categories depending upon the phases of contamination: Slurry phase bioremediation and Solid phase bioremediation.

Slurry phase bioremediation

Slurry phase bioremediation works by blending contaminated soil or water into a slurry and then treating it using slurry phase bioreactors. This method is faster and more effective than any other bioremediation technique, degradation rates of organic pollutants are higher than *in-situ* bioremediation, it provides more contact efficiency of microbes with nutrients and pollutants, more than one electron acceptors can be used, and it gives complete control over the process by making adjustments to various parameters such as pH, temperature, aeration, *etc*., as per the requirement [34, 35]. Even though the equipment, excavation, and construction costs are high, it is still more effective than other physicochemical remediation processes. Its applications range from remediation of polychlorinated biphenyls (PCBs), Polycyclic Aromatic Hydrocarbons (PAHs), total petroleum hydrocarbons (TPH), and pesticide contamination [35].

Bioreactors

A bioreactor is a vessel designed to convert polluted raw material *via* biological processes into a specific product under a controlled environment. It provides optimum conditions for microbial growth. The bioreactor can be operated in different modes depending on the type of pollution removal and installation cost: batch, fed-batch, continuous, multistage, and sequencing [7]. They are utilized to effectively treat heavy metals and volatile compounds, including benzene, toluene, BTEX, pesticides, *etc*, as the operating parameters can be managed [10]. Bioreactors can also be used for biostimulation and bioaugmentation using Genetically Engineered Microorganisms (GEM).

Aerobic Granular Sludge Membrane Bioreactors (AGMBR), similar to membrane bioreactors, are found effective in wastewater treatment with a percentage removal of 92% for Total Organic Contaminants (TOC), 76% for Total Nitrogen (TN), and 82% for Total Phosphorus (TP) [36]. Osmotic membrane bioreactors (OMBR) have shown excellent results in removing 90-95% of TOC and 100% of phosphate [37].

Solid phase bioremediation

Solid phase bioremediation is an *ex-situ* bioremediation technique that involves the extraction of polluted soil and forming it into heaps. Organic waste, such as animal dung, leaves, agricultural waste, household waste, industrial waste, *etc*, is added to the heaps. Air circulation is maintained for ventilation and microbial growth. This technique requires more space and time to achieve the desired result [7].

Bio Piling

Also known as biocells, bioheaps, or biomounds, it is an advanced technology that involves piling the contaminated surface area and supplying nutrients, aeration through a piping system, and irrigation to support microbial growth. The stimulated microbes facilitate the degradation of excavated pollutants aerobically [33]. This type of bioremediation has gained much attention due to its cost-effectiveness, flexibility to adjust the temperature, pH, humidity, nutrient availability, aeration if required, and less space utilization. Bio-piling has widely been used to treat volatile compounds, petroleum hydrocarbons, low molecular weight pollutants, and harsh cold environments [5]. The introduction of a heating system reduces remediation time and increases microbial reactivity. It has previously showcased effective treatment of diesel contamination of soil in the sub-Antarctic region by removing about 93% of TPH in a year [3].

Windrows

Windrows focuses on the periodic turning of heaped polluted soil to accelerate the bioremediation process by enhancing the degradation rates of hydrocarbonoclastic bacteria through assimilation, biotransformation, and mineralization. Microbial reactivity can also be increased by adjusting aeration, adding water and nutrients, and ensuring uniform distribution of pollutants [7]. Windrows were found to have a higher rate of hydrocarbon removal than bio-pile treatment [5]. One such example of successful remediation is of the oil spill sites in the Ogoniland communities in the Niger Delta, where the topsoil of 15-20cm depth was piled to form windrows with a height of 30-40cm [38].

Landfarming

Landfarming is a simple and low-cost bioremediation technique employed in agricultural settings. It requires no pre-assessment of the contaminated soil and minimal supervision, and it can be used to remediate large volumes of soil. After the contaminated soil is extracted, it can be spread over a prepared bed, and tilling is performed occasionally until all the pollutants are degraded [18]. Generally, landfarming is considered an *ex-situ* bioremediation technique, but sometimes, if the treatment is done on-site, it is regarded as an *in-situ* bioremediation technique. Enhanced landfarming uses fertilizers and hydrocarbon-degrading microbes in bioaugmentation and employs different irrigation procedures [39]. This technique is widely used to treat polycyclic and aliphatic hydrocarbons and PCB-contaminated sites [40].

Biofiltration

Biofiltration is an *ex-situ* bioremediation method used to treat contaminated air and water through biological filters, which consist of filter media with microorganisms immobilized on it. The filtration process starts by absorbing air and water contaminants through the biofilm or cellular membrane. The contaminants are transferred to the bed media, where the microorganisms use them as a carbon source, increasing in number and forming colonies. These colonies perform their normal metabolic activities and degrade the contaminants. In the end, the purified water and air are expelled [41]. Various microbes utilized for biofiltration of Volatile Organic Compounds (VOC), BTEX, and air pollutants such as NO_x, SO_2, CO_2, *etc*, are *P.putida, Bacillus subtilis, Pseudomonas fluorescens, Alcaligenes xylosoxidans, Pseudomonas aeruginosa, Candida subhashii, Fusarium solani,* and so on, with 80-90% removal efficiency [42].

FUTURE PERSPECTIVES IN BIOREMEDIATION

Biotechnology and genetic engineering in bioremediation processes have remarkable potential to fulfill the current needs for a sustainable, efficient, and effective solution to eradicate environmental pollution.

Genetically Engineered Microorganisms (GEM)

Advancements in biotechnology have enabled researchers to modify bacteria at the gene level to target specific pollutants, extend the substrate spectrum, include recalcitrant compounds, and increase catabolic activity and stability [18]. Gene editing methods, such as CRISPR-Cas, TALENs, and ZFNs, have produced genetically modified *P. putida* KT2440, *Sphingobium japonicum,* and *Pseudomonas* sp. WBC-3, *Pteris vittata, Arabidopsis helleri,etc*, gave successful bioremediation and phytoremediation results. Unlike the low-throughput ZFNs and TALENs, CRISPR-Cas is a next-generation genetic manipulation technique that researchers widely accept [6, 42].

Bioremediation by Nano Materials

The application of nanoparticles in bioremediation is occasionally termed Nanoremediation. Nanoparticles can serve as carriers for bacteria or be employed directly for cleanup purposes. It is a cheap and fast process that can be used for large-scale cleanup of contaminated sites [5]. Nanoremediation has recently been utilized for wastewater treatment and recycling due to its high functionality, enhanced catalysis, and reactivity. Industrial effluents have been treated to provide water with reduced toxic compounds, heavy metals, and other organic and inorganic contaminants, as reported by Salem *et al.* [43]. As compared to conventional techniques, nano-particles have shown targeted removal of contaminants such as dyes including methylene blue, eriochrome black T, methyl orange; chlorinated compounds, and heavy metal ions like vanadium (V), chromium (Cr), lead (Pb), nickel (Ni), copper (Cu), zinc (Zn), *etc* [44].

Bioinformatics in Bioremediation

Bioinformatics tools and databases are involved in bioremediation to analyze the chemical structure and composition of xenobiotic compounds, catalytic enzymes, degradation pathways, and protein expression capacity of microbes. Research and review articles can be viewed using literature databases such as PubMed and SCOPUS. Biodegradative databases are used to obtain information about the degradation of xenobiotic compounds. Some examples of biodegradative databases are PMBD (Plastics Microbial Biodegradation Database), OxDBase (A database of Biodegradative oxygenases), EAWAG-BBD (Biocatalysis/ Biodegradation Database), BioCyc, BBD, *etc*. Pathway prediction tools utilize knowledge-based and machine learning-based algorithms to identify the degradation mechanisms of microbes, such as BNICE (Biochemical Network Integrated Computational Explorer), EAWAG-Pathway Prediction System (PPS), *etc* [45].

CONCLUSION

Bioremediation provides a leading sustainable approach to address the urgent problem of environmental contamination triggered by industrial processes, intensive agriculture, and chemical misuse. The detrimental impact of pollutants, including heavy metals, hydrocarbons, pesticides, and microplastics, on ecosystems and human health raises the urgent need for innovative remediation strategies. Bioremediation harnesses the natural metabolic processes of microorganisms and plants to detoxify various environmental pollutants.

In-situ techniques, such as biostimulation, bioaugmentation, biosparging, bioventing, and phytoremediation, enable pollutant breakdown directly at the contaminated site, reducing operational costs and environmental disturbances. Meanwhile, *ex-situ* techniques, including biofiltration, landfarming, bio piles, windrows, and bioreactors, offer enhanced control over the polluted area and are often better suited for deeply polluted or geographically challenging sites. Despite being more resource-intensive, these methods are highly adaptable and effective when used strategically.

The integration of advanced technologies, particularly genetic engineering and nanotechnology, has significantly expanded the potential of bioremediation. Genetically Engineered Microorganisms (GEMs) and nanoparticles have shown remarkable efficiency in breaking down stubborn contaminants and providing targeted pollutant removal. Furthermore, the development of bioinformatics tools has accelerated research by mapping metabolic pathways, identifying degradation mechanisms, and optimizing microbial strains for more effective outcomes. With continuous research advancements, refinement, and proper regulation, bioremediation can address any contaminated site and convert it into thriving ecosystems.

REFERENCES

[1] Priya AK, Muruganandam M, Kumar A, *et al*. Recent advances in microbial-assisted degradation and remediation of xenobiotic contaminants; challenges and future prospects. J Water Process Eng 2024; 60: 105106.
[http://dx.doi.org/10.1016/j.jwpe.2024.105106]

[2] Hu F, Wang P, Li Y, *et al*. Bioremediation of environmental organic pollutants by *Pseudomonas aeruginosa*: Mechanisms, methods and challenges. Environ Res 2023; 239(Pt 1): 117211.
[http://dx.doi.org/10.1016/j.envres.2023.117211] [PMID: 37778604]

[3] Bala S, Garg D, Thirumalesh BV, *et al*. Recent strategies for bioremediation of emerging pollutants: a review for a green and sustainable environment. Toxics 2022; 10(8): 484.
[http://dx.doi.org/10.3390/toxics10080484] [PMID: 36006163]

[4] Mokrani S, Houali K, Yadav KK, *et al*. Bioremediation techniques for soil organic pollution: Mechanisms, microorganisms, and technologies - A comprehensive review. Ecol Eng 2024; 207: 107338.

[http://dx.doi.org/10.1016/j.ecoleng.2024.107338]

[5] Alori ET, Gabasawa AI, Elenwo CE, Agbeyegbe OO. Bioremediation techniques as affected by limiting factors in soil environment. Front Soil Sci 2022; 2: 937186.
[http://dx.doi.org/10.3389/fsoil.2022.937186]

[6] Kuppan N, Padman M, Mahadeva M, Srinivasan S, Devarajan R. A comprehensive review of sustainable bioremediation techniques: eco friendly solutions for waste and pollution management. Waste Manag Bull 2024 Sep; 2(3): 154–71.
[http://dx.doi.org/10.1016/j.wmb.2024.07.005]

[7] Azubuike CC, Chikere CB, Okpokwasili GC. Bioremediation techniques–classification based on site of application: principles, advantages, limitations and prospects. World J Microbiol Biotechnol 2016; 32(11): 180.
[http://dx.doi.org/10.1007/s11274-016-2137-x] [PMID: 27638318]

[8] Majone M, Verdini R, Aulenta F, *et al.* In situ groundwater and sediment bioremediation: barriers and perspectives at European contaminated sites. N Biotechnol 2015; 32(1): 133-46.
[http://dx.doi.org/10.1016/j.nbt.2014.02.011] [PMID: 24607450]

[9] Hu Z, Chan CW. *In-situ* bioremediation for petroleum contamination: A fuzzy rule-based model predictive control system. Eng Appl Artif Intell 2015; 38: 70-8.
[http://dx.doi.org/10.1016/j.engappai.2014.10.019]

[10] Sales da Silva IG, Gomes de Almeida FC, Padilha da Rocha e Silva NM, Casazza AA, Converti A, Asfora Sarubbo L. Soil Bioremediation: Overview of Technologies and Trends. Energies 2020; 13(18): 4664.
[http://dx.doi.org/10.3390/en13184664]

[11] Balland-Bolou-Bi C, Brondeau F, Jusselme MD. Can Natural Attenuation be Considered as an Effective Solution for Soil Remediation? In: Mustafa A, Naveed M, eds. Soil Contamination - Recent Advances and Future Perspectives. London: IntechOpen; 2023. p. 1-17. Available from: https://www.intechopen.com/chapters/84766.

[12] Abo-Alkasem MI, Hassan NH, Abo Elsoud MM. Microbial bioremediation as a tool for the removal of heavy metals. Bull Natl Res Cent 2023; 47(1): 31.
[http://dx.doi.org/10.1186/s42269-023-01006-z]

[13] Wu C, Li F, Yi S, Ge F. Genetically engineered microbial remediation of soils co-contaminated by heavy metals and polycyclic aromatic hydrocarbons: Advances and ecological risk assessment. J Environ Manage 2021; 296: 113185.
[http://dx.doi.org/10.1016/j.jenvman.2021.113185] [PMID: 34243092]

[14] Janssen DB, Stucki G. Perspectives of genetically engineered microbes for groundwater bioremediation. Environ Sci Process Impacts 2020; 22(3): 487-99.
[http://dx.doi.org/10.1039/C9EM00601J] [PMID: 32095798]

[15] Sanjana M, Prajna R, Katti US, Kavitha RV. Bioremediation–the recent drift towards a sustainable environment. Environ Sci Adv 2024; 3(8): 1097-110.
[http://dx.doi.org/10.1039/D3VA00358B]

[16] Aljabri M. Recent advances in pesticide bioremediation: integrating microbial, phytoremediation, and biotechnological strategies—a comprehensive review. Environ Pollut Bioavailat 2025; 37(1): 2554173.
[http://dx.doi.org/10.1080/26395940.2025.2554173]

[17] Romantschuk M, Lahti-Leikas K, Kontro M, *et al.* Bioremediation of contaminated soil and groundwater by *in situ* biostimulation. Front Microbiol 2023; 14: 1258148.
[http://dx.doi.org/10.3389/fmicb.2023.1258148] [PMID: 38029190]

[18] Sharma S, Singh G, Mehta P, Ranout AS. Bioremediation As A Strategy To Combat Soil Pollution. In: Futuristic Trends in Biotechnology. Vol 3. IIP Series; 2023. p. 105-14.

[http://dx.doi.org/10.58532/V3BJBT11P4CH1]

[19] Omokhagbor Adams G, Tawari Fufeyin P, Eruke Okoro S, Ehinomen I. Bioremediation, biostimulation and bioaugmention: a review. Int J Environ Bioremediat Biodegrad 2020; 3(1): 28-39.
[http://dx.doi.org/10.12691/ijebb-3-1-5]

[20] Ma H, Zhao Y, Yang K, Wang Y, Zhang C, Ji M. Application oriented bioaugmentation processes: Mechanism, performance improvement and scale-up. Bioresour Technol 2022; 344(Pt B): 126192.
[http://dx.doi.org/10.1016/j.biortech.2021.126192] [PMID: 34710609]

[21] Khodabakhshi Soureshjani M, Zytner RG. Developing a Robust Bioventing Model. Math Comput Appl 2023; 28(3): 76.
[http://dx.doi.org/10.3390/mca28030076]

[22] Praveen R, Nagalakshmi R. Review on bioremediation and phytoremediation techniques of heavy metals in contaminated soil from dump site. Mater Today Proc 2022; 68: 1562-7.
[http://dx.doi.org/10.1016/j.matpr.2022.07.190]

[23] Odoh CK, Zabbey N, Sam K, Eze CN. Status, progress and challenges of phytoremediation - An African scenario. J Environ Manage 2019; 237: 365-78.
[http://dx.doi.org/10.1016/j.jenvman.2019.02.090] [PMID: 30818239]

[24] Wei Z, Van Le Q, Peng W, et al. A review on phytoremediation of contaminants in air, water and soil. J Hazard Mater 2021; 403: 123658.
[http://dx.doi.org/10.1016/j.jhazmat.2020.123658] [PMID: 33264867]

[25] Yu F, Tang S, Shi X, et al. Phytoextraction of metal(loid)s from contaminated soils by six plant species: A field study. Sci Total Environ 2022; 804: 150282.
[http://dx.doi.org/10.1016/j.scitotenv.2021.150282] [PMID: 34798760]

[26] Kristanti RA, Ngu WJ, Yuniarto A, Hadibarata T. Rhizofiltration for removal of inorganic and organic pollutants in groundwater: a review. Biointerface Res Appl Chem. 2021; 11(4): 12326–47.
[http://dx.doi.org/10.33263/BRIAC114.1232612347]

[27] Bakshe P, Jugade R. Phytostabilization and rhizofiltration of toxic heavy metals by heavy metal accumulator plants for sustainable management of contaminated industrial sites: A comprehensive review. J Hazard Mater Adv 2023; 10: 100293.
[http://dx.doi.org/10.1016/j.hazadv.2023.100293]

[28] Bashir Z, Raj D, Selvasembian R. A combined bibliometric and sustainable approach of phytostabilization towards eco-restoration of coal mine overburden dumps. Chemosphere 2024; 363: 142774.
[http://dx.doi.org/10.1016/j.chemosphere.2024.142774] [PMID: 38969231]

[29] Kafle A, Timilsina A, Gautam A, Adhikari K, Bhattarai A, Aryal N. Phytoremediation: Mechanisms, plant selection and enhancement by natural and synthetic agents. Environ Adv 2022; 8: 100203.
[http://dx.doi.org/10.1016/j.envadv.2022.100203]

[30] Sharma M, Rawat S, Rautela A. Phytoremediation in sustainable wastewater management: an eco-friendly review of current techniques and future prospects. AQUA—Water Infrastructure. Ecosystems and Society 2024; 73(9): 1946-75.
[http://dx.doi.org/10.2166/aqua.2024.427]

[31] Babu SMOF, Hossain MB, Rahman MS, et al. Phytoremediation of toxic metals: a sustainable green solution for clean environment. Appl Sci (Basel) 2021; 11(21): 10348.
[http://dx.doi.org/10.3390/app112110348]

[32] Paul O, Jasu A, Lahiri D, Nag M, Ray RR. In situ and ex situ bioremediation of heavy metals: the present scenario. J Environ Eng Landsc Manag 2021; 29(4): 454-69.
[http://dx.doi.org/10.3846/jeelm.2021.15447]

[33] Maglione G, Zinno P, Tropea A, Mussagy CU, Dufossé L, Giuffrida D, Mondello A. Microbes' role in environmental pollution and remediation: a bioeconomy focus approach. AIMS Microbiol. 2024;

10(3): 723-55.
[http://dx.doi.org/10.3934/microbiol.2024033]

[34] Wang F, Sun J, Pang R, Xiao X, Wang X, Lou H. Bio-slurry-based biodegradation technology for organically contaminated soils: current work and future directions. J Environ Chem Eng 2024; 12(2): 112033.
[http://dx.doi.org/10.1016/j.jece.2024.112033]

[35] Sun J, Wang F, Jia X, Wang X, Xiao X, Dong H. Research progress of bio-slurry remediation technology for organic contaminated soil. RSC Advances 2023; 13(15): 9903-17.
[http://dx.doi.org/10.1039/D2RA06106F] [PMID: 37034448]

[36] Truong HTB, Bui HM. Potential of aerobic granular sludge membrane bioreactor (AGMBR) in wastewater treatment. Bioengineered 2023; 14(1): 2260139.
[http://dx.doi.org/10.1080/21655979.2023.2260139] [PMID: 37732563]

[37] Aftab B, Khan SJ, Maqbool T, Hankins NP. Heavy metals removal by osmotic membrane bioreactor (OMBR) and their effect on sludge properties. Desalination 2017; 403: 117-27.
[http://dx.doi.org/10.1016/j.desal.2016.07.003]

[38] Mafiana MO, Bashiru MD, Erhunmwunsee F, Dirisu CG, Li SW. An insight into the current oil spills and on-site bioremediation approaches to contaminated sites in Nigeria. Environ Sci Pollut Res Int 2021; 28(4): 4073-94.
[http://dx.doi.org/10.1007/s11356-020-11533-1] [PMID: 33188631]

[39] Yap HS, Zakaria NN, Zulkharnain A, Sabri S, Gomez-Fuentes C, Ahmad SA. Bibliometric analysis of hydrocarbon bioremediation in cold regions and a review on enhanced soil bioremediation. Biology (Basel) 2021; 10(5): 354.
[http://dx.doi.org/10.3390/biology10050354] [PMID: 33922046]

[40] Nayak P, Solanki H. Impact of agriculture on environment and bioremediation techniques for improvisation of contaminated site. Int Assoc Biol Comput Digest 2022; 1: 145-56.
[http://dx.doi.org/10.56588/iabcd.v1i1.29]

[41] Pachaiappan R, Cornejo-Ponce L, Rajendran R, Manavalan K, Femilaa Rajan V, Awad F. A review on biofiltration techniques: recent advancements in the removal of volatile organic compounds and heavy metals in the treatment of polluted water. Bioengineered 2022; 13(4): 8432-77.
[http://dx.doi.org/10.1080/21655979.2022.2050538] [PMID: 35260028]

[42] Isha AS, Ali S, Khalid A, Naseer IA, Raza H, Chang Y-C. Bioremediation of Smog: Current Trends and Future Perspectives. Processes (Basel) 2024; 12(10): 2266.
[http://dx.doi.org/10.3390/pr12102266]

[43] Salem SS, Fouda A. Green synthesis of metallic nanoparticles and their prospective biotechnological applications: an overview. Biol Trace Elem Res 2021; 199(1): 344-70.
[http://dx.doi.org/10.1007/s12011-020-02138-3] [PMID: 32377944]

[44] Kapoor RT, Salvadori MR, Rafatullah M, Siddiqui MR, Khan MA, Alshareef SA. Exploration of microbial factories for synthesis of nanoparticles–a sustainable approach for bioremediation of environmental contaminants. Front Microbiol 2021; 12: 658294.
[http://dx.doi.org/10.3389/fmicb.2021.658294] [PMID: 34149647]

[45] Arora PK, Kumar A, Srivastava A, Garg SK, Singh VP. Current bioinformatics tools for biodegradation of xenobiotic compounds. Front Environ Sci 2022; 10: 980284.
[http://dx.doi.org/10.3389/fenvs.2022.980284]

CHAPTER 6

Microbial Communities in Environmental Engineering

Arunima Singh[1], Sonali Ranjan[1], Suparna Bardhan[2], Anupam Jayas[2], Yogesh Kumar Vishwakarma[1] and R. S. Singh[1,*]

[1] *Department of Chemical Engineering and Technology, Indian Institute of Technology, Banaras Hindu University, Varanasi, Uttar Pradesh, India*

[2] *Department of Botany, Institute of Science, Banaras Hindu University, Varanasi, Uttar Pradesh, India*

Abstract: Microbial communities play an important role in environmental engineering by providing natural solutions for pollution control, waste treatment, and resource recovery. This chapter will explore the complex dynamics of microbes in engineered systems, which are involved in the complex biogeochemical processes, specifically those engaged during the breakdown of toxic organic pollutants to less harmful substances and the cycling of nutrients in the ecosystem. The main focus involves the composition, functions, and interactions of the microbial colonies. For instance, in the field of wastewater treatment, bacteria like *Nitrosomonas* and *Nitrobacter* play a vital role in processes like nitrification; in addition to that, other microbes like *Pseudomonas aeruginosa* and *Burkholderia pseudomallei* are efficient in degrading hydrocarbons. Some microbes act as heavy metal reducers, like *Bacillus sp., Acinetobacter schindleri,* and *Rhodococcus sp.*, which have the potential to reduce Chromium (VI). Other than these, some microbes, such as *Citrobacter amalonaticus* and *Enterobacter cloacae,* are also effective in degrading different pollutants. Environmental microbiome engineering is one of the emerging strategies in mitigating climate change, which is basically modified to enhance ecosystem functions.

Keywords: Environmental engineering, Heavy metals, Microbial communities, Microbial dynamics, Waste treatment.

INTRODUCTION

One of the best ways to describe environmental engineering is the wide range of problems that its practitioners deal with. In general, environmental engineers provide solutions and systems at the human-environment interface. Historically,

---

* **Corresponding author R. S. Singh:** Department of Chemical Engineering and Technology, Indian Institute of Technology (BHU), Varanasi, Uttar Pradesh, India; E-mail: rssingh.che@itbhu.ac.in

Rajneesh Kumar, Ram Sharan Singh & Maulin P. Shah. (Eds.)
All rights reserved-© 2026 Bentham Science Publishers

this work concentrated on water supply and wastewater treatment, utilizing the field's foundations in public health protection and sanitation system design. As the field's focus expanded to encompass the mitigation of pollution in air, water, and soil in the 1970s, the name "environmental engineering" supplanted the earlier term, "sanitary engineering." At about the same time, the field's design methodology changed from emphasizing engineered treatment systems to a stronger focus on ecological processes and concepts. The field has lately broadened to include efforts like green manufacturing and sustainable urban design, as well as toxins and chemical exposures from products and materials [1].

Microbial community-based processes have existed for almost a century [2 - 4]. Anaerobic sludge digestion and activated sludge (AS) treatment of wastewater are two established methods that were successfully used even before their microbiological underpinnings were known [5]. Engineering bioreactors that are efficient, durable, and immune to shocks requires knowledge of the microbial populations that fundamentally govern bioreactor performance [6].

The recognition of ecological and microbiological concepts in environmental engineering has made significant advancements feasible. One excellent example is biological nutrient removal (BNR), which allows for the total removal of Nitrogen (N) and Phosphorus (P) from wastewater by cycling the microbial population through a number of aerobic, anoxic, and anaerobic phases [2, 7]. Three separate kinds of bacteria are chosen using three different stages: those that can oxidise NH_3-N to NO_3-N, decrease NO_3-N to N_2 gas, and store excess PO_4-P, respectively [8].

Environmental engineering and microbial ecology are inextricably linked. The principles and instruments of microbial ecology serve as the foundation for environmental engineering process management, and these processes offer fascinating environments that advance the principles and instruments of microbial ecology [8]. Microbial communities are clusters of diverse microorganisms that have the ability to self-organize and self-sustain. Microbes are widely distributed in nature, and their capacity to convert different types of contaminants into nutrients or other useful products allows them to contribute to the goal of a sustainable ecosystem [9]. Microbial communities offer a variety of benefits when maintained effectively in an environmental engineering setting.

It is estimated that there are ~10^{18} microbes in a Wastewater Treatment Plant (WWTP) [10], known to be less diverse than those in sediments, soil, and sea [11]. The organisms in AS consist of bacteria, archaea, eukaryotes (fungi, algae, protozoa, and metazoa), and viruses (*e.g.*, bacteriophages). Bacteria are the main components of the AS community. Alive or metabolically active bacterial cells revealed by cultivation-independent studies are typically around 80% of the total count of cells [12]. Knowledge of microbial communities is important in understanding the stability and performance of bioreactors. For instance, in full-scale AS bioreactors, the removal efficiencies of influent Chemical oxygen demand and Total nitrogen were strongly correlated with the abundances of genes involved in carbon and nitrogen cycling, respectively [13]. In anaerobic Sequential Batch Reactors (SBRs), the structure of the microbial population was linked to the effectiveness of phenol degradation [14]. Higher diversity in the AS microbial community increased resistance to toxic shock loadings [15]. Metabolic diversity, or the number of usable carbon sources, was favourably correlated with the diversity indices of bacteria that accumulate polyhydroxyalkanoates (PHA) [16].

The process of bioremediation uses biological mechanisms, mostly carried out by certain wild-type or designed microbes, to create energy and biomass while lowering the concentration of environmental pollutants to a harmless level. The primary bio-remediators are bacteria, fungi, and microalgae, which are extensively dispersed throughout the biosphere and are known for their rapid reproduction rate and adaptability to various environmental circumstances. By using their enzymatic abilities to modulate the breakdown and conversion of toxins, these organisms may be used alone or in a consortium to reclaim the original natural environment and stop further contamination.

The exploitation of coal, petroleum, and metal resources has seriously harmed soil surface microecology, particularly the microbial community, under the strain of rapidly growing industrialization and enormous energy demand, rendering the ecosystem vulnerable. Owing to their superior nitrogen-fixing capabilities, as well as their capacity to impact plant development and the conversion of soil nutrients and organic matter, soil microbial diversity aids in the preservation and restoration of ecosystem processes. The foundation of sustained repair for damaged soil is the microbial community's recovery [17 - 19]. Additionally, plastic is a significant contributor to environmental pollution due to its complicated nature and extensive use of plastic polymer. However, plastic can be partially removed by bioremediation and encourages sustainability in the environment. To recycle the plastics, the microbial enzymes created during the bioremediation process can also turn them into fuel oil [20].

Molecular biological methods for examining microbial communities in bioreactors and other engineered systems have produced incredible findings that link process stability to diversity and dynamics. Given that large-scale ecosystems are frequently more difficult to maintain than constructed systems, and because interactions occur among engineered environments and other ecosystems, the former might be used to explain some ecological problems that haven't been solved. For example, the process stability of methanogenic bioreactors with distinct trophic groups seems to be influenced by the variety of functional groups within each trophic level and also by the ways in which these functional groups work in concert. Along with this, microbial ecologists and environmental engineers can investigate conditions, processes, and interactions in engineered environments, which in turn allow them to make the ecological engineering of bioreactor design and operation more feasible [21].

The present chapter describes the various engineering approaches to waste treatment using microbial communities, including bacteria, fungi, and several algae. These can be treated using *in-situ* and *ex-situ* treatment methods, depending on the adaptability and economic feasibility as mentioned. Many modern methods are still evolving, which have various advantages in terms of efficiency, but there are several limitations that have been addressed.

ENGINEERING APPROACHES TO WASTE TREATMENT USING MICROBIAL COMMUNITIES

The majority of waste is produced by human activity. The world's rapid and unanticipated changes in livelihood and development have complicated the garbage that is produced. The constant production of dangerous contaminants from numerous industries worldwide causes the biosphere to deteriorate quickly. The advancement of agricultural methods and the quick expansion of healthcare facilities produce a lot of agricultural and biomedical waste, which has a negative impact on environmental health. Solid trash, liquid waste, and gaseous waste are the three primary categories of waste [22]. Human, animal, and plant health are all negatively impacted by these contaminants. Additionally, they cause the microbial community to be destroyed in both terrestrial and aquatic environments, necessitating remediation. Although several remediation techniques, including chemical and physical techniques, have been used for many years, their shortcomings and difficulties have encouraged the adoption of bioremediation as an alternative. Utilizing biological agents like microorganisms and plants, bioremediation aims to eliminate or mitigate the impacts of environmental contaminants [23]. Microbes are utilized more often than plants, mostly due to their quick development and ease of manipulation, which improves their potential as bioremediation agents [24]. Several bacterial, fungal, and algal species have been used to remove different types of environmental contaminants.

There are several major categories into which the field of bioremediation may be divided. For instance, *in-situ* or *ex-situ* media can be treated with bioremediation technology. *In situ*, groundwater and soils are treated on-site, without removal, *via in-situ* procedures. This strategy is beneficial as material handling expenses and some environmental effects may be decreased. On the other hand, the capacity to regulate or alter the existing chemical and physical environment could be limited by *in situ* processes. For example, at several locations, aerobic bioventing has been utilized to remediate fuel-contaminated subsurface soils *via* the *In Situ* Bioremediation (ISB) method. During aerobic bio-venting, in order to promote aerobic hydrocarbon metabolism, the air is usually pumped beneath the surface. *Ex-situ* procedures entail moving the polluted material to a cleanup location. The processes of composting and land treatment are examples of *ex-situ* procedures. In these procedures, soils are dug up, blended with amendments, and then run in a way that promotes the degradation of the harmful pollutants [25]. The distinct steps involved in the bioremediation process are mentioned in Fig. (**1**).

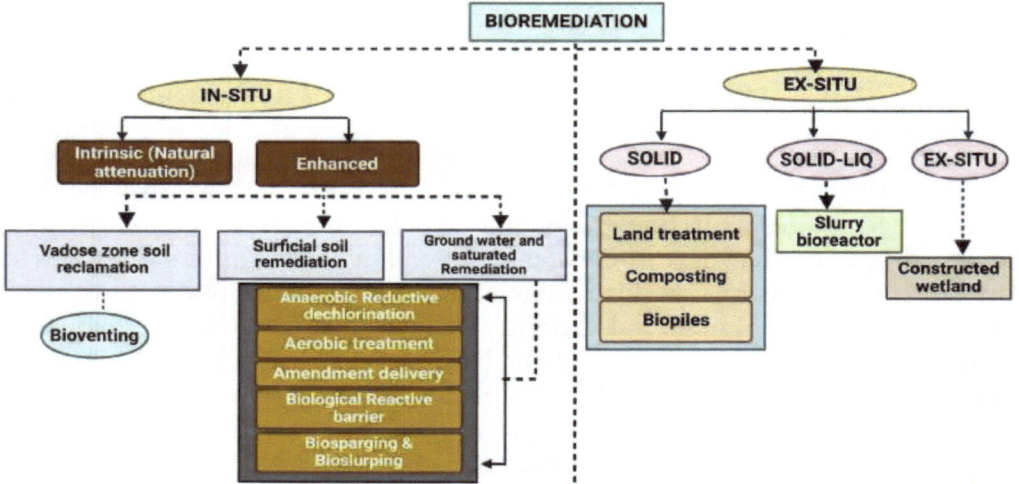

Fig. (1). Distinct steps involved in bioremediation.

In-situ Bioremediation

Since the mid-1970s, engineered *in situ* bioremediation has been a popular approach for decontaminating polluted areas. It has several benefits over excavation or solely chemical/physical methods [26]. ISB technology allows remediation engineers to preserve site function and availability while striking a balance between cost and efficacy. There are other potential hazards, nevertheless, which might arise from poor application design, incorrect installation, or insufficient site characterization. The degradation process and the pollutants to be broken down determine whether ISB may be used in oxic or anoxic geochemical settings. An effective remediation method for chlorinated solvents in groundwater with little disturbance at the ground surface is anaerobic ISB. On the other hand, it may have potentially harmful effects in some situations [27]. *In situ*, bioremediation may be divided into two main categories: enhanced and intrinsic. Both use natural processes, either with (intrinsic) or enhanced, to break down pollutants.

Intrinsic In-situ Bioremediation

Intrinsic bioremediation uses natural mechanisms to break down pollutants without changing the existing environment or adding modifications. Intrinsic bioremediation may be involved in Monitored Natural Attenuation (MNA) sites. The National Research Council (NRC) and Environmental Pollution Act (EPA) define MNA sites as "biodegradation, dispersion, dilution, sorption, volatilization, radioactive decay, and chemical or biological stabilization, transformation, or destruction of contaminants" [28]. A comprehensive site evaluation and the generation of a conceptual model of the site are necessary for MNA implementation. Implementing MNA at a location may be made easier by determining the extent of a stable or diminishing plume, site-specific, risk-based choices based on numerous lines of evidence. Monitored Natural Attenuation (MNA) is a relatively passive remediation process, which means that no new materials are added to the contaminated zone. However, its implementation necessitates active, ongoing monitoring, which must be incorporated into a site's design plan [29].

Enhanced In-situ Bioremediation

Vadose Zone Soil Remediation

Vadose zone contamination by different contaminants is a global issue, and treatment without excavation is often practiced due to economic or technological limitations (Höhener & Ponsin, 2014). Even though the basic biological processes utilized in situ bioremediation could happen spontaneously, numerous sites will need assistance in order to be cleaned up. For instance, the right conditions for a particular microbial activity or increased clearance rates can be developed by adding organic substrates, nutrients, or air [30].

Bioventing

The main in situ biological technique relevant to the unsaturated zone is bioventing. It is the aerobic degradation process of waste substances. Different solid wastes produced from oil reservoirs during the extraction of petroleum and petrol are treated by bioventing. In order to speed up the cleanup process, oxygen and nutrients like phosphorus and nitrogen are introduced into the polluted location [31]. Depending on the modifications utilized, it can be classified as aerobic, metabolic, or anaerobic.

Surficial Soil Remediation

When pollution is superficial, the soil may be treated in situ using techniques similar to composting or land treatment. Variations of these techniques include tilling shallow soils and adding nutrients to enhance aeration and bioremediation. This procedure has similarities with land farming and composting, except for not excavating the soil.

Ground Water and Saturated Soil Remediation

Anaerobic Reductive Dechlorination

Numerous locations where groundwater has been tainted with chlorinated solvents, such as trichloroethylene (TCE) or tetrachloroethene (PCE), have employed anaerobic reductive dechlorination. This method involves delivering organic substrates to the subsurface for fermentation. Hydrogen is produced as a byproduct of fermentation, and an anaerobic environment is created in the region that has to be remediated. Chlorine atoms are successively extracted from chlorinated solvents by a second microbial population using the hydrogen [32]. Sequential dechlorination would occur as follows if PCE were broken down by reductive dechlorination: PCE would undergo conversion to TCE, followed by 1,2-dichloroethane (DCE), Vinyl Chloride (VC), and/or dichloroethane [33].

Aerobic Treatment

Much like bioventing, enhanced *in situ* aerobic groundwater bioremediation techniques can be employed when aerobically degradable pollutants, like fuels, are present in anaerobic areas of an aquifer. In these cases, air or other oxygen sources are pumped into the aquifer close to the pollution. As the oxygenated water moves through the contaminated area, it helps the native bacteria to potentially break down the pollutants [34].

Amendment Delivery

In situ, groundwater treatment, both methods, anaerobic and aerobic, can be set up to include groundwater recirculation or direct air or aqueous stream injection. Direct injection involves directly injecting additives into the aquifer, such as nutrients, oxygen sources, or organic substrates. For instance, a gas such as oxygen might be sparged into the aquifer. It is possible to inject lactate or hydrogen peroxide as a liquid stream; however, hydrogen peroxide should be used cautiously since it may have disinfecting properties. Both liquids and gases are occasionally introduced. In the groundwater recirculation setup, groundwater is extracted, modified as necessary, and reinjected into the aquifer. Extracting groundwater at one elevation, modifying it in the subsurface, and re-injecting it into another height is another way to recirculate below the earth's surface [35].

Biological Reactive Barriers

Biological reactive barriers include an active bioremediation zone that is established in the zone of contamination. Depending on the pollutant of interest and site requirements, these barriers can be built to take advantage of either aerobic or anaerobic processes. A bioremediation zone is created by excavating a trench and filling it with sand that has been pre-mixed with ingredients rich in nutrients, oxidants, or reductants. An alternative is to inject amendments or recirculate altered groundwater at the toe of the plumes to create a bioremediation curtain [36]. Passing through a permeable reactive barrier (PRB) causes contaminants to biodegrade [37].

Biosparging

It is a method of treating trash from locations that contain petroleum products such as lubricating oil, petrol, and diesel. This technique raises the oxygen content by injecting air beneath the groundwater under pressure. Proper air pressure regulation is necessary to prevent the release of volatile particles into the environment, which causes air pollution [38].

Bioslurping

Free products floating on the water table can be effectively removed by bioslurping, often called multi-phase extraction. The two remedial techniques of vacuum-enhanced free-product recovery and bioventing are combined in bioslurping [39, 40]. Light, nonaqueous-phase liquids (LNAPL) are extracted from the water table and capillary fringe by vacuum-enhanced free-product recovery, whereas bioventing promotes aerobic bioremediation of polluted soils in situ [40]. The vertical range of bioslurping is restricted to 25 feet below the earth since this technique cannot lift impurities deeper than 25 feet [41]. A height-adjustable bioslurping tube is placed inside a screened section at the water table after being lowered into a groundwater well. When the bioslurping tube is vacuumed, the free product is "slurped" up the tube and into an oil-water separator or trap for additional processing. LNAPL flow from surrounding places towards the bioslurping well is encouraged when the LNAPL is removed, since it lowers the LNAPL elevation. The bioslurping tube starts to draw vapours from the unsaturated zone when the fluid level in the bioslurping well decreases in response to the vacuum extraction of LNAPL. By encouraging the mobility of soil gases, this vapour extraction improves aerobic biodegradation and aeration [42].

Ex-situ Bioremediation

The term "*ex-situ*" encompasses a broad range of technologies, each with varying levels of sophistication, but they are all characterized by the utilization of excavated soil. This technique has a drawback because of the additional expense involved and the potential for contamination to spread during excavation and transportation. However, in many situations, the latter risk is mitigated by treating the area of interest on-site at a treatment facility. Furthermore, *ex-situ* technologies have many advantages that make them competitive, the most significant of which is the potential for improved remediation process control (even with the most basic technological solutions, such as biopiles), as the contained reaction environment is easier to manage and the treatment process is more predictable than an *in situ* environment. Consequently, improving mass transfer, optimizing operating parameters, boosting biodegradation kinetics, and adding more pretreatments to boost process efficiency are feasible.

Solid Phase Treatment

Land Treatment

Another name for land treatment is land farming. It is a bioremediation technique in which soil amendments are combined with polluted soil and then tilled into the ground. Enhancing native biodegradative bacteria to aerobically degrade pollutants is the primary goal [25].

Composting

A diverse population of microorganisms facilitates the aerobic decomposition process known as composting, which has been widely used for a variety of waste types through the metabolic activity of microbial consortia. Composting has been used to stabilise and transform organic waste into a more stable and safe form that can be utilised in various agricultural practices [38]. It is a waste management technique that is both inexpensive and ecologically sound. Humus and plant nutrients are the primary byproducts of composting, whereas the byproducts include carbon dioxide, water, and heat [43]. This process involves a variety of microorganisms, including bacteria, actinomycetes, yeasts, and fungi. Mesophilic, thermophilic, and cooling and maturation phases are the three stages of composting. The kinds of composting organic matter (OM) and the process's efficiency or efficacy, which is determined by the level of agitation and aeration, are the two parameters that control how long the composting stages last [25].

Biopiles

It is a hybrid method that combines composting with land farming. Both aerobic and anaerobic microbes can flourish in this technique as it provides an ideal environment for their growth. With the help of biodegradation, bio-piles are used to reduce the quantities of petroleum elements [25]. Biopiles and vessel systems are more technically advanced constructed composting solutions; they are more costly but offer greater process control, which leads to increased efficiency and a smaller volume of treated material [44].

Solid–Liquid Mix Phase Treatment

Materials like slurries and sludges make up the solid-liquid mixture. Below is a discussion of one technology for handling such mixes.

Slurry Bioreactors

Slurry bioreactors are used to treat solid or semi-solid wastes such as sludge, soil, and sediments. Due to their high cost, slurry bioreactors are probably going to be utilized for more challenging treatment procedures. Waste is screened to get rid of big particles and trash, and then is usually combined with water in a tank or other container until the solids are suspended in the liquid phase. If required, further particle size reduction can be achieved either before (by pulverizing and/or screening the wastes) or after (by using a shearing mixer) the addition of water. In addition to increased interaction between pollutants and bacteria that can break down those pollutants and speed up mass transfer rates, suspension and mixing of the solids can occur [45]. Tanks or lined lagoons are used for mixing. Tanks are often used for mechanical mixing. 10–30% solids by weight are typical for slurries [46]. Lagoons commonly employ aeration using submerged aerators or spargers, which can be paired with mechanical mixing to get the intended effects. To enhance handling properties and microbial breakdown rates, nutrients and other additives, including neutralising agents, surfactants, dispersants, and co-metabolites (such as phenol and pyrene), may be added. The bioreactor can be seeded with native bacteria or microorganisms, or they can be added continually to maintain appropriate biomass levels. The matrix, pollutant type, and concentration all affect residence time in the bioreactor [47]. When the dry-weight concentrations of contaminants reach the acceptable values, the slurry is dewatered. The slurry is usually dewatered by gravity using a clarifier. Depending on the properties of the slurry and financial concerns, further dewatering equipment may be employed [48].

Liquid Phase Treatment

Constructed Wetlands

Constructed wetlands are linked to biological processes such as phytoremediation, which is based on plants, and bioremediation, which is based on microbes. The organic materials in the water undergoing treatment are broken down and/or absorbed by microbes affixed to plant surfaces, plant litter, and the wetland substrate [49]. Through biological, chemical, and physical processes influenced by plants and their roots (*i.e.*, the rhizosphere), phytoremediation uses plants to remove, transfer, stabilise, or destroy contaminants. These processes include degradation, extraction through accumulation in plant roots, shoots, or leaves, metabolism of contaminants, and immobilisation of contaminants at the interface between roots and soil [50].

TYPE OF MICROBIAL COMMUNITIES USED IN ENVIRONMENTAL ENGINEERING

Various microbial communities are used nowadays in the application of environmental engineering using bacteria, fungi, and algae, which is discussed in the following section. Fig. (**2**) shows the diverse roles of microbes used in various engineering approaches

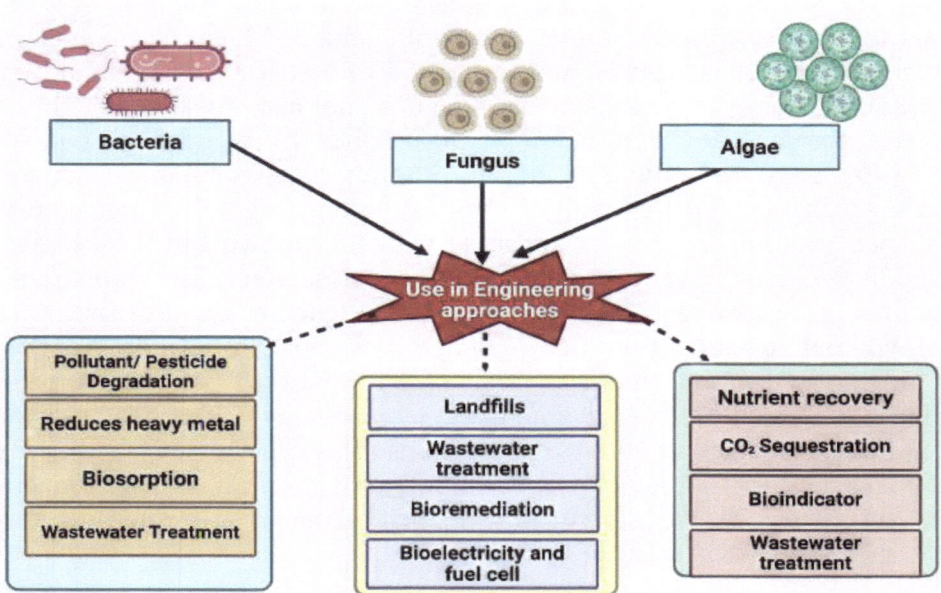

Fig. (2). Diverse roles of microbes used in engineering approaches.

Bacterial Communities in Different Engineering Approaches

The microbial communities, like bacteria, play a vital role in environmental engineering, especially in wastewater treatment, control of pollution, and recovery of resources [51, 52]. Their capacity to adapt in diverse environments, metabolic variability, and also to form biofilms make them suited for various environmental applications [53]. Table 1 represents different types of bacterial community helpful in environmental engineering.

Table 1. Different types of microbial communities helpful in environmental engineering.

Microbial Community	Species	Role in Environmental Engineering	Reference
Bacteria	*Escherichia coli*	Used for biomonitoring in waterbodies	[57]
	Pseudomonas sp.	Helps to degrade different organic compounds like Benzene, Toluene, xylene, *etc.*	[117]
	Bacillus subtilis	Biodegradation of crude oil	[118]
	Streptomyces spp.	Degradation of Petroleum and Naphthalene	[119]
	Rhodococcus erythropolis	Biodegradation of Phenol	[120]
Fungi	*Phanerochaete chrysosporium*	Decolourisation and Bioremediation of Textile Effluent	[121]
	Trametes versicolor	Bioremediation of phenol-containing wastewater	[122]
	Aspergillus niger	Helpful in the removal of azo dye from aqueous solution	[123]
	Penicillium chrysogenum	Detoxification of heavy metals	[124]
Algae	*Spirulina* sp.	Treatment of aquaculture wastewater	[125]
	Chlorella vulgaris	Production of Biofuel	[126]
	Cystoseira compressa	Bioaccumulation of heavy metals	[127]
	Scenedesmus obliquus	Helpful in the treatment of wastewater generated from a poultry farm	[128]

Degradation of Pollutants

Bacteria have the capability to degrade the environmental pollutant, specifically plastics and petroleum hydrocarbons. *Ideonella sakaiensis* can degrade polyethylene terephthalate within a period of 6 weeks. The biodegradation of plastic is regarded as one of the most effective ways, and during the process, bacteria produce a few enzymes like esterase, cutinase, lipase, *etc.*, which help them to break down the polymers [54, 55].

The study found that over 79 genera, like those of *Pseudomonas, Streptococcus, Mycobacterium, etc.*, have the capability to degrade the petroleum hydrocarbon chain by utilising it as a carbon source [56]. Apart from this, bacteria like the species of *Coliform, E. coli, Streptococcus, Pseudomonas, Vibrio, etc.* are also used as an indicator of pollution and one of the environmental monitoring parameters of the water body [57].

Heavy Metal Reducers

Heavy metal is one of the significant environmental and health risks that needs an effective remediation strategy. The study found that sulfate-reducing bacteria (SRB) like *Desulfovibrio vulgaris* are effective in the removal of heavy metals like Chromium (Cr), Copper (Cu), Manganese (Mn), Nickel (Ni), and Zinc (Zn) by precipitating the metals as sulfides [58]. SRB bacteria have the capability to live in high concentrations of metal, which makes them more suitable for treating acid mine drainage as well as highly contaminated environments [59].

Biosorption

It is another process that involves the heavy metal binding to the metal-binding protein of the bacterial cell wall [60]. The study found that bacteria like *Pseudomonas fluorescens* and *Bacillus safensis* have efficiency of around 84% and 72% in chromium degradation with the help of this process [61]. It has been studied that bacterial biomass is recognized to be a safe, cost-effective, and environmentally friendly method for the removal of heavy metals [62].

Wastewater Treatment

The wide variety of microorganisms and their function in wastewater treatment can be best illustrated in two biological treatment units—the anaerobic digester and the AS process. At municipal wastewater treatment facilities, the most popular aerobic biological treatment unit is the AS process. This process involves both eukaryotes (metazoans and protozoa) and prokaryotes (bacteria). Based on respiration, biological processes take place in both anoxic and aerobic environments. At municipal wastewater treatment facilities, the anaerobic digester is the most often utilised anaerobic biological treatment unit. The organism solely involved is prokaryotes [63].

Decomposition of Organic Matter: Through metabolic processes, bacteria break down organic contaminants in wastewater, lowering the Biochemical Oxygen Demand (BOD).

For instance, the most prevalent bacteria are *Escherichia coli* and *Pseudomonas aeruginosa,* which break down organic molecules, including sewage and food waste, into simpler chemicals [64].

Nutrient Removal: In order to eliminate nutrients like nitrogen and phosphorus, which can cause eutrophication in water bodies, certain types of microbes are essential.

For instance, Nitrifying Bacteria: *Nitrosomonas* species produce nitrites (NO_2^-) from ammonia (NH_3), whereas *Nitrobacter* species produce nitrates (NO_3^-) from nitrites [65].

Denitrifying Bacteria:*Pseudomonas stutzeri* and *Bacillus* species lower the amount of nitrogen in water by converting nitrates into nitrogen gas (N_2), which is then discharged into the atmosphere [66].

The well-known Phosphate Accumulating Organism (PAO) *Candidatus Accumulibacter* efficiently lowers the phosphorus levels in the effluent by absorbing phosphate during aerobic circumstances and releasing it under anaerobic conditions [66, 67].

Different microbial systems have been created for wastewater treatment, which relies heavily on microbial populations. Metaproteomics provides information on how these communities work. For example, Lacerda *et al.* (2007) identified more than 100 differentially expressed proteins by using metaproteomics to investigate how a bacterial population responded to cadmium exposure in a bioreactor [68]. Similarly, Wilmes *et al.* investigated the molecular processes behind improved biological phosphorus removal (IBPR), discovering important proteins associated with metabolic stability and IBPR-specific pathways, especially those derived from *Candidatus Accumulibacter, phosphatis* [69, 70]. More than 700 proteins, including those involved in the fatty acid cycle and denitrification, have been discovered using non-gel-based techniques. Research by Park *et al.* (2008) [71] and others emphasises how microbial interactions and extracellular proteins play a part in sludge digestion. All these findings have greatly improved our knowledge of the roles played by microbes in wastewater treatment systems [66].

Pesticide Degradation

The bacterial population is very effective at breaking down pesticides. The genus Bacillus, along with *Klebsiella, Flavobacterium, Acinetobacter, Aerobacter, Alcaligenes, Neisseria, Sphingomonas, Burkholderia, Pseudomonas, Micrococcus,* and *Arthrobacter,* are among the significant groups [72 - 75].

New Molecular Techniques used in Bacteria to Study Communities

The new advancement in molecular techniques has specifically enhanced the bacterial application in the field of environmental engineering. Those techniques help to study deeper and gain a better understanding of microbial communities in the processes involved in the environment. Some of the techniques are:

Culture-Independent Method: This approach includes metagenomics as well as metaproteomics, which bypasses the culturing of bacteria and allows for studying the view of bacterial diversity in a comprehensive manner and also their interactions [76, 77].

Omics Technologies: This technology includes a collection of modern molecular biology techniques which is used to examine the roles and interactions of different molecules that constitute the organism's cell [78]. Genomics, transcriptomics, proteomics, and metabolomics help to analyse the genes, proteins, metabolites, and also cellular components at a global level [79, 80].

Quantitative PCR (qPCR): This technique is also known as real-time PCR. This method is important for detecting the sensitive as well as rapid analysis of gene abundance and expression in any complex environment [81, 82].

Applications

The various applications of bacterial communities in environmental engineering are as follows:

Pollution Degradation: Bacteria are mostly used as catalysts to degrade the pollutants with specialized molecular techniques after proper identification and optimization of the process [83, 84].

Environmental Monitoring: The modern techniques advancement helps in the rapid examination of changes in the environment, especially by studying microbial responses, which is important for proper management of the environment as well as remediation [85].

Wastewater Treatment: Bacterial communities are widely used to treat wastewater and help to manage the industrial as well as municipal wastes because it is cost-effective and an efficient way to decontaminate water [86, 87].

Fungal Communities in Different Engineering Approaches

Fungal communities play a crucial role in environmental engineering, especially in sustainable technology development and bioremediation, as they possess a unique feature of interacting with other organisms and have distinct enzymatic activity [88, 89]. The following sections will signify the contributions made by the fungi in this field. Table 1 depicts different types of fungal communities helpful in environmental engineering.

Wastewater Treatment

In Wastewater Treatment Plants (WWTPs), fungi constitute a significant group of microbial communities. Through a variety of intracellular and extracellular enzymes and sorption processes, fungi help break down environmental organic chemicals, including proteins, complex carbohydrates, lipids, aromatic hydrocarbons, pharmaceutical compounds, heavy metals, and chemicals that disrupt hormones. Furthermore, they have been related to the stabilisation and denitrification of aggregates of AS cells; however, their higher biomass is connected to operational issues including bulking, foaming, and membrane biofouling [90 - 93].

Sanitary Landfill

According to a study [94], fungi are primarily capable of breaking down cellulose and lignin, two essential steps in the composting of green waste. Several fungal strains with strong tyrosinase, peroxidase, and endogluconase enzyme activity were found in the study by [95] and demonstrated a significant role in the breakdown of rye straw. According to [96], who tracked the development of fungal communities during the composting of wheat straw, cellulolytic fungi were both very numerous in the mature compost and ubiquitous throughout the whole composting process. Several studies have also documented the biodegradation of plastic by fungi [97].

Bioremediation

Fungi have strong metabolic processes and can produce a wide range of extracellular enzymes, which makes them a potent agent in this field and works efficiently in degrading toxic pollutants [98, 99]. They use various enzymes like catalases, laccases, peroxidases, *etc.*, to treat various organic pollutants by detoxifying them, and hence this makes fungi an environmentally friendly way to address pollution [100]. The study found that *Pleurotus ostreatus* efficiently decolourises the dyes from textiles as well as degrades the phenolic compounds [101]. The advancement in microbial techniques will uplift the natural ability of wild strains of natural fungi, which in turn will help in providing improved degradation efficiency of pollutants.

Bioelectricity and Fuel Cell

The generation of bioelectricity by fungi is an emerging field that includes biodegradation by fungi and also energy production [102]. The enzyme, fungal laccase, which is produced by Monodictys castaneae, has enhanced electricity production and degraded phenolic compounds. This efficiency is achieved because the enzyme acts as a biocatalyst [103]. *Saccharomyces cerevisiae* is used in (MFCs) as they have the ability to electron transfer using mediators or directly, which increases the electricity generation [104].

Algal Communities in Different Engineering Approaches

Algae are considered to be one of the efficient organisms in the field of environmental engineering, which offers both natural and sustainable solutions to various challenges that humankind is facing in daily life. These organisms have the capacity to absorb the heavy metals as well as toxins, which in turn clean up the polluted areas [105]. Apart from that, they have the ability to absorb CO_2 from the surrounding environment, which is helpful in combating climate change, and also these communities can be used as bioindicators of the environment [106]. Table 1 depicts different types of algal communities helpful in environmental engineering.

CO_2 Sequestration

The sequestration of CO_2 by algae is one of the encouraging approaches for mitigating climate change by reducing the level of atmospheric carbon dioxide [107]. Photosynthesis in algae is one of the important processes by which they can convert the atmospheric $CO2$ into organic compounds. The factors that influence the process are light availability, algal type, pH, and temperature [108, 109]. The study highlighted that algal species like *Spirulina, Chlorella, Hematococcus,* and *Dunaliella* are efficient in CO_2 sequestration [109].

Bioindicators

Algal communities are widely used as bioindicators in the environment for their high sensitivity to variation in the quality of water, so they can be used to understand the presence and extent of pollutants in water bodies [110]. A recent study found that algae species like *Cystoseira compressa* and *Ericaria mediterranea* are capable of detecting heavy metal contamination in marine environments [111]. Apart from these, other algal species like *Sargassum, Padina,* and *Turbinaria* are used for detecting the cadmium levels in marine waters [55]. The use of algae for bioindicators is a cost-effective and efficient means for monitoring the environment due to their metabolic processes and life span.

Algae have shown great promise in addressing environmental issues and advancing sustainability. Problems, including pollution, energy crises, and waste management, have been brought on by resource exploitation and rapid population increase. The promise of algae for carbon neutrality, wastewater treatment, sustainable agriculture, and other applications is highlighted by developments in phycology research [112].

Nutrient Recovery

In wastewater treatment, the use of microalgae-bacteria consortia for nutrient recovery is a successful strategy that improves the removal of contaminants while also recovering valuable resources. For instance, it has been demonstrated that when the growth-promoting bacterium *Azospirillum brasilense* is co-immobilized with the microalga *Chlorella vulgaris*, the latter can efficiently remove phosphorus and ammonium ions from synthetic wastewater. The efficiency of nutrient removal is increased by this combination, which also makes it easier to recover the nutrients for possible future use in agricultural applications [113].

Biofuel Production

The high lipid content of algae, which may be turned into biodiesel, and their capacity to thrive in a variety of conditions have made them an appealing alternative. Due to their versatility in cultivation—including non-arable land and wastewater—algae are a desirable alternative for the generation of sustainable biofuel [114]. The benefits of algae include high biomass yields and quick growth rates, which may be used to create biofuels and mitigate some of the environmental issues related to conventional energy sources [115].

Biofuel Production from Heavy Metal-Contaminated Wastewater

Wastewater-based algal biofuel production solves wastewater management issues while providing a sustainable energy source. Because of their high lipid content and versatility, microalgae such as *Chlorella, Nannochloropsis*, and *Scenedesmus* are perfect for biofuel. By supplying vital nutrients like phosphate and nitrogen, wastewater minimises environmental effects like eutrophication, lowers expenses, and lessens the demand for fertilisers [116].

MODERN APPROACH TO WASTE TREATMENT

Every activity done by humans is responsible for some amount of waste generation, and effluents from houses, agricultural, and industrial operations produce wastewater streams that contain organic contaminants that are difficult to remove and need proper disposal. These pollutants contain toxic complex compounds, which cause severe damage to the environment and health [129, 130]. Additionally, to adhere to the increasingly strict environmental [131] regulations, the concentration of pollutants in the stream where effluents are discharged must be maintained at a specific minimal level [132]. One of the main causes of the depletion of nutrients (such as nitrogen and phosphorus) in ocean water is wastewater discharge. Wastewater discharge releases approximately 1 million tonnes of phosphorus and 7.7 million tonnes of nitrogen into natural water basins annually. It degrades the environment and results in eutrophication [132]. This presents a significant challenge to the management of water distribution in the various parts of the world [133, 134]. To fulfil the current demand for potable water, several regulations, guidelines, and techniques pertaining to the treatment, discharge, and recycling of wastewater into water bodies have been developed and put into practice globally over time. However, the effectiveness of those rules is always in question, and industries are always finding ways not to comply with their rules and regulations [133].

Traditional remediation techniques relied on chemical and physical dispersants, which would eventually endanger the environment's health [128, 135]. Current methods are predicated on the premise that bioremediation is the only practical way to treat hazardous waste. The process by which a material undergoes biological degradation is the present techniques are predicated on the premise that bioremediation is the only practical way to treat hazardous waste. The process by which a material undergoes biological degradation is biodegradation. It is a natural process that occurs unhindered by humans. The designed process of using biological means such as bacteria, algae, fungi, *etc.*, to break down a contaminant [128, 136]. The development of genetically engineered organisms for ecological remediation of contaminants in soil and water requires the design of new metabolic pathways, modification of existing removal pathways to eliminate deposition of toxic substances, increase bioavailability of pollutants, and improve catalytic capacity of microorganisms [137, 138].

Bacteria are highly effective at breaking down environmental contaminants. The ability of certain bacterial strains to break down various contaminants, such as nitro-aromatics, chloro-aromatics, and polycyclic aromatic compounds, has been found to be useful for bioremediation of contaminated areas [139]. Toxic substances can't be removed by natural bacteria, or they can't do so well. Through the interdisciplinary application of molecular methods, genomics, bioinformatics, and microbiology, efforts have been made to boost the potential of these local bacteria to remove pollutants by altering their genetic makeup, a process known as Genetically Modified Organisms (GMOs) [140, 141].

Genetic engineering can also be referred to as recombinant DNA technology; Genetic engineering involves modifying the microbial cell's ability to kill, enhance protein production, and other metabolic activities of the cell to achieve the desired effect [142, 143]. GMOs possess characteristics of many native bacteria because the genes are transferred from these bacteria [142, 144, 145]. These modified bacteria can be effectively used for bioremediation, as they involve the removal of toxins and pollutants from the environment.

Bacteria, fungi, and algae have a higher ability to oxidize pollutants [142]. The use of recombinant DNA and genetic engineering in microorganism breeding has increased dramatically, producing a significant number of bacteria with improved pollutant-degrading capabilities [146]. Table 2 showcases different Genetically Modified Organisms and their role in environmental engineering. Bacteria that have been genetically engineered include *Pseudomonas putida*, *Escherichia coli*, *Bacillus idriensis*, *Mycobacterium marinum*, *etc.*, like bacteria, plants have also been modified to increase tolerance, ability to detoxify heavy metals, and also to improve biomass and plant growth in contaminated areas [142, 147].

Table 2. Genetically modified organism and their role in environmental engineering.

Type	Genetically Modified Organisms	Function	References
Bacteria	*Pseudomonas fluorescens* HK44	Bioremediation indicator	[163]
	Mesorhizobium huakuii sub sp. Rengei B3	Enhance the accumulation of Cd^{2+}	[164]
	Pseudomonas putida KT2440	Rapid and high-yield production of vanillin from ferulic acid	[165]
	Ralstonia eutropha	A rhizosymbiont offers the advantage of removing the metal bound on the rhizobacteria while harvesting the plants	[164]
	E. coli JM1	Ability to decolorize azo dyes	[166]
	Sphingomonas desiccabilis	Bacteria methylate inorganic arsenic into less toxic organoarsenicals.	[167]
	Spingomonas paucimobilis UT26XEGM	Degrade methyl-parathion and γ-HCH simultaneously and the activity of self-destruction after 3-methylbenzoate induction	[168]
	Achromobacter sp. AO22	Biosensor for detecting copper bioavailability (hence potential toxicity) in a soil bacterial background	[169]
	Pseudomonas sp. BF1–3	Degrades organophosphorus pesticide chlorpyrifos	[170]
	Sphingobium indicum B90A	xenobiotic degradation	[171]

Type	Genetically Modified Organisms	Function	References
Fungi	Saccharomyces cerevisiae mleA ML01	Prevent the formation of noxious biogenic amines produced by lactic acid bacteria in wine.	[172]
	Saccharomyces cerevisiae ML01	Able to convert malic acid into lactic acid during alcoholic fermentation	[173]
	Recombinant Phanerochaete chrysosporium	Stable production of ligninolytic enzymes	[174]
	Pycnoporus sanguineus MUCL 41582	ligninolytic peroxidases, Dye-Decolorizing Peroxidases (DyPs), and Heme-Thiolate Peroxidases (HTPs)	[175]
	Aspergillus niger phyA2	Phytase production, Degradation of phytate to release bound phosphorus	[176]
	Trichoderma reesei Xyr1	Plant biomass conversion into renewable biofuel and chemicals.	[177]
	Aspergillus niger xlnR	Plant biomass conversion into renewable biofuel and chemicals	[177]
	Aspergillus nidulans gaaR	Pectinases: Utilization of pectin-rich feedstock in biofuel production	[178]
Algae	Nannochloropsis salina NsbHLH2	Industrial production of biodiesel	[181]
	Nannochloropsis salina AtWRI1	Increased lipid production	[182]
	Pseudochoricystis ellipsoidea G418	Synthesize and accumulate a significant amount of aliphatic hydrocarbons intracellularly.	[183]
	Scenedesmus obliquus GH2 Transgenic	Phytoremediation	[184]

The ability of GMOs to endure ecological stressors and, once their intended function is fulfilled, to have an appropriate mechanism for removing them from the area of action are prerequisites for the successful use of these organisms [148]. Many studies have been done on the genetic modification of plants and their function in remediation of contaminated soil, water, plants, and bacteria have been continuously used for bioremediation purposes [149]. Modifying the organism so that it can tolerate and thrive in heavily polluted areas would allow the genetically modified strain to sustain itself in an environment with a native microbial community.

The extent of bioremediation depends on the genes responsible for the elimination and growth of microbial populations under adverse stress conditions [135, 142]. Therefore, selecting and designing a suitable microbial strain with a rapid growth rate and effective remediation properties without any ecological risks will remain a challenging step towards achieving a safe and reliable ecosystem [150, 151].

The processes depend on interactions between microbes and contaminants, such as heavy metals. These interactions with contaminants can be facilitated by bacteria, fungi, lichens, archaea, protozoans, and algae [142, 152, 153].

Using Genetically Modified Organisms

Common and wild microbial strains are less strong and moderate degraders of poisons and wastage; hence, bioaugmentation came into the situation by presenting heritarily built microbial strains into the contaminated locales for productive and quick degradation [154]. Thus, GMOs form a broad category that can include all plants, animals, or microorganisms, while the term "genetically engineered microorganisms" (GEM) refers explicitly to microorganisms (*i.e.*, bacteria or fungi, including yeast) [138]. A foreign gene was inserted into the genome of these four organisms with the ability to remove toxic heavy metals and other pollutants that get accumulated in the ecosystem [155]. These genetic modifications include a number of methods used to alter the genetic structure of domesticated plants and animals to achieve a desired result (such as substitution, hybridization, and mutagenesis). While genetic engineering is a form of gene editing that involves intentionally making a targeted change to achieve a specific result in the genetic sequence of a plant, animal, or microorganism [156].

Microorganisms have the capacity to adapt to different natural conditions and show assorted hereditary reactions to particular harmful substances. This versatility can be improved by consolidating fitting qualities into the microorganisms. Various enhancements in quality integration inside microorganisms have been made possible by fast progress in recombinant DNA innovation in conjunction with complex atomic science strategies [157]. The use of bacteria for heavy metal remediation is a promising approach, supported by advancements in molecular tools like directed evolution and ecological design to engineer bacterial strains and enhance enzyme production for pollutant removal, leveraging their high oxidation potential and modifiable metabolic pathways [158, 159]. The detailed systematic illustration of the mechanism involved in GMO production is shown in Fig. (**3**).

Recent advances in fungal engineering and mycoremediation have improved their effectiveness in heavy metal remediation, with fungi distinguished by their adaptability, dispersal capacity, and enzyme production. Capable of thriving in harsh environments and industrial WWTPs, fungi such as *Aspergillus flavus* and *Aspergillus niger* have shown potential for reducing heavy metal toxicity. Genetic engineering has further improved fungal strains by enabling them to mineralize contaminants and develop mutants with specialized enzymes, making fungi a valuable tool for wastewater and soil remediation. Fungi para-microalgae are effective in absorbing heavy metals like copper, zinc, and manganese for metabolic and enzymatic activities, while demonstrating protective mechanisms against toxic metals such as lead, mercury, and arsenic. These mechanisms include metal immobilization, gene regulation, chelation, and antioxidant enzyme production, which help neutralize free radicals. Algal species like *Chlorella* sp. and *Chlamydomonas* sp. have been engineered for enhanced pollutant removal, a process known as phycoremediation. Advances in genetic engineering, such as introducing metallothionein genes into Chlorophyta *C.reinhardtii*, have improved their resilience and efficiency in mitigating environmental contamination.

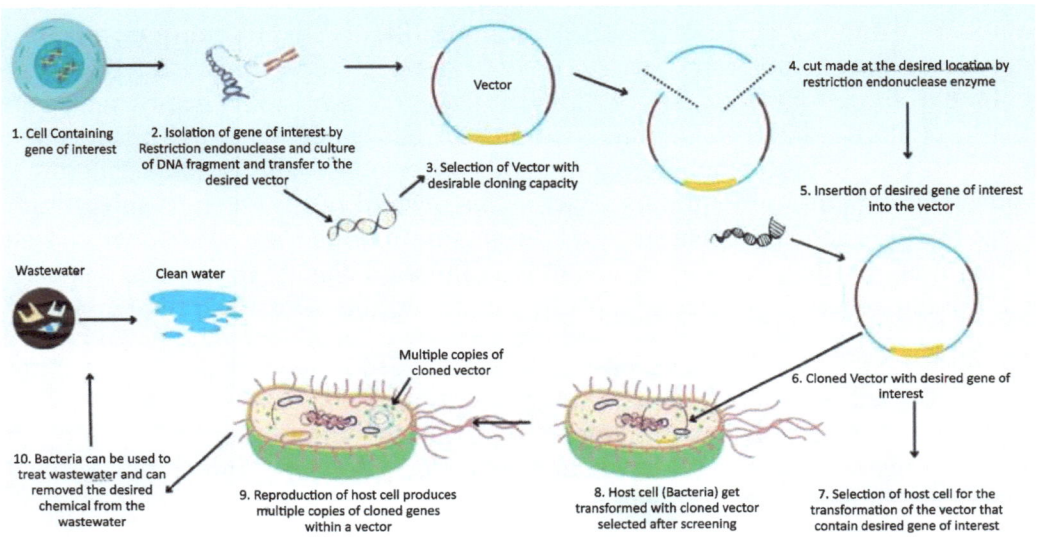

Fig. (3). Systematic illustration of the mechanism involved in GMO production.

Using a Hybrid System

Hybrid techniques have gained increased attention due to the drawbacks of conventional water treatment methods and the restrictions of single-stage membrane filtration (*e.g.*, Reverse Osmosis). The researchers demonstrated that combining processes reduces their flaws and improves system performance compared to using only one process [160]. In order to save fuel, recover energy, and improve overall system efficiency, a hybrid energy system often combines two or more energy sources or methods through appropriate energy conversion techniques [161]. This approach combines physical, chemical, and/or biological treatment modalities to enable the effective elimination of different Emerging Contaminants (ECs). Utilizing biodegradation mechanisms, biological treatment technologies are the most popular methods for removing ECs. High molecular weight molecules are broken down into smaller molecules by microorganisms, including bacteria, fungi, and algae, during the biodegradation process [162].

LIMITATIONS

The use of microbes in environmental engineering has the following limitations, as discussed in the following section.

Use of Proteomics in Microbes Identification

Very few laboratories are currently employing proteomics to address environmental issues, despite the fact that it has already been effectively established as a useful method for identifying the functional molecules of diverse signalling pathways in biomedical research. Without a doubt, this technology is expensive, necessitates highly specialised facilities, and requires knowledgeable personnel to conduct the tests. It is obvious that more affordable proteomics technologies must be created for environmental remediation. However, efforts to use proteomics research in environmental biotechnology labs are currently underway [185].

Analytic Computational Tools

A large amount of information about genes, proteins, and metabolites is provided by Systems Biology; however, adequate and efficient computational methods are needed for improved gene, pathway, and metabolite annotation [186]. Due to the high expenses of sample processing and the need for specialised equipment, SB has not been utilised extensively for bioremediation up to this moment [185]. The initial database of its kind to include data on a vast number of microbial biodegradation reactions and pathways for resistant chemicals is the EAWAG-BBD, which was first created by the University of Minnesota in the United States and is hosted by EAWAG (Swiss Federal Institute of Aquatic Science and Technology) [187]. There are lists of 219 pathways, 1503 reactions, 1396 chemicals, 993 enzymes, and 543 different kinds of microbiological entries in the most recent version of this database, which was updated on May 10, 2017. MetaRuter, another openly accessible online resource, provides comprehensive information on a range of biological substances, enzymes, reactions, and organisms associated with biodegradation and bioremediation processes [188].

Risks of Genetically Engineered Organisms

The most significant problem is the unpredictability of the consequences for other environmental life forms besides the intended one. Because the genetic material in GMOs is viable, there is a risk of genetic contamination, which might have a detrimental impact on the ecosystem as a whole if it spreads horizontally or vertically to analogous or compatible species [189]. In addition to these problems, perceived risks to the integrity and inherent worth of species and ecosystems raised ethical concerns about horizontal gene transfer from GMOs to other organisms [190].

CONCLUSION

The nature of microorganisms is distinct and frequently unforeseen. Microorganisms are a useful tool for resolving a variety of environmental engineering issues. A remarkable evolution of research and creative instruments has been developed to give an effective approach to protecting our planet, as well as current ways of biological waste treatment and environmental monitoring. The scientific and reliable use of microorganisms is inevitably irresistible, substantially contributing to waste management, sustainability, and the shift to a circular bioeconomy. The study emphasizes the importance of microorganisms in environmental engineering, particularly their use in waste and pollution remediation. In addition, the role of microbes in *in-situ* and *ex-situ* bioremediation methods, either by themselves or in conjunction with other strategies, is to clean up polluted areas. However, additional research is required to fully grasp this subject's advantages and constraints.

REFERENCES

[1] Robert S. Environmental Engineering for the 21st Century. Addressing Grand Challenges 2019; (Jan): 1-121.
[http://dx.doi.org/10.17226/25121]

[2] Rittmann BE, McCarty, Perry L. "Environmental biotechnology: principles and applications," Accessed: Jan 03, 2025 Available from: https://cir.nii.ac.jp/crid/1130003901570290944

[3] Grady Jr CPL, Daigger GT, Lim HC. Biological wastewater treatment. 2nd ed. New York: Marcel Dekker, Inc; 1999.

[4] Metcalf L, Eddy HP. Wastewater engineering: treatment, disposal, and reuse. 4th ed. New York: McGraw-Hill; 2004. Available from: https://library.wur.nl/WebQuery/titel/1979505

[5] Rittmann BE. Microbial ecology to manage processes in environmental biotechnology. Curr Opin Biotechnol 2006 Jun; 17(3): 261–6.
[http://dx.doi.org/10.1016/j.tibtech.2006.04.003]

[6] Smith SK, Weaver JE, Ducoste JJ, de los Reyes III FL. Microbial community assembly in engineered bioreactors. Water Res 2024 May 15; 255: 121495.

[7] Filipe CDM, Grady Jr CPL. Biological wastewater treatment, revised and expanded. 2nd ed. Boca Raton (FL): CRC Press; 1998.
[http://dx.doi.org/10.1201/9780849306730]

[8] Wagner M, Loy A. Bacterial community composition and function in sewage treatment systems. Curr Opin Biotechnol. 2002; 13(3): 218–227.
[http://dx.doi.org/10.1016/S0958-1669(02)00315-4]

[9] Kaur J, Gosal SK. Biotransformation of Pollutants: A Microbiological Perspective. Rhizobiont in

Bioremediation of Hazardous Waste. 2021; pp. 151-62.
[http://dx.doi.org/10.1007/978-981-16-0602-1_8]

[10] Woodcock S, Curtis TP, Head IM, Lunn M, Sloan WT. Taxa–area relationships for microbes: the unsampled and the unseen. Ecol Lett 2006; 9(7): 805-12.
[http://dx.doi.org/10.1111/j.1461-0248.2006.00929.x] [PMID: 16796570]

[11] Curtis TP, Sloan WT, Scannell JW. Estimating prokaryotic diversity and its limits. Proc Natl Acad Sci USA 2002; 99(16): 10494-9.
[http://dx.doi.org/10.1073/pnas.142680199] [PMID: 12097644]

[12] Seviour RJ, Nielsen PH. Microbial communities in activated sludge plants. In: Blackall LL, Seviour RJ, Schlieper L, eds. Microbial Ecology of Activated Sludge. London: IWA Publishing; 2010. p. 95-126. Available from: https://vbn.aau.dk/en/publications/microbial-communities-in-activated-sludge-plants/

[13] Xia Y, Wang X, Wen X, Ding K, Zhou J, Yang Y, *et al.* Overall functional gene diversity of microbial communities in three full-scale activated sludge bioreactors. Appl Microbiol Biotechnol. 2014; 98(16): 7233–42.
[http://dx.doi.org/10.1007/s00253-014-5791-7]

[14] Rosenkranz F, Cabrol L, Carballa M, *et al.* Relationship between phenol degradation efficiency and microbial community structure in an anaerobic SBR. Water Res 2013; 47(17): 6739-49.
[http://dx.doi.org/10.1016/j.watres.2013.09.004] [PMID: 24083853]

[15] Saikaly PE, Oerther DB. Diversity of dominant bacterial taxa in activated sludge promotes functional resistance following toxic shock loading. Microb Ecol 2011; 61(3): 557-67.
[http://dx.doi.org/10.1007/s00248-010-9783-6] [PMID: 21153808]

[16] Yang C, Zhang W, Liu R, *et al.* Phylogenetic diversity and metabolic potential of activated sludge microbial communities in full-scale wastewater treatment plants. Environ Sci Technol 2011; 45(17): 7408-15.
[http://dx.doi.org/10.1021/es2010545] [PMID: 21780771]

[17] Barberán A, McGuire KL, Wolf JA, *et al.* Relating belowground microbial composition to the taxonomic, phylogenetic, and functional trait distributions of trees in a tropical forest. Ecol Lett. 2015 Dec; 18(12): 1397–405.
[http://dx.doi.org/10.1111/ele.12536]

[18] Salama ES, Jeon BH, Wang J, Abou-Shanab RAI, Xiong JQ. Editorial: Microbial advances towards sustainable environment: Microbiome structure & integrated technologies. Front Microbiol 2022; 13: 971696.
[http://dx.doi.org/10.3389/fmicb.2022.971696] [PMID: 35923399]

[19] Peay KG, Baraloto C, Fine PVA. Strong coupling of plant and fungal community structure across western Amazonian rainforests. ISME J. 2013 Sep; 7(9): 1852–61.
[http://dx.doi.org/10.1038/ismej.2013.66]

[20] Tamoor M, Samak NA, Jia Y, *et al.* Potential Use of Microbial Enzymes for the Conversion of Plastic Waste Into Value-Added Products: A Viable Solution. Front Microbiol 2021; 12: 777727.
[http://dx.doi.org/10.3389/fmicb.2021.777727] [PMID: 34917057]

[21] Briones A, Raskin L. Diversity and dynamics of microbial communities in engineered environments and their implications for process stability. Curr Opin Biotechnol 2003 Jun; 14(3): 270–6.
[http://dx.doi.org/10.1016/S0958-1669(03)00065-X]

[22] Mondal S, Palit D. Effective Role of Microorganism in Waste Management and Environmental Sustainability. In: Jhariya MK, Banerjee A, Meena RS, Yadav DK, eds. Sustainable Agriculture, Forest and Environmental Management. Singapore: Springer; 2019. p. 485–515.
[http://dx.doi.org/10.1007/978-981-13-6830-1_14]

[23] Bala S, Garg D, Thirumalesh BV, *et al.* Recent strategies for bioremediation of emerging pollutants: A

[24] Sivaperumal P, Kamala K, Rajaram R. Bioremediation of Industrial Waste Through Enzyme Producing Marine Microorganisms. Adv Food Nutr Res 2017; 80: 165-79.
[http://dx.doi.org/10.1016/bs.afnr.2016.10.006] [PMID: 28215325]

[25] Naseem M, Syab S, Akhtar S, *et al.* Ex-Situ and In-Situ bioremediation strategies and their limitations for solid waste management: a mini-review. J Qual Assur Agric Sci. 2023; 3(1): 28–31.

[26] Höhener P, Ponsin V. In situ vadose zone bioremediation. Curr Opin Biotechnol. 2014 Jun; 27: 1–7.
[http://dx.doi.org/10.1016/j.copbio.2013.08.018]

[27] Hou D, O'Connor D. Green and sustainable remediation: concepts, principles, and pertaining research. In: Hou D, ed. Sustainable Remediation of Contaminated Soil and Groundwater: Materials, Processes, and Assessment. Amsterdam: Elsevier; 2020. p. 1-17.
[http://dx.doi.org/10.1016/B978-0-12-817982-6.00001-X]

[28] National Research Council. Natural attenuation for groundwater remediation [Internet]. Washington (DC): National Academies Press; 2000 [cited 2025 Mar 04]. Available from: https://nap.nationalacademies.org/catalog/9792/natural-attenuation-for-groundwaer-remediation

[29] Krupka KM, Martin WJ. Subsurface Contaminant Focus Area: Monitored Natural Attenuation (MNA)—Programmatic, Technical, and Regulatory Issues. Richland (WA): Pacific Northwest National Laboratory; 2001 Jul. Available from: https://www.pnnl.gov/main/publications/external/technical_reports/pnnl-13569.pdf
[http://dx.doi.org/10.2172/965699]

[30] Höhener P, Ponsin V. *In situ* vadose zone bioremediation. Curr Opin Biotechnol 2014; 27: 1-7.
[http://dx.doi.org/10.1016/j.copbio.2013.08.018] [PMID: 24863890]

[31] Lee MD, Swindoll CM. Bioventing for in situ remediation. Hydrological sciences journal. 1993 Aug 1; 38(4): 273-82.
[http://dx.doi.org/10.1080/02626669309492674]

[32] Bioremediation E. Principles and Practices of Enhanced Anaerobic Bioremediation of Chlorinated Solvents Accessed: Jan 13, 2025 2004. Available from: https://www.frtr.gov/matrix-2019/documents/Enhanced-In-Situ-Reductive-Dechlorinated-for-Groundwater/2004-Principles-and-Practices-of-Enhanced-Anaerobic-Bioremediation-of-Chlorinated-Solvents.pdf

[33] Tiedje JM, Quensen JF III, Chee-Sanford J, Schimel JP, Boyd SA. Microbial reductive dechlorination of PCBs. Biodegradation 1993-1994; 4(4): 231-40.
[http://dx.doi.org/10.1007/BF00695971] [PMID: 7764920]

[34] Show K-Y, Lee D-J. Anaerobic Treatment Versus Aerobic Treatment. In: Lee D-J, Jegatheesan V, Pandey A, eds. Biological Treatment of Industrial Effluents. Curr Dev Biotechnol Bioeng. Amsterdam: Elsevier; 2017. p. 205–30.
[http://dx.doi.org/10.1016/B978-0-444-63665-2.00008-4]

[35] Muller KA, Johnson CD, Bagwell CE, Truex MJ. Methods for Delivery and Distribution of Amendments for Subsurface Remediation: A Critical Review. Ground Water Monit Remediat 2021; 41(1): 46-75.
[http://dx.doi.org/10.1111/gwmr.12418]

[36] Environmental Protection Agency. Engineered approaches to in situ bioremediation of chlorinated solvents: fundamentals and field applications [Internet]. Washington (DC): U.S. Environmental Protection Agency; 2000 [cited 2026 Jan 14]. Available from: https://nepis.epa.gov/Exe/ZyPURL.cgi?Dockey=10002UIJ.TXT

[37] Upadhyay S. and Sinha A. Role of microorganisms in Permeable Reactive Bio-Barriers (PRBBs) for environmental clean-up: A review, Global NEST Journal 2018; 20: 1-12. Available from: journal.gnest.org Available from:

https://journal.gnest.org/sites/default/files/Submissions/gnest_02525/gnest_02525_proof.pdf

[38] García-Gómez A, Bernal MP, Roig A. Organic matter fractions involved in degradation and humification processes during composting. Compost Sci Util. 2005; 13(2): 127–35.
[http://dx.doi.org/10.1080/1065657X.2005.10702229]

[39] DiLandro AC, Chappell TM, Panchani PN. *et al.* A chemical application method with underwater dissection to improve anatomic identification of cadaveric foot and ankle structures in podiatric education. J Am Podiatr Med Assoc. 2013 Sep; 103(5): 378–82. Available from: https://japmaonline.org/view/journals/apms/103/5/1030387.xml

[40] Sharma J. Advantages and limitations of in situ methods of bioremediation. Recent Adv Biol Med. 2019; 6: 32–7. Available from: https://www.academia.edu/65194280/Advantages_and_Limitations_of_In_Situ_Methods_of_Bioremediation

[41] Kuppusamy S, Palanisami T, Megharaj M, Venkateswarlu K, Naidu R. In-Situ Remediation Approaches for the Management of Contaminated Sites: A Comprehensive Overview. Rev Environ Contam Toxicol 2016; 236: 1-115.
[http://dx.doi.org/10.1007/978-3-319-20013-2_1] [PMID: 26423073]

[42] Miller RK. Ground-Water Remediation Technology Analysis Center. Technology overview report. TO-96-03; 1996. Available from: https://scholar.google.com/scholar?hl=en&as_sdt=0%2C5&q=Miller%2C+R.+K.+%281996%29.+Ground-Water+Remediation+Technology+Analysis+Center.+Technology+Overview+Report.+TO-96-03.&btnG=

[43] Abbasi SA, Ramasamy EV, Gajalakshmi S, Khan FI, Abbasi N. A waste management project involving engineers and scientists of a university, a voluntary (nongovernmental) organization, and lay people—a case study. In: Proceedings of International Conference on Transdisciplinarity. Zurich: Swiss Federal Institute of Technology; 2000 Feb. p. 1-3. Available from: https://scholar.google.com/scholar?cluster=909070609715642076&hl=en&oi=scholarr

[44] Tomei MC, Daugulis AJ. Ex Situ Bioremediation of Contaminated Soils: An Overview of Conventional and Innovative Technologies. Crit Rev Environ Sci Technol 2013; 43(20): 2107-39.
[http://dx.doi.org/10.1080/10643389.2012.672056]

[45] Evans BS, Dudley CA, Klasson KT. Sequential anaerobic-aerobic biodegradation of PCBs in soil slurry microcosms. In: Wyman CE, Davison BH, eds. Seventeenth Symposium on Biotechnology for Fuels and Chemicals. ABAB Symposium, vol 57/58. Totowa (NJ): Humana Press; 1996. p. 885–94.
[http://dx.doi.org/10.1007/978-1-4612-0223-3_83]

[46] FRTR Technology Screening Matrix | Federal Remediation Technologies Roundtable (FRTR). Accessed: Jan 14, 2025 [Online] Available from: https://frtr.gov/matrix/

[47] U.S. Environmental Protection Agency. Engineering bulletin: slurry biodegradation [Internet]. Washington (DC): U.S. Environmental Protection Agency; 1990 [cited 2025 Jan 14].

[48] Robles-González IV, Fava F, Poggi-Varaldo HM. A review on slurry bioreactors for bioremediation of soils and sediments. Microb Cell Fact 2008; 7(5): 5.
[http://dx.doi.org/10.1186/1475-2859-7-5]

[49] U.S. Environmental Protection Agency, Office of Wetlands. A handbook of constructed wetlands: a guide to creating wetlands for: agricultural wastewater, domestic wastewater, coal mine drainage, stormwater. In the Mid [Internet]. Washington (DC): U.S. Environmental Protection Agency; 1995 [cited 2025 Mar 04]. Available from: https://sub.cehum.org/bitstream/CEHUM2018/1287/1/Davis.%20A%20Handbook%20of%20Constructed%20Wetlands%2C%20Volume%201%2C%20General%20Considerations.pdf

[50] National Research Council. Contaminants in the subsurface: source zone assessment and remediation [Internet]. Washington (DC): National Academies Press; 2005 [cited 2025 Mar 04].
[http://dx.doi.org/10.17226/11146]

[51] Dhanker R, Khatana K, Verma K, *et al.* An integrated approach of algae-bacteria mediated treatment

[52] Sharma P, Pandey AK, Kim S, Singh SP, Chaturvedi P, Varjani S. Critical review on microbial community during in-situ bioremediation of heavy metals from industrial wastewater. Environ Technol Innov 2021 Nov; 24: 101826.
[http://dx.doi.org/10.1016/j.eti.2021.101826]

of industries generated wastewater: optimal recycling of water and safe way of resource recovery. Biocatal Agric Biotechnol 2023 Nov; 54: 102936.
[http://dx.doi.org/10.1016/j.bcab.2023.102936]

[53] Flemming HC, Wingender J, Szewzyk U, Steinberg P, Rice SA, Kjelleberg S. Biofilms: an emergent form of bacterial life. Nat Rev Microbiol 2016; 14: 563–575.
[http://dx.doi.org/10.1038/nrmicro.2016.94]

[54] Mallick P, Misra J. Degradation of plastics causing pollution using bacteria for improvement of freshwater fish cultivation. Sustain Environ Eng Sci 2021; 93: 175-80.
[http://dx.doi.org/10.1007/978-981-15-6887-9_20]

[55] Ho MTG, Bantoto-Kinamot V. Sargassum, Padina and Turbinaria as bioindicators of cadmium in Bais Bay, Negros Oriental. Palawan Sci 2021; 13(1): 90–8.

[56] Xu X, Liu W, Tian S, et al. Petroleum Hydrocarbon-Degrading Bacteria for the Remediation of Oil Pollution Under Aerobic Conditions: A Perspective Analysis. Front Microbiol 2018; 9: 2885.
[http://dx.doi.org/10.3389/fmicb.2018.02885] [PMID: 30559725]

[57] Sumampouw KJ, Risjani Y. Bacteria as indicators of environmental pollution: review. Int J Ecosyst 2014; 4(6): 251–8.
[http://dx.doi.org/10.5923.j.ije.20140406.03.html]

[58] Cabrera G, Pérez R, Gómez JM, Ábalos A, Cantero D. Toxic effects of dissolved heavy metals on Desulfovibrio vulgaris and Desulfovibrio sp. strains. J Hazard Mater 2006 Jul 31; 135(1-3): 40–6.
[http://dx.doi.org/10.1016/j.jhazmat.2005.11.058]

[59] Martins M, Faleiro ML, Barros RJ, Veríssimo AR, Barreiros MA, Costa MC. Characterization and activity studies of highly heavy metal resistant sulphate-reducing bacteria to be used in acid mine drainage decontamination. J Hazard Mater 2009 Jul 30; 166(2-3): 706–13.
[http://dx.doi.org/10.1016/j.jhazmat.2008.11.088]

[60] Shamim S. Biosorption of heavy metals. In: Derco J, Vrana B, eds. Biosorption. London: InTech; 2018. p. 1–25.
[http://dx.doi.org/10.5772/intechopen.72099]

[61] Kalaimurugan D, Balamuralikrishnan B, Durairaj K, et al. Isolation and characterization of heavy-metal-resistant bacteria and their applications in environmental bioremediation. Int J Environ Sci Technol 2020; 17(3): 1455-62.
[http://dx.doi.org/10.1007/s13762-019-02563-5]

[62] Priyadarshanee M, Das S. Biosorption and removal of toxic heavy metals by metal tolerating bacteria for bioremediation of metal contamination: a comprehensive review. J Environ Chem Eng 2021 Feb; 9(1): 104686.
[http://dx.doi.org/10.1016/j.jece.2020.104686]

[63] Gerardi MH. Wastewater bacteria. Hoboken (NJ): John Wiley & Sons, Inc; 2006.
[http://dx.doi.org/10.1002/0471979910]

[64] Zheng C, Zhao L, Zhou X, Fu Z, Li A. Treatment technologies for organic wastewater. In: Al-Malek SAR, eds. Water Treatment [Internet]. London: InTech; 2013 [cited 2025 Dec 7].
[http://dx.doi.org/10.5772/52665]

[65] Yao Q, Peng DC. Nitrite oxidizing bacteria (NOB) dominating in nitrifying community in full-scale biological nutrient removal wastewater treatment plants. AMB Express 2017; 7(1): 25.
[http://dx.doi.org/10.1186/s13568-017-0328-y] [PMID: 28116698]

[66] Wang D, Kong L, Li Y, Xie Z. Wang DZ, Kong LF, Li YY, Xie ZX. Environmental microbial

community proteomics: status, challenges and perspectives. Int J Mol Sci. 2016; 17(8): 1275.
[http://dx.doi.org/10.3390/ijms17081275]

[67] Rout PR, Shahid MK, Dash RR, *et al.* Nutrient removal from domestic wastewater: A comprehensive review on conventional and advanced technologies. J Environ Manage 2021; 296: 113246.
[http://dx.doi.org/10.1016/j.jenvman.2021.113246] [PMID: 34271353]

[68] Lacerda CM, Reardon KF. Environmental proteomics: applications of proteome profiling in environmental microbiology and biotechnology. Brief Funct Genomics Proteomics. 2009 Jan; 8(1): 75–87.
[http://dx.doi.org/10.1093/bfgp/elp005]

[69] Lacerda CMR, Choe LH, Reardon KF. Metaproteomic analysis of a bacterial community response to cadmium exposure. J Proteome Res 2007 Feb 7; 6(3): 1145–52.
[http://dx.doi.org/10.1021/pr060477v]

[70] Wilmes P, Andersson AF, Lefsrud MG, *et al.* Community proteogenomics highlights microbial strain-variant protein expression within activated sludge performing enhanced biological phosphorus removal. ISME J 2008 Aug; 2(8): 853–64.
[http://dx.doi.org/10.1038/ismej.2008.38]

[71] Wilmes P, Wexler M, Bond PL. Metaproteomics provides functional insight into activated sludge wastewater treatment. PLoS ONE 2008 Mar 19; 3(3): e1778.
[http://dx.doi.org/10.1371/journal.pone.0001778]

[72] Park C, Novak JT, Helm RF, Ahn Y-O, Esen A. Evaluation of the extracellular proteins in full-scale activated sludges. Water Res 2008 Aug; 42(14): 3879–89.
[http://dx.doi.org/10.1016/j.watres.2008.05.014]

[73] Mamta, Rao RJ, Wani KA. Bioremediation of pesticides under the influence of bacteria and fungi. In: Singh S, Srivastava K, editors. Handbook of Research on Uncovering New Methods for Ecosystem Management through Bioremediation. Hershey (PA): IGI Global; 2015. p. 22. https://www.igi-global.com/gateway/chapter/135089

[74] van Herwijnen R, van de Sande BF, van der Wielen FWM, Springael D, Govers HAJ, Parsons JR. Influence of phenanthrene and fluoranthene on the degradation of fluorene and glucose by Sphingomonas sp. strain LB126 in chemostat cultures. FEMS Microbiol Ecol 2003; 46(1): 105-11.
[http://dx.doi.org/10.1016/S0168-6496(03)00202-2]

[75] Mohammed A I. Isolation of Pesticide Degrading Microorganisms from soil Asian Business Review 2014; 5(4): 164-8.

[76] Bose S, Kumar P S, Vo D N, Rajamohan N, Saravanan R. Microbial degradation of recalcitrant pesticides: a review. Environmental Chemistry Letters 2021; 19(4): 3209-28.
[http://dx.doi.org/10.1007/s10311-021-01236-5]

[77] Ranjard L, Poly F, Nazaret S. Monitoring complex bacterial communities using culture-independent molecular techniques: application to soil environment. Res Microbiol 2000 Apr; 151(3): 167–77.
[http://dx.doi.org/10.1016/S0923-2508(00)00136-4]

[78] Rastogi G, Sani RK. Molecular techniques to assess microbial community structure, function, and dynamics in the environment. In: Ahmad I, Ahmad F, Pichtel J, eds. Microbes and Microbial Technology. New York (NY): Springer; 2011. p. 23-45.
[http://dx.doi.org/10.1007/978-1-4419-7931-5_2]

[79] Schneider MV, Orchard S. Omics technologies, data and bioinformatics principles. Methods Mol Biol 2011; 719: 3-30.
[http://dx.doi.org/10.1007/978-1-61779-027-0_1] [PMID: 21370077]

[80] Li C, Xia F, Zhang Y, Chang CC, Wei D, Wei L. Molecular biological methods in environmental engineering. Water Environ Res 2017; 89(10): 942-59.
[http://dx.doi.org/10.2175/106143017X15023776270197] [PMID: 28954649]

[81] Goel R, Kotay S M, Butler C S, Torres C I, Mahendra S. Molecular biological methods in environmental engineering Water Environment Research 2011; 83(10): 10.
[http://dx.doi.org/10.2175/106143011X13075599869092]

[82] Gedalanga P, Kotay S M, Sales C M, Butler C S, Goel R, Mahendra S. Novel applications of molecular biological and microscopic tools in environmental engineering Water Environment Research 2013; 85(10): 10.
[http://dx.doi.org/10.2175/106143013X13698672321742]

[83] Mahendra S, Gedalanga P, Kotay S M, Torres C I, Butler C S, Goel R. Advancements in molecular techniques and applications in environmental engineering Water Environment Research 2012; 84(10): 10.
[http://dx.doi.org/10.1002/j.1554-7531.2012.tb00236.x]

[84] Narayanan M, Ali SS, El-Sheekh M. A comprehensive review on the potential of microbial enzymes in multipollutant bioremediation: mechanisms, challenges, and future prospects. J Environ Manage 2023 May 15; 334: 117532.
[http://dx.doi.org/10.1016/j.jenvman.2023.117532]

[85] Kumari S, Das S. Bacterial enzymatic degradation of recalcitrant organic pollutants: catabolic pathways and genetic regulations. Environ Sci Pollut Res Int 2023; 30(33): 79676-705.
[http://dx.doi.org/10.1007/s11356-023-28130-7] [PMID: 37330441]

[86] Kumar A, Bisht B, Joshi V, Dhewa T. Review on bioremediation of polluted environment: A management tool Int J Environ Sci 2011; 1(6): 1079-93.

[87] Dutta D, Arya S, Kumar S. Industrial wastewater treatment: current trends, bottlenecks, and best practices. Chemosphere 2021 Dec; 285: 131245.
[http://dx.doi.org/10.1016/j.chemosphere.2021.131245]

[88] Priya AK, Pachaiappan R, Kumar PS, Jalil AA, Vo D-VN, Rajendran S. The war using microbes: a sustainable approach for wastewater management. Environ Pollut 2021 Apr 15; 275: 116598.
[http://dx.doi.org/10.1016/j.envpol.2021.116598]

[89] Rao MA, Scelza R, Acevedo F, Diez MC, Gianfreda L. Enzymes as useful tools for environmental purposes. Chemosphere 2014; 107: 145–62.
[http://dx.doi.org/10.1016/j.chemosphere.2013.12.059]

[90] Deveau A, Bonito G, Uehling J, *et al.* Bacterial–fungal interactions: ecology, mechanisms and challenges. FEMS Microbiol Rev 2018 May; 42(3): 335–52.
[http://dx.doi.org/10.1093/femsre/fuy008]

[91] Guest R, Smith D W. A potential new role for fungi in a wastewater MBR biological nitrogen reduction system J Environ Eng Sci 2002; 1(6): 433-7.
[http://dx.doi.org/10.1139/S02-037]

[92] Harms H, Schlosser D, Wick L. Untapped potential: exploiting fungi in bioremediation of hazardous chemicals. Nat Rev Microbiol 2011 Mar; 9(3): 177–92.
[http://dx.doi.org/10.1038/nrmicro2519]

[93] Zhang H, Feng J, Chen S, *et al.* Disentangling the drivers of diversity and distribution of fungal community composition in wastewater treatment plants across spatial scales. Front Microbiol 2018; 9(JUN): 1291.
[http://dx.doi.org/10.3389/fmicb.2018.01291] [PMID: 29967600]

[94] Assress HA, Selvarajan R, Nyoni H, *et al.* Diversity, co-occurrence and implications of fungal communities in wastewater treatment plants. Sci Rep 2019 Oct 1; 9(1): 14056.
[http://dx.doi.org/10.1038/s41598-019-50624-z]

[95] Wang K, Mao H, Li X. Functional characteristics and influence factors of microbial community in sewage sludge composting with inorganic bulking agent. Bioresour Technol 2018; 249: 527–35.
[http://dx.doi.org/10.1016/j.biortech.2017.10.034]

[96]   Ye R, Xu S, Wang Q, Fu X, Dai H, Lu W. Fungal diversity and its mechanism of community shaping in the milieu of sanitary landfill. Front Environ Sci Eng 2021; 15(1): 77.
[http://dx.doi.org/10.1007/s11783-020-1370-6]

[97]   Zhang X, Zhong Y, Yang S, *et al.* Diversity and dynamics of the microbial community on decomposing wheat straw during mushroom compost production. Bioresour Technol 2014; 170: 183-95.
[http://dx.doi.org/10.1016/j.biortech.2014.07.093] [PMID: 25129234]

[98]   Moore-Kucera J, Cox SB, Peyron M, *et al.* Native soil fungi associated with compostable plastics in three contrasting agricultural settings. Appl Microbiol Biotechnol 2014; 98(14): 6467-85.
[http://dx.doi.org/10.1007/s00253-014-5711-x] [PMID: 24797311]

[99]   El-Gendi H, Saleh AK, Badierah R, Redwan EM, El-Maradny YA, El-Fakharany EM. A comprehensive insight into fungal enzymes: structure, classification, and their role in mankind's challenges. J Fungi 2022; 8(1): 23.
[http://dx.doi.org/10.3390/jof8010023]

[100] Gupta S, Wali A, Gupta M, Annepu SK. Fungi: An Effective Tool for Bioremediation. In: Singh DP, Singh HB, Prabha R, eds Plant-Microbe Interactions in Agro-Ecological Perspectives 2017; 2: 593-606.
[http://dx.doi.org/10.1007/978-981-10-6593-4_24]

[101] Deshmukh R, Khardenavis AA, Purohit HJ. Diverse Metabolic Capacities of Fungi for Bioremediation. Indian J Microbiol 2016; 56(3): 247-64.
[http://dx.doi.org/10.1007/s12088-016-0584-6] [PMID: 27407289]

[102] Díaz-Godínez G. Fungi as important agents in water bioremediation. J Environ Biol 2024 Nov; 45(6) :i-iii.
[http://dx.doi.org/10.22438/jeb/45/6/Ed-1]

[103] Umar A, Mubeen M, Ali I, *et al.* Harnessing fungal bio-electricity: a promising path to a cleaner environment. Front Microbiol 2024; 14: 1291904.
[http://dx.doi.org/10.3389/fmicb.2023.1291904] [PMID: 38352061]

[104] Moubasher H, Tammam A, Saleh M. Enhancing electricity generation using fungal laccase-based microbial fuel cell. J Microbiol Biotechnol Food Sci 2024; 14(2): e9703.
[http://dx.doi.org/10.55251/jmbfs.9703]

[105] Sarma H, Bhattacharyya PN, Jadhav DA, *et al.* Fungal-mediated electrochemical system: prospects, applications and challenges. Curr Res Microb Sci 2021; 2: 100041.
[http://dx.doi.org/10.1016/j.crmicr.2021.100041]

[106] Zeraatkar AK, Ahmadzadeh H, Talebi AF, Moheimani NR, McHenry MP. Potential use of algae for heavy metal bioremediation, a critical review. J Environ Manage 2016; 181: 817–31.
[http://dx.doi.org/10.1016/j.jenvman.2016.06.059]

[107] Singh UB, Ahluwalia AS. Microalgae: a promising tool for carbon sequestration. Mitig Adapt Strategies Glob Change 2013; 18(1): 73-95.
[http://dx.doi.org/10.1007/s11027-012-9393-3]

[108] Paul V, Chandra Shekharaiah PS, Kushwaha S, Sapre A, Dasgupta S, Sanyal D. Role of Algae in CO2 Sequestration Addressing Climate Change: A Review. Smart Innovation, Systems and Technologies 2020; 161: 257-65.
[http://dx.doi.org/10.1007/978-981-32-9578-0_23]

[109] Kumar K, Nag Dasgupta C, Nayak B, Lindblad P, Das D. Development of suitable photobioreactors for CO2 sequestration addressing global warming using green algae and cyanobacteria. Bioresour Technol. 2011; 102(8): 4945-4953.
[http://dx.doi.org/10.1016/j.biortech.2011.01.054]

[110] Alami AH, Alasad S, Ali M, Alshamsi M. Investigating algae for CO_2 capture and accumulation and

simultaneous production of biomass for biodiesel production. Sci Total Environ 2021; 759: 143529. [http://dx.doi.org/10.1016/j.scitotenv.2020.143529]

[111] Gökçe D. Algae as an Indicator of Water Quality. In: Thajuddin N, Dhanasekaran D, eds. Algae - Organisms for Imminent Biotechnology [Internet]. London: InTech; 2016 [cited 2025 Dec 6]. [http://dx.doi.org/10.5772/62916]

[112] Pagana I, Nava V, Puglia GD, et al. Cystoseira compressa and Ericaria mediterranea: Effective Bioindicators for Heavy- and Semi-Metal Monitoring in Marine Environments with Rocky Substrates. Plants 2024; 13(4): 530. [http://dx.doi.org/10.3390/plants13040530] [PMID: 38498557]

[113] Meeranayak UFJ, Nadaf RD, Toragall MM, Nadaf U, Shivasharana CT. The role of algae in sustainable environment: a review. J Algal Biomass Utln 2020; 11(2): 28–34.

[114] Mata TM, Martins AA, Caetano NS. Microalgae for biodiesel production and other applications: a review. Renew Sust Energ Rev. 2010; 14(1): 217–32.

[115] Mata TM, Martins AA, Caetano NS. Microalgae for biodiesel production and other applications: a review. Renew Sust Energ Rev 2010; 14(1): 217–32. [http://dx.doi.org/10.1016/j.rser.2009.07.020]

[116] Demirbas A. Political, economic and environmental impacts of biofuels: a review. Appl Energy 2009; 86(Suppl 1): 108–17. [http://dx.doi.org/10.1016/j.apenergy.2009.04.036]

[117] Kwakye JM, Ekechukwu DE, Ogbu AD. Challenges and opportunities in algal biofuel production from heavy metal-contaminated wastewater. Int J Eng Res Dev 2024 Jul; 20(7): 236–47.

[118] Di Martino C, López NI, Raiger Iustman LJ. Isolation and characterization of benzene, toluene and xylene degrading Pseudomonas sp. selected as candidates for bioremediation. Int Biodeterior Biodegradation 2012; 67: 15–20. [http://dx.doi.org/10.1016/j.ibiod.2011.11.004]

[119] Parthipan P, Preetham E, Machuca LL, Rahman PKSM, Murugan K, Rajasekar A. Biosurfactant and degradative enzymes mediated crude oil degradation by bacterium Bacillus subtilis A1. Front Microbiol 2017; 8(FEB): 193. [http://dx.doi.org/10.3389/fmicb.2017.00193] [PMID: 28232826]

[120] Ferradji FZ, Mnif S, Badis A, et al. Naphthalene and crude oil degradation by biosurfactant producing Streptomyces spp. isolated from Mitidja plain soil (North of Algeria). Int Biodeterior Biodegradation. 2014; 86(C) :300–8. [http://dx.doi.org/10.1016/j.ibiod.2013.10.003]

[121] Prieto M, Hidalgo A, Rodríguez-Fernández C, Serra J, Llama M. Biodegradation of phenol in synthetic and industrial wastewater by Rhodococcus erythropolis UPV-1 immobilized in an air-stirred reactor with clarifier. Appl Microbiol Biotechnol 2002; 58(6): 853-60. [http://dx.doi.org/10.1007/s00253-002-0963-2] [PMID: 12021809]

[122] Asamudo N, Daba A, Ezeronye O. Bioremediation of textile effluent using Phanerochaete chrysosporium. Afr J Biotechnol. 2013; 4(13). https://www.ajol.info/index.php/ajb/article/view/71767

[123] Ryan D, Leukes W, Burton S. Improving the bioremediation of phenolic wastewaters by Trametes versicolor. Bioresour Technol 2007; 98(3): 579–87. [http://dx.doi.org/10.1016/j.biortech.2006.02.001]

[124] Mahmoud MS, Mostafa MK, Mohamed SA, Sobhy NA, Nasr M. Bioremediation of red azo dye from aqueous solutions by Aspergillus niger strain isolated from textile wastewater. J Environ Chem Eng 2017 Feb; 5(1): 547–54. Available from: https://www.sciencedirect.com/science/article/pii/S2213343716304742

[125] Xu X, Xia L, Zhu W, Zhang Z, Huang Q, Chen W. Role of Penicillium chrysogenum XJ-1 in the detoxification and bioremediation of cadmium. Front Microbiol 2015; 6(DEC): 1422.

[http://dx.doi.org/10.3389/fmicb.2015.01422] [PMID: 26733967]

[126] Cardoso LG, Duarte JH, Costa JAV, *et al*. *Spirulina sp.* as a Bioremediation Agent for Aquaculture Wastewater: Production of High Added Value Compounds and Estimation of Theoretical Biodiesel. BioEnergy Res 2021; 14(1): 254-64.
[http://dx.doi.org/10.1007/s12155-020-10153-4]

[127] Peter AP, Tan X, Lim JY, Chew KW, Koyande AK, Show PL. Environmental analysis of Chlorella vulgaris cultivation in large scale closed system under waste nutrient source. Chem Eng J 2022; 433: 134254. Available from: https://www.sciencedirect.com/science/article/pii/S1385894721058277

[128] Benfares R, Seridi H, Belkacem Y, Inal A. Heavy metal bioaccumulation in brown algae Cystoseira compressa in Algerian coasts, Mediterranean Sea. Environ Processes 2015; 2(3): 429–39.
[http://dx.doi.org/10.1007/s40710-015-0075-5]

[129] Oliveira AC, Barata A, Batista AP, Gouveia L. *Scenedesmus obliquus* in poultry wastewater bioremediation. Environ Technol 2019; 40(28): 3735-44.
[http://dx.doi.org/10.1080/09593330.2018.1488003] [PMID: 29893195]

[130] Ojha N, Karn R, Abbas S, Bhugra S. Bioremediation of industrial wastewater: a review. In: IoP conference series: earth and environmental science 2021 Jun 1 (Vol. 796, No. 1, p. 012012). IOP Publishing.

[131] Samaksaman U, Peng T-H, Kuo J-H, Lu C-H, Wey M-Y. Thermal treatment of soil co-contaminated with lube oil and heavy metals in a low-temperature two-stage fluidized bed incinerator. Appl Therm Eng 2016; 93: 131–8.
[http://dx.doi.org/10.1016/j.applthermaleng.2015.09.024]

[132] Ansari FA, Singh P, Guldhe A, Bux F. Microalgal cultivation using aquaculture wastewater: Integrated biomass generation and nutrient remediation. Algal Res 2017 Jan; 21: 169–77.
[http://dx.doi.org/10.1016/j.algal.2016.11.015]

[133] Gogate PR, Pandit AB. A review of imperative technologies for wastewater treatment I: oxidation technologies at ambient conditions. Adv Environ Res 2004; 8(3-4): 501–51.
[http://dx.doi.org/10.1016/S1093-0191(03)00032-7]

[134] Aditya L, Mahlia TMI, Nguyen LN, Vu HP, Nghiem LD. Microalgae-bacteria consortium for wastewater treatment and biomass production. Sci Total Environ 2022; 838(Pt 1): 155871.
[http://dx.doi.org/10.1016/j.scitotenv.2022.155871] [PMID: 35568165]

[135] Sharma N, Singh A, Batra N. Modern and Emerging Methods of Wastewater Treatment. In: Achal V, Mukherjee A, eds. Ecological Wisdom Inspired Restoration Engineering. Singapore: Springer; 2018. p. 223–47.
[http://dx.doi.org/10.1007/978-981-13-0149-0_13]

[136] Qu X, Alvarez P. Applications of nanotechnology in water and wastewater treatment Water Research 2013; 47: 3931-46.
[http://dx.doi.org/10.1016/j.watres.2012.09.058]

[137] Singh M, Pant G, Hossain K. Green remediation: Tool for safe and sustainable environment — a review. App Water Sci 2017; 7(6): 2629-35.
[http://dx.doi.org/10.1007/s13201-016-0461-9]

[138] Ramrakhiani L, Ghosh S, Majumdar S. Surface Modification of Naturally Available Biomass for Enhancement of Heavy Metal Removal Efficiency, Upscaling Prospects, and Management Aspects of Spent Biosorbents: A Review. Appl Biochem Biotechnol 2016; 180(1): 41-78.
[http://dx.doi.org/10.1007/s12010-016-2083-y] [PMID: 27097928]

[139] Azad MAK, Amin L, Sidik NM. Genetically engineered organisms for bioremediation of pollutants in contaminated sites. Chin Sci Bull 2014; 59(8): 703-14.
[http://dx.doi.org/10.1007/s11434-013-0058-8]

[140] Liu L, Bilal M, Duan X, Iqbal HMN. Mitigation of environmental pollution by genetically engineered

bacteria — current challenges and future perspectives. Sci Total Environ. 2019; 667 :444–54.
[http://dx.doi.org/10.1016/j.scitotenv.2019.02.390]

[141] Kumar S, Dagar VK, Khasa YP, Kuhad RC. Genetically modified microorganisms (GMOs) for bioremediation. In: Biotechnology for environmental management and resource recovery. India: Springer 2013: (pp. 191-218).

[142] Hussain I, Aleti G, Naidu R, *et al.* Microbe and plant assisted-remediation of organic xenobiotics and its enhancement by genetically modified organisms and recombinant technology: A review. Sci Total Environ 2018; 628-629: 1582-99.
[http://dx.doi.org/10.1016/j.scitotenv.2018.02.037] [PMID: 30045575]

[143] Pant G, Garlapati D, Agrawal U, Prasuna RG, Mathimani T, Pugazhendhi A. Biological approaches practised using genetically engineered microbes for a sustainable environment: A review. J Hazard Mater 2021; 405: 124631.
[http://dx.doi.org/10.1016/j.jhazmat.2020.124631] [PMID: 33278727]

[144] Saravanan A, Kumar PS, Ramesh B, Srinivasan S. Removal of toxic heavy metals using genetically engineered microbes: Molecular tools, risk assessment and management strategies. Chemosphere 2022; 298: 134341.
[http://dx.doi.org/10.1016/j.chemosphere.2022.134341] [PMID: 35307383]

[145] Abdel-Mawgoud AM, Markham KA, Palmer CM, Liu N, Stephanopoulos G, Alper HS. Metabolic engineering in the host Yarrowia lipolytica. Metab Eng 2018; 50: 192–208.
[http://dx.doi.org/10.1016/j.ymben.2018.07.016]

[146] Velkov VV. Stress-induced evolution and the biosafety of genetically modified microorganisms released into the environment. J Biosci 2001; 26(5): 667-83.
[http://dx.doi.org/10.1007/BF02704764] [PMID: 11807296]

[147] Paul D, Pandey G, Jain RK. Suicidal genetically engineered microorganisms for bioremediation: Need and perspectives. BioEssays 2005; 27(5): 563-73.
[http://dx.doi.org/10.1002/bies.20220] [PMID: 15832375]

[148] Imron MF, Kurniawan SB, Ismail NI, Abdullah SRS. Future challenges in diesel biodegradation by bacteria isolates: A review. J Clean Prod 2020; 251: 119716.
[http://dx.doi.org/10.1016/j.jclepro.2019.119716]

[149] Erdem H. The effects of biochars produced in different pyrolsis temperatures from agricultural wastes on cadmium uptake of tobacco plant. Saudi J Biol Sci 2021 Jul; 28(7): 3965–71.
[http://dx.doi.org/10.1016/j.sjbs.2021.04.016]

[150] Rafeeq H, Afsheen N, Rafique S, *et al.* Genetically engineered microorganisms for environmental remediation. Chemosphere 2023; 310: 136751.
[http://dx.doi.org/10.1016/j.chemosphere.2022.136751] [PMID: 36209847]

[151] Das S, Jean J-S, Kar S, Chou M-L, Chen C-Y. Screening of plant growth-promoting traits in arsenic-resistant bacteria isolated from agricultural soil and their potential implication for arsenic bioremediation. J Hazard Mater 2014 May 15 ;272: 112–20.
[http://dx.doi.org/10.1016/j.jhazmat.2014.03.012]

[152] Dong D, Sun H, Qi Z, Liu X. Improving microbial bioremediation efficiency of intensive aquacultural wastewater based on bacterial pollutant metabolism kinetics analysis. Chemosphere. 2021 Feb; 265: 129151.
[http://dx.doi.org/10.1016/j.chemosphere.2020.129151]

[153] Xia X, Wu S, Zhou Z, Wang G. Microbial Cd(II) and Cr(VI) resistance mechanisms and application in bioremediation. J Hazard Mater 2021 Jan 5; 401: 123685.
[http://dx.doi.org/10.1016/j.jhazmat.2020.123685]

[154] Wang F, Dong W, Zhao Z, *et al.* Heavy metal pollution in urban river sediment of different urban functional areas and its influence on microbial community structure. Sci Total Environ 2021; 778:

146383.
[http://dx.doi.org/10.1016/j.scitotenv.2021.146383]

[155] Sharma P, Pandey AK, Udayan A, Kumar S. Role of microbial community and metal-binding proteins in phytoremediation of heavy metals from industrial wastewater. Bioresour Technol 2021; 326: 124750.
[http://dx.doi.org/10.1016/j.biortech.2021.124750]

[156] Yadav A, Chowdhary P, Kaithwas G, Bharagava RN. Toxic Metals in the Environment: Threats on Ecosystem and Bioremediation Approaches. In: Das S, Dash HR, eds. Handbook of Metal-Microbe Interactions and Bioremediation. 1st ed. Boca Raton (FL): CRC Press; 2017. p. 14.
[http://dx.doi.org/10.1201/9781315153353-11]

[157] Pieper DH, Reineke W. Engineering bacteria for bioremediation. Curr Opin Biotechnol. 2000; 11(3): 262-70.
[http://dx.doi.org/10.1016/S0958-1669(00)00094-X]

[158] Jacob JM, Karthik C, Saratale RG, *et al.* Biological approaches to tackle heavy metal pollution: A survey of literature. J Environ Manage. 2018; 217: 56-70.
[http://dx.doi.org/10.1016/j.jenvman.2018.03.077]

[159] Saradadevi GP, Das D, Mangrauthia SK, *et al.* Genetic, epigenetic, genomic and microbial approaches to enhance salt tolerance of plants: a comprehensive review. Biol. 2021; 10(12): 1255.
[http://dx.doi.org/10.3390/biology10121255]

[160] Vymazal J. Horizontal sub-surface flow and hybrid constructed wetlands systems for wastewater treatment. Ecol Eng 2005 Dec 1; 25(5): 478-490.
[http://dx.doi.org/10.1016/j.ecoleng.2005.07.010]

[161] Yang K, Zhu L, Zhao Y, Wei Z, Chen X, Yao C. A novel method for removing heavy metals from composting system: The combination of functional bacteria and adsorbent materials. Bioresour Technol 2019 Dec; 293: 122095.
[http://dx.doi.org/10.1016/j.biortech.2019.122095]

[162] Ejraei A, Aroon MA, Ziarati Saravani A. Wastewater treatment using a hybrid system combining adsorption, photocatalytic degradation and membrane filtration processes. J Water Process Eng 2019; 28: 45-53.
[http://dx.doi.org/10.1016/j.jwpe.2019.01.003]

[163] Abdullah MO. Introduction to Applied Energy. In: Abdullah MO, ed. Applied Energy. 1st ed. Boca Raton (FL): CRC Press; 2012. p. 36.
[http://dx.doi.org/10.1201/b12758-8]

[164] Dhangar K, Kumar M. Tricks and tracks in removal of emerging contaminants from the wastewater through hybrid treatment systems: A review. Sci Total Environ 2020; 738: 140320.
[http://dx.doi.org/10.1016/j.scitotenv.2020.140320] [PMID: 32806367]

[165] Trögl J, Chauhan A, Ripp S, Layton A C, Kuncová G, Sayler G S. *Pseudomonas fluorescens* HK44: Lessons Learned from a Model Whole-Cell Bioreporter with a Broad Application History Sensors 2012; 12(2): 1544-71.
[http://dx.doi.org/10.3390/s120201544]

[166] Singh JS, Abhilash PC, Singh HB, Singh RP, Singh DP. Genetically engineered bacteria: An emerging tool for environmental remediation and future research perspectives. Gene 2011; 480(1-2): 1-9.
[http://dx.doi.org/10.1016/j.gene.2011.03.001] [PMID: 21402131]

[167] Graf N, Altenbuchner J. Genetic engineering of Pseudomonas putida KT2440 for rapid and high-yield production of vanillin from ferulic acid. Appl Microbiol Biotechnol 2014; 98(1): 137-49.
[http://dx.doi.org/10.1007/s00253-013-5303-1] [PMID: 24136472]

[168] Jin R, Yang H, Zhang A, Wang J, Liu G. Bioaugmentation on decolorization of C.I. Direct Blue 71 by using genetically engineered strain Escherichia coli JM109 (pGEX-AZR). J Hazard Mater 2009;

163(2-3): 1123-8.
[http://dx.doi.org/10.1016/j.jhazmat.2008.07.067] [PMID: 18755538]

[169] Chen J, Qin J, Zhu YG, de Lorenzo V, Rosen BP. Engineering the soil bacterium Pseudomonas putida for arsenic methylation. Appl Environ Microbiol 2013; 79(14): 4493-5.
[http://dx.doi.org/10.1128/AEM.01133-13] [PMID: 23645194]

[170] Lan WS, Lu TK, Qin ZF, et al. Genetically modified microorganism Spingomonas paucimobilis UT26 for simultaneously degradation of methyl-parathion and γ-hexachlorocyclohexane. Ecotoxicology 2014; 23(5): 840-50.
[http://dx.doi.org/10.1007/s10646-014-1224-8] [PMID: 24648032]

[171] Ng SP, Palombo EA, Bhave M. Identification of a copper-responsive promoter and development of a copper biosensor in the soil bacterium *Achromobacter* sp. AO22. World J Microbiol Biotechnol 2012; 28(5): 2221-8.
[http://dx.doi.org/10.1007/s11274-012-1029-y] [PMID: 22806045]

[172] Barman DN, Haque MA, Islam SMA, Yun HD, Kim MK. Cloning and expression of ophB gene encoding organophosphorus hydrolase from endophytic Pseudomonas sp. BF1-3 degrades organophosphorus pesticide chlorpyrifos. Ecotoxicol Environ Saf 2014; 108: 135-41.
[http://dx.doi.org/10.1016/j.ecoenv.2014.06.023] [PMID: 25062445]

[173] Sangwan N, Verma H, Kumar R, et al. Reconstructing an ancestral genotype of two hexachlorocyclohexane-degrading Sphingobium species using metagenomic sequence data. ISME J. 2014; 8(2): 398–408.
[http://dx.doi.org/10.1038/ismej.2013.153]

[174] Husnik JI, Volschenk H, Bauer J, Colavizza D, Luo Z, van Vuuren HJJ. Metabolic engineering of malolactic wine yeast. Metab Eng 2006; 8(4): 315-23.
[http://dx.doi.org/10.1016/j.ymben.2006.02.003] [PMID: 16621641]

[175] Vaudano E, Costantini A, Garcia-Moruno E. An event-specific method for the detection and quantification of ML01, a genetically modified Saccharomyces cerevisiae wine strain, using quantitative PCR. Int J Food Microbiol 2016; 234: 15-23.
[http://dx.doi.org/10.1016/j.ijfoodmicro.2016.06.017] [PMID: 27367966]

[176] Coconi-Linares N, Magaña-Ortíz D, Guzmán-Ortiz DA, Fernández F, Loske AM, Gómez-Lim MA. High-yield production of manganese peroxidase, lignin peroxidase, and versatile peroxidase in Phanerochaete chrysosporium. Appl Microbiol Biotechnol 2014; 98(22): 9283-94.
[http://dx.doi.org/10.1007/s00253-014-6105-9] [PMID: 25269601]

[177] Knop D, Yarden O, Hadar Y. The ligninolytic peroxidases in the genus Pleurotus: divergence in activities, expression, and potential applications. Appl Microbiol Biotechnol 2015; 99(3): 1025-38.
[http://dx.doi.org/10.1007/s00253-014-6256-8] [PMID: 25503316]

[178] Zhou X, Hui E, Yu XL, et al. Development of a rapid immunochromatographic lateral flow device capable of differentiating phytase expressed from recombinant aspergillus niger phy A2 and genetically modified corn. J Agric Food Chem 2015; 63(17): 4320-6.
[http://dx.doi.org/10.1021/acs.jafc.5b00188] [PMID: 25901899]

[179] Jiang Y, Duarte AV, van den Brink J, et al. Enhancing saccharification of wheat straw by mixing enzymes from genetically-modified Trichoderma reesei and Aspergillus niger. Biotechnol Lett 2016; 38(1): 65-70.
[http://dx.doi.org/10.1007/s10529-015-1951-9] [PMID: 26354856]

[180] Alazi E, Knetsch T, Di Falco M, et al. Inducer-independent production of pectinases in Aspergillus niger by overexpression of the D-galacturonic acid-responsive transcription factor gaaR. Appl Microbiol Biotechnol 2018; 102(6): 2723-36.
[http://dx.doi.org/10.1007/s00253-018-8753-7] [PMID: 29368217]

[181] Kang NK, Jeon S, Kwon S, et al. Effects of overexpression of a bHLH transcription factor on biomass and lipid production in Nannochloropsis salina. Biotechnol Biofuels 2015; 8(1): 200.

[http://dx.doi.org/10.1186/s13068-015-0386-9] [PMID: 26628914]

[182] Kang NK, Kim EK, Kim YU, *et al.* Increased lipid production by heterologous expression of AtWRI1 transcription factor in *Nannochloropsis salina*. Biotechnol Biofuels 2017; 10(1): 231.
[http://dx.doi.org/10.1186/s13068-017-0919-5] [PMID: 29046718]

[183] Imamura S, Hagiwara D, Suzuki F, Kurano N, Harayama S. Genetic transformation of Pseudochoricystis ellipsoidea, an aliphatic hydrocarbon-producing green alga. J Gen Appl Microbiol 2012; 58(1): 1-10.
[http://dx.doi.org/10.2323/jgam.58.1] [PMID: 22449745]

[184] Peña Salamanca EJ, Madera-Parra CA, Avila-Williams CA, Rengifo-Gallego AL, Ascúntar Ríos D. Phytoremediation Using Terrestrial Plants. In: Ansari AA, Gill SS, Gill R, Lanza GR, Newman L, eds. Phytoremediation Management of Environmental Contaminants, Volume 2. Cham: Springer; 2015. p. 305-19.
[http://dx.doi.org/10.1007/978-3-319-10969-5_25]

[185] Dangi AK, Sharma B, Hill RT, Shukla P. Bioremediation through microbes: systems biology and metabolic engineering approach. Crit Rev Biotechnol. 2019; 39(1): 79–98.
[http://dx.doi.org/10.1080/07388551.2018.1500997]

[186] Pavlopoulos GA, Malliarakis D, Papanikolaou N, Theodosiou T, Enright AJ, Iliopoulos I. Visualizing genome and systems biology: technologies, tools, implementation techniques and trends, past, present and future. GigaScience. 2015; 4(1): s13742–015–0077–2.
[http://dx.doi.org/10.1186/s13742-015-0077-2]

[187] Gao J, Ellis LBM, Wackett LP. The University of Minnesota Biocatalysis/Biodegradation Database: improving public access. Nucleic Acids Res. 2010; 38(suppl_1): D488–91.
[http://dx.doi.org/10.1093/nar/gkp771]

[188] Pazos F, Guijas D, Valencia A, De Lorenzo V. MetaRouter: bioinformatics for bioremediation. Nucleic Acids Res 2004; 33(Database issue): D588-92.
[http://dx.doi.org/10.1093/nar/gki068] [PMID: 15608267]

[189] Prakash D, Verma S, Bhatia R, Tiwary BN. Risks and Precautions of Genetically Modified Organisms. ISRN Ecol 2011; 2011: 1-13.
[http://dx.doi.org/10.5402/2011/369573]

[190] Keese, Paul. "Risks from GMOs due to horizontal gene transfer." Environmental biosafety research 7.3 (2008): 123-149.

CHAPTER 7

Climate Change and Microbial Processes

Preeti Mehta Kakkar[1,*]**, Aindree Lohumi**[1]**, Diya Saha**[1] **and Shruti Khanna Ahuja**[2,*]

[1] *Department of Biotechnology, Amity Institute of Biotechnology, Amity University, Noida, Uttar Pradesh, India*

[2] *Centre for Medical Biotechnology, Amity Institute of Biotechnology, Amity University, Noida, Uttar Pradesh, India*

Abstract: Climate change poses a significant existential threat to all life forms on Earth, with far-reaching impacts on ecosystems globally. Microorganisms play an important role in biogeochemical cycles such as carbon, nitrogen, and phosphorus. They are the most abundant life forms in the world. They contribute significantly to greenhouse gas generation and consumption, as well as serving as important pathogens. The responses of microbial communities to climate change are intricate and multifaceted, influencing ecosystem functionality, biodiversity, and the global distribution of microbes. This chapter examines the impact of climate change on microbial communities, specifically how temperature, carbon dioxide levels, and ocean acidification affect diversity. Additionally, it discusses the implications of altered microbial dynamics on human health, agriculture, and the environment while also highlighting future research directions and potential strategies for harnessing microbial processes to mitigate environmental challenges.

Keywords: Biogeochemical cycles, Carbon cycle, Climate change, Greenhouse gases, Microbial communities, Ocean acidification.

INTRODUCTION

This climate change issue is one that concerns us greatly today and gives long-reaching repercussions, not just on ecosystems, economics, and health [1, 2]. It forms an increasingly formidable threat to life on Earth by upsetting natural processes and compromising ecological stability around the world [3]. The rising

---

[*]**Corresponding authors Preeti Mehta Kakkar and Shruti Khanna Ahuja:** Department of Biotechnology, Amity Institute of Biotechnology, Amity University, Noida, Uttar Pradesh, India; and Centre for Medical Biotechnology, Amity Institute of Biotechnology, Amity University, Noida, Uttar Pradesh, India;
E-mails: pmkakkar@amity.edu, skahuja@amity.edu

Rajneesh Kumar, Ram Sharan Singh & Maulin P. Shah. (Eds.)
All rights reserved-© 2026 Bentham Science Publishers

greenhouse gas levels in the atmosphere have never been observed, with these gases elevating global temperature levels and forcing severe deviations in climatic patterns, which put huge stresses on the ecosystems [4, 5]. These changes are reshuffling biodiversity, water resources, and food security, radically changing the ecosystems. Rising temperatures with increased CO_2 levels, ocean acidification, altered frequency, intensity, timing, and duration of climate extremes are affecting not just the structure and function but also species ranges, productivity, and resilience [3]. This overview discusses the global implications of climate change on biodiversity, food security, and water availability, emphasizing the complex problems that ecosystems face as they try to adapt to these rapid changes.

Microorganisms perform vital functions in these ecosystems, ensuring the balance of life on earth. Although they are very small, they play a crucial role in the major biogeochemical cycles of carbon, nitrogen, and phosphorus. This is because these cycles influence life by controlling nutrient supply, decomposing organic materials, and regulating greenhouse gas in the environment [3, 4]. Microbes are important for soil health, plant productivity, and the stability of aquatic food webs. Their impacts thus extend into human health and agriculture, altering pathogen dynamics and crop nutrient availability [4, 5]. Despite their vital relevance, the interaction between climate change and microbial activities is often overlooked in broader studies in climate science, allowing major gaps in our perception of how ecosystems respond to climatic changes [4, 5]. While CO_2 levels and temperatures across the globe are increasingly becoming high, microbial communities adapt in numerous unpredictable manners, influencing ecosystem functionality, resilience, and the global spread of microbial life [4, 6, 7]. Microbial processes are intrinsically linked to the Earth's climate systems, with their activities having a direct impact on greenhouse gas fluxes, carbon storage, and nutrient cycles.

Microorganisms play important roles in processes like methane generation and oxidation, both of which have a direct impact on atmospheric warming potential. Soil microbial communities mediate the storage of carbon, acting as both sources and sinks of CO_2 depending on the environment [5, 7]. Oceanic bacteria, particularly phytoplankton, play a major role in carbon fixation within aquatic habitats. They provide the base for the food chain in the ocean and regulate global carbon dynamics [5, 6]. Rising climate-induced stressors are increasingly threatening these vital services: warmer oceans, ocean acidification, and oxygen depletion. Such stressors disrupt the precarious balance in which microorganisms operate; this upsets the carbon-regulating services provided and threatens ecosystem stability [4]. Climate change impacts the diversity and functionality of microbes in diverse settings such as terrestrial, aquatic, and extreme settings [4 - 6]. Rising temperatures in terrestrial ecosystems change the composition and activity of soil microbial communities with impacts on soil fertility and carbon storage [5, 6].

Changes in temperature and nutrient availability impact microbial populations in aquatic settings with significant implications for the nutrient cycle and primary production [1, 8]. Climate change threatens the disruption of ecological niches in microorganisms that are well-suited to harsh conditions, thereby potentially leading to the loss of unique microbial functions. Changes that are critical and crucial include those that are expected to have far-reaching impacts on ecosystem resilience, agricultural output, and natural resources' sustainability [7, 9]. This chapter is meant to fill the knowledge gap in understanding the complex interactions between climate change and microbiological activity [3, 6, 9]. It will study how changes in environmental parameters, such as temperature, CO_2 levels, and ocean pH, affect microbial diversity, functionality, and interspecies interactions across various ecosystems. The chapter will also consider the larger consequences of these microbial changes for human health, agricultural sustainability, and environmental management [3 - 5].

For instance, shifts in microbial populations can exacerbate the spread of plant and animal diseases, interrupt agricultural nutrient cycles, and alter water quality, all of which directly impact food security and public health [7, 10]. Moreover, studying how microbes respond to environmental shifts may uncover valuable insights about ecosystem resilience and adaptability in the context of climate change [8, 10]. This chapter not only explores the effects of climate change on microbial processes but also examines potential ways to leverage microbial capacities to address environmental concerns. Novel approaches such as bioengineering, carbon sequestration, and ecosystem restoration can be used to mitigate and adapt to climate change through microbial processes. For instance, enhancing microbial activity in soils can improve carbon storage, lower greenhouse gas emissions, and increase agricultural output [4, 5, 10]. Similarly, promoting the proliferation of phytoplankton in marine habitats may enhance oceanic sequestration of carbon; this may offset some CO_2 increase impacts [3, 6].

With the development of microbial biotechnology, solutions to several climate issues are prospective for solutions, from cleaning an environment that is dirty back to developing biofuels that are sustainable [1, 4]. From this perspective, the functionality of microorganisms as both indicators and agents underscores complexity in their interaction with its habitat. Microbial reactions, as indicators, can give early warning signs of ecological disturbances, allowing scientists to monitor ecosystem health and forecast future changes [7, 10]. As agents, microorganisms actively regulate biogeochemical cycles and maintain ecosystem stability, emphasizing their potential to contribute to global sustainability initiatives [5, 6, 9, 10]. Recognizing and incorporating microbial activities into climate adaptation and mitigation measures is critical for dealing with the numerous problems presented by a constantly changing planet. This chapter attempts to provide an overview of how environmental shifts like increased temperatures, higher levels of CO_2, and ocean acidification influence the functionality of microbial activity, as well as interspecies interactions in diverse habitats.

It will explore the way in which alterations in environmental parameters such as temperature, CO_2, and pH of the ocean impact the diversity, functioning, and interspecies interactions of microbes across a range of habitats [7, 9, 10]. Moreover, it examines how shifting microbial dynamics influence important ecosystem services, like soil fertility, water quality, and plant production [2, 4]. This chapter provides insight into the broader implications of these changes in microbes for human health, agricultural sustainability, and environmental management. Besides, knowing the resilience and adaptability processes of microbes could help conservationists develop approaches to safeguard ecosystem functionality from changing climate circumstances. It examines the most recent studies to show how microbial responses to climate change might restructure ecosystem dynamics, affect nutrient cycles, and influence ecosystem resilience [4, 6].

Finally, this chapter points out that microbes are of great significance in the struggle against climate change. If we investigate their responses to environmental changes and creative approaches to taking advantage of their capabilities, we can uncover new avenues for the improvement of ecosystem resilience and sustainability [6, 9, 11]. The more dire the impact of climate change becomes, the more vital it will be to understand and utilize microbial activities in safeguarding biodiversity, food, and water, and making a greener future possible for all earth's species [4, 5].

MICROBIAL DIVERSITY AND CLIMATE CHANGE

Microbial Biodiversity in Terrestrial and Aquatic Ecosystems

Microorganisms are among the most diverse and abundant life forms on earth, thriving in almost any environment, from fertile soils to harsh locations such as hydrothermal vents and polar ice [1, 10]. Microbial communities—which include bacteria, archaea, fungi, and protists—play critical roles in terrestrial ecosystems by cycling nutrients, decomposing organic materials, and developing beneficial connections with plants [5, 7, 10]. These bacteria enhance soil fertility and encourage plant growth through symbiotic interactions and convert organic matter into nutrients that other creatures can conveniently consume [9, 10]. Microbial diversity is crucial in all aquatic habitats, freshwater or marine. The planktonic organisms include bacteria, archaea, and algae, which contribute to the global carbon cycle and oxygen production. Photosynthetic microorganisms, like cyanobacteria and algae, not only produce oxygen but also capture atmospheric carbon dioxide. Microbial communities recycle nutrients, digest organic debris, and clean water in freshwater ecosystems [9]. For instance, microbial biofilms contribute to contaminant degradation, thereby protecting water quality [7, 9]. They contribute, in a delicate equilibrium of the microorganisms' effect, to the wellness and productivity of both aquatic and terrestrial ecosystems.

Earth's diverse ecosystem processes, plant productivity, and nutrient cycling are particularly critical to support human well-being. With ongoing global population growth, substantial increases in plant production and intensified land use will be required to meet future food and fibre demands. Understanding the factors influencing plant production and nutrient cycling, especially in the context of a changing environment, is important for preserving and managing both natural and human-dominated ecosystems [5, 7, 10].

We propose that soil microbial diversity is critical in maintaining ecosystem multifunctionality by facilitating litter decomposition and organic matter mineralization processes that facilitate the flow of matter and energy between aboveground and belowground communities [12]. A rapidly increasing body of evidence suggests that the relationship between biodiversity, including microbes and plants, and ecosystem functioning is more linear than saturating [10, 12]. This means that microbial diversity loss due to the worldwide environmental changes caused, for example, by altered land use patterns, nitrogen enrichment, or climate change, would further impair microbes' capacity for sustaining multiple ecosystem functions [5].

Despite this awareness, empirical evidence for a positive relationship between microbial diversity and multifunctionality in terrestrial ecosystems remains scarce [1, 4]. The relative importance of microbial diversity as compared to abiotic properties in soil, climate, and plant species richness remains poorly understood for driving ecosystem functioning [10, 13]. This lack of knowledge constrains our ability to predict changes in multifunctionality under ongoing global environmental changes and to develop effective management and conservation strategies.

Impacts of Climate Change on Microbial Diversity

Temperature Variations

Temperature has a profound impact on microbial activity and diversity (Fig. **1**). As the globe's temperatures increase, microbial metabolic rates, community structures, and geographic distributions all shift [4, 5]. For instance, in terrestrial ecosystems, heat accelerates decomposition of organic matter through saprotrophic fungi and bacteria, which leads to the liberation of carbon stored in soil [4]. However, prolonged hot or high temperature events can stress microbial populations, thereby lowering diversity and function. Such disruptions can impede nitrogen cycling and even decrease soil fertility over time. Thawing of permafrost in polar regions exposes ancient, dormant microbial communities [4, 10, 13].

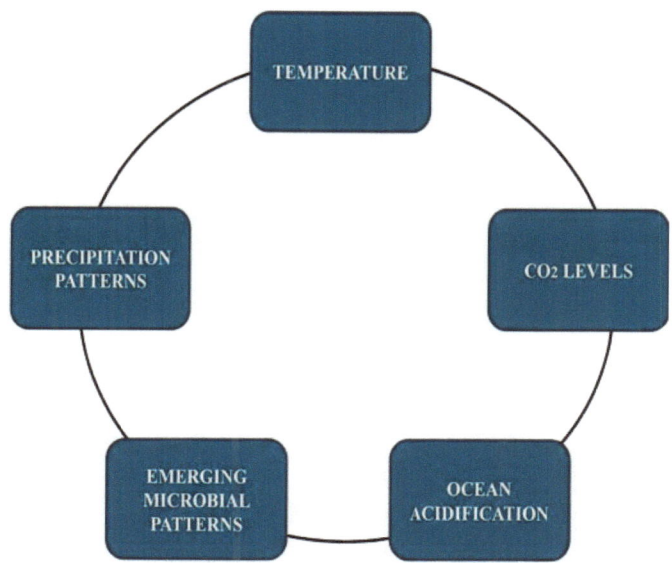

Fig. (1). Impacts of climate change on microbial diversity.

Once activated, these microbes continue to break down frozen organic matter, releasing greenhouse gases in the form of methane and carbon dioxide [4, 5, 13]. This process accelerates the pace of climate change and gives it a kind of feedback, which further increases the strength of influence over microbial ecology [10, 13]. Similarly, stratification arising from changes in temperature over aquatic habitats influences the stratification of microbial populations over their habitat. Warmer waters often select for thermophilic, fast-growing microorganisms that break microbial networks [11, 14].

There are many adaptive strategies used by microbes, *i.e.*, while respiration by bacteria increases with warmer temperatures, this same increase tends to result in smaller bacterial cells [11, 13, 15]. Viruses have evolved to encode cold shock response proteins, enabling them to survive at lower temperatures, in the cooler oceans. Changes in the microbial community composition have been documented in both warmer and drier soils and in warming oceans, where they might alter carbon and nutrient cycling in these ecosystems [14, 16, 17]. Viruses that infect soil and aquatic microbes also have an influence on microbial activity and community structure. For example, cyanophages, which are viruses that infect cyanobacteria, can affect carbon dioxide fixation during infection and may thereby enhance the availability of greenhouse gases in the atmosphere. However, there is still a need for more research to better understand the virome's role in microbial adaptation to climate change [13, 17].

Rising CO_2 Levels

Higher levels of atmospheric CO_2 have several impacts on microbial populations. Elevated CO_2 enhances plant growth by altering the type and amount of root exudates, which are substances produced by plant roots [5, 15, 18]. This then influences the soil microbial populations, affecting activities such as nutrient cycling and carbon storage (Fig. **1**). Increased carbon inputs have the potential to enhance microbial activities but also disrupt the balance of microbial community structures, which compromise ecosystem stability. High CO_2 levels stimulate the growth of some microorganisms, including cyanobacteria, in aquatic systems [4, 5, 13].

Even though these microbes are very important primary producers, an overabundance of them can lead to Hazardous Algal Blooms (HAB) [6, 19]. HABs disrupt the aquatic food chains by decreasing oxygen levels and releasing poisons that kill other organisms. These events not only threaten microbial diversity but also have a chain of effects on the broader ecosystems [18, 20].

Ocean Acidification and Marine Microbiomes

Ocean acidification, caused by seawater's absorption of excess CO_2, affects the marine microbial communities greatly [4, 21]. The subsequent drop in seawater pH affects the physiological and metabolic processes of numerous marine microorganisms. Calcifying microorganisms are very vulnerable as they help create reefs and form carbonates [2, 19, 21]. The decline of these species poses a significant threat to coral reefs, which are critical biodiversity hotspots.

Furthermore, ocean acidification affects nitrogen-fixing cyanobacteria, which play a vital role in sustaining the nutrient cycle of marine environments [4, 21]. Changes in their quantity and activity could disrupt the availability of nitrogen and thus create an imbalance in the marine ecosystem [3]. Acidic conditions may also favor the growth of toxic microorganisms, such as Vibrio species, which are poisonous to both marine life and human beings (Fig. **1**). These changes reflect the broader ecological and health risks associated with changes in microbial dynamics [3, 21]. Enrichment of oceans with anthropogenic CO_2 and acidification will have deep impacts on marine biogeochemistry and microbes. In a high-CO_2 world, fluxes involving heterotrophic bacteria and the microbial loop are thought to shift [9, 21]. Microbially mediated processes are fundamental to the functioning of marine ecosystems. Although marine microbes are known to be extremely plastic in response to pH shifts, the actual effects on microbially mediated processes are still not well established [9, 19, 20]. These microbes contribute to over 50% of global primary production and are essential for major biogeochemical cycles. The high abundance and diversity of marine microbes are fundamental in controlling Earth's climate through controlling CO_2 and CH_4 [2, 9, 21].

Altered Precipitation Patterns

Fluctuations in precipitation due to climatic change have an intensive influence on microbial ecosystems [22]. Heavy rainfall and floods erode the soil, leach the nutrients, and alter the structural variations of the microbial communities. Drought decreases soil moisture, thereby stressing microbial communities and lowering their metabolic activities [2, 21]. These disturbances can cause impairment of microbial functions like fixing of nitrogen and decomposition of organic matter [4, 22]. In aquatic ecosystems, altered precipitation changes the flow of water and input of nutrients, thus modifying the dynamics of microbial communities. Excessive runoff can transport pollutants and organic matter into the water bodies, leading to blooms of microbes that consume dissolved oxygen and kill aquatic life. Understanding how microbial communities respond to these changes is critical for maintaining water quality and ecosystem health [3, 23].

This multi-factor compound effect on the low-altitude grassland microorganisms may indicate increased availability of water in the soil because of increased precipitation that can reduce some of the warming impacts on the soil's biomass, diversity, activity, and rates of turnover for bacteria [2, 12, 23]. Soil water is an important determinant of plant productivity and microorganisms within soil in all semiarid ecosystems, as well as alpine steppes [20, 24]. The availability of soil water in water-limited ecosystems is highly dependent on the balance between temperature and precipitation changes, thus resulting in diverse microbial responses. Warming typically represses microbial growth and reduces both the soil microbial biomass and its activity in semiarid grasslands [12, 22, 24]. However, increased precipitation usually enhances both the microbial biomass and diversity, directly or indirectly, through an increase in plant productivity and carbon allocation [3, 4].

Emerging Microbial Pathogens

Climate change facilitates the spread of microbial diseases by creating the necessary conditions [25, 26]. Increased temperature and disturbed ecosystems allow the spread of disease-causing microorganisms into new territories and hence increase the risks of epidemics [26]. For instance, rising waters have been related to an increased amount of pathogenic Vibrio species that could harm marine organisms and hence pose health hazards for human beings (Fig. 1). Similarly, changes in the terrestrial ecosystems may favour the survival and spread of soil-borne diseases, which threaten agricultural production and food security [20, 24, 26].

The last two decades have brought out several emergent multidrug-resistant species: MRSA (Methicillin-Resistant Staphylococcus aureus), extensively resistant tuberculosis, as well as drug-resistant ones, vancomycin-resistant enterococci, extended-spectrum β-lactamase-producing *E. coli*, and carbapenem-producing Gram-negative bacteria, all of which emerged and expanded to hospital-wide transmission among hospital-acquired patients and population and became emerging issues to be put up with heavy concern [25, 26]. Of course, handling the above-emerging critical issue falls outside this review [25].

MICROBIAL INFLUENCE ON BIOGEOCHEMICAL CYCLES

The Earth's biogeochemical cycles are complex systems supporting life, and microbes are the unsung heroes [8, 27]. They orchestrate the transformation and movement of essential components, such as carbon, nitrogen, and phosphorus. Everything, from climate regulation to food production, is influenced.

Carbon Cycle: Production and Consumption of Greenhouse Gases

Microorganisms play a critical role in the global carbon cycle by facilitating the generation and consumption of greenhouse gases, such as carbon dioxide and methane [6]. Decomposer microbes in terrestrial ecosystems degrade organic matter, emitting CO_2 as a byproduct of respiration [8, 18]. This mechanism is essential for nutrient recycling but adds to atmospheric carbon levels. On the other hand, some bacteria take CO_2 from the atmosphere through photosynthesis and carbon sequestration. Natural carbon sinks, such as soil bacteria and sea phytoplankton, fix CO_2 and store it in organic forms. In the oceans, phytoplankton play a significant role in the biological carbon pump [6, 18, 27]. These microorganisms convert CO_2 into organic matter by using sunlight, thus forming the basis of the marine food chain. Upon their death, their carbon-rich remains fall to the ocean's depths, thus sequestering carbon for centuries. This process highlights the importance of protecting marine ecosystems since any disturbance can propagate across global carbon storage systems [6, 18].

Role of Microorganisms in Carbon Sequestration

Soils and sedimentary bacteria play a crucial role in the storage of carbon, with profound implications for combating climate change [18]. For instance, mycorrhizal fungi form symbiotic associations with plant roots, stretching the reach of the latter further into the soil to trap and store carbon. It also enhances the uptake of plant nutrients, thus maintaining a healthier ecosystem [27, 28].

Microbial communities in wetlands and peatlands inhibit decomposition, which allows for long-term carbon storage [18, 28]. However, such ecosystems are vulnerable: changing global temperatures and moving water levels threaten their stability and the potential release of enormous amounts of stored carbon into the atmosphere [8, 18, 27]. An example is thawing Arctic permafrost, leading to increased microbial activity with increased production of CO_2 and methane, a powerful greenhouse gas, and making the climate worse.

Nitrogen and Phosphorus Cycles: Agricultural and Environmental Impacts

Nitrogen and phosphorus are crucial nutrients for all living species, and microbes play a central role in their cycling [3, 23]. Nitrogen-fixing bacteria and archaea, such as those in legume roots, convert atmospheric nitrogen into plant-usable forms. Such natural fertilization increases the yield of agriculture while reducing dependence on synthetic fertilizers [29]. Denitrifying bacteria are responsible for bringing nitrogen back into the atmosphere, which completes the nitrogen cycle. While useful, this process may also emit nitrous oxide (N_2O), a greenhouse gas. Microbes have an equally important role in phosphorus cycling [3, 20].

Some bacteria degrade organic phosphorus molecules, while others solubilize inorganic phosphorus, making it available to plants [3, 29]. This microbial activity is essential for both food production and ecological health. Human activities, such as excessive fertilizer use and climate change, interrupt natural cycles [23]. Excess nutrients in water bodies lead to eutrophication, a process in which algae growth depletes oxygen levels and endangers aquatic life. Meanwhile, evolving climatic patterns worsen nutrient imbalances, posing challenges for both farmers and conservationists [20, 29].

Methane Production and Microbial Regulation

Methane is a potent greenhouse gas that microorganisms both produce and control [5]. Methanogenic archaea are well adapted to anaerobic environments like wetlands, rice paddies, and ruminant digestive tracts, where they produce methane as a byproduct of metabolism [3, 30]. These emissions make a significant contribution to global warming, especially as agriculture and permafrost thawing increase methane-producing settings [22, 30].

On the other hand, methanotrophic bacteria are natural mitigators, absorbing methane before it escapes into the atmosphere. Such microorganisms are crucial for balance, although their activity is affected by environmental factors. For instance, warming temperatures in Arctic wetlands enhance methanogenic activity but challenge methanotrophs, thereby increasing methane emissions [3].

Why does Microbial Stewardship Matter?

Understanding and safeguarding microbial functions is an important approach in addressing global problems, such as climate change, food security, and ecosystem health [5]. Efforts to assist microbial communities, such as sustainable farming and wetland restoration, may enhance their beneficial impacts. As caretakers of these secret partners, humans may use microbial activities to build a more resilient and balanced Earth [3, 5, 30].

These invisible microorganisms are the ones responsible for an immense impact on the environment we view and dwell in [5, 30]. The recognition of their contributions and vulnerabilities has helped us to better negotiate interwoven issues of environmental conservation and human well-being.

CLIMATE CHANGE AND MICROBIAL INTERACTIONS

A climate change study regarding microbial interactions in the climate helps understand differences in ecosystems, agriculture, and public health. Microorganisms, often invisible to the naked eye, have the greatest impact on the habitat in which they dwell [19]. As the climate increases, microbial populations take centre stage as active players in changes that are taking place across the world [24, 31]. This chapter explores how bacteria interact in many ecosystems, from plants and animals to harsh settings, and how these interactions are affected by a changing climate [19, 22, 32].

Microbe-Plant Interactions: Agricultural and Forest Ecosystems

In agricultural and forest ecosystems, microorganisms and plants create intricate, mutually beneficial connections [5, 32]. Climate change, including rising temperatures, altered rainfall patterns, and higher CO_2 levels, is affecting these interactions in predictable and unanticipated ways.

Soil Microbes and Plant Health

These soil microorganisms include bacteria, fungi, and protozoa. This helps in the cycle of nutrients, allows for the proper growth of a plant, and increases a plant's ability to not contract diseases. Agricultural application of microbes improves plant uptake of vital nutrients from the soil, like nitrogen, phosphorus, and potassium [32]. It helps plants against infections and pest attacks [5, 32]. As the climate warms, changes in soil moisture and temperature can affect microbial activity and variety, affecting plant health. Some useful microorganisms may thrive, while others may be challenged and become less effective, resulting in reduced crop yield [5, 22].

Freshly dead plants and living plants release carbon into the soil as rhizodeposits, which act as food for soil microbes and aid in their reproduction [5]. In return, microbes supply essential nutrients such as nitrogen and phosphorus through processes like nitrogen fixation and the mineralization of organic matter [22, 23, 28]. Many microbes can fix atmospheric nitrogen, but a few of them specialize in the improvement of nutrients for crops [5, 28, 32]. For example, rhizobia associated with legumes enable the fixation of nitrogen, whereas arbuscular mycorrhizal fungi extend their hyphae into the soil to improve phosphorus acquisition. The other benefit provided by these fungi is root growth enhancement, further increasing phosphorus uptake [31, 32]. To continue their functioning activity and nutrient-providing functions, the soil microbes require regular replenishments of organic matter rich in carbon [5, 22].

Microbes in Forests

In the forests, plants have relied more on the mycorrhizal fungus for exchanging nutrients with the soil, especially in nutrient-poor areas. The fungi make symbiotic relationships with tree roots, thereby increasing the uptake of water and minerals, especially in drought conditions. However, climate change is likely to change these relationships [32]. For instance, the changes in precipitation patterns may affect soil moisture, thus affecting both trees and their fungal companions. Moreover, warmer temperatures may enhance the development of pathogens such as root rot fungi that are harmful to forest health [11]. The supply of growth-limiting nutrients such as nitrogen, through mobilization from organic matter and conversion to mineral forms accessible to plants, relies heavily on microbial symbionts [32]. In temperate and boreal forests, mycorrhizal fungi and nitrogen-fixing bacteria provide up to 80% of the nitrogen and 75% of the phosphorus taken up by plants.

Impact on Agricultural Productivity

Rising global temperatures are causing pests and pathogens to expand their ranges, further stressing crops [32, 24]. For instance, warm weather encourages the dispersal of fungal diseases, including species of Fusarium, that destroy crucial crops such as wheat and maize. Help microbes may fail to adapt to changing environments, weakening the crop's own defences. These dynamics represent the necessity to regulate the health of microorganisms in agriculture for sustainability in climate change [24].

Plant growth-promoting microbes and arbuscular mycorrhizae are widely used for the promotion of plant growth and crop yields both under normal and stress conditions [32, 24]. These microbes influence various physiological parameters in response to external stimuli and improve plant growth through multiple mechanisms. These include the production of plant growth regulators, synthesis of various metabolites, and the conversion of atmospheric nitrogen into ammonia, among others, through direct and indirect pathways [24]. These microbes also provide resistance against pathogens by activating ISR and SAR. The interaction of plants with microbes is an essential factor in enhancing plant growth and controlling diseases, especially in a changing environment [32, 24]. This contributes to more sustainable agricultural practices without compromising ecosystem functionality.

Microbe-Animal and Human Pathogens: Health Implications

The complex interplay between diseases, commensal microbes, and changes in the environment due to climate change is increasingly influencing human and animal health [19, 33].

1. Pathogens now exist in a Warmer World. Global warming and extreme climatic factors are more suitable for pathogens such as bacteria [19, 23]. Some of the waterborne diseases that have been found to infect people include malaria, dengue fever caused by mosquito vectors, among others, as their ranges have expanded through rising temperatures. Other such diseases include cholera, which is on the rise because of flooding and modifications in the water systems [19, 22, 23].
2. In humans, the microbiome—the collection of bacteria that live in our bodies—is essential for digestion, immunity, and overall health. Climate change has the potential to disturb this delicate balance [19, 22]. Increased temperatures, for example, may promote gut dysbiosis, in which dangerous bacteria outcompete good microbes, resulting in illnesses such as irritable bowel syndrome (IBS) [19, 25]. Extreme weather events, such as floods, can

also have an influence on sanitation, allowing germs to spread and disturb the microbiome, contributing to diseases [23].

3. Climate change also increases the danger of zoonotic infections, which are passed from animals to humans [22, 23]. Climate change has affected wildlife movement patterns and habitats, increasing human contact with possible disease vectors. Emerging zoonotic diseases such as Lyme disease, Ebola, and new coronaviruses are all associated with environmental changes aggravated by climate change [19, 25, 33].

Microbial Interactions in Extreme Environments

The most extraordinary microbial communities can be found in some of Earth's most extreme conditions, where life appears to be impossible. Such habitats are becoming more relevant as scientists investigate life's resilience and the potential consequences of climate change [11].

Extremophiles and Their Adaptation

Extremophiles are microorganisms that flourish in severe environments like boiling hot springs, deep-sea vents, and acidic lakes [14]. These microorganisms have developed distinct biochemical pathways that allow them to withstand high salinity, strong pressure, and extreme temperatures [24]. Thermophiles thrive in temperatures above 100°C, whilst Halophiles thrive in saline surroundings. Understanding how these species adapt to severe environments can shed light on resilience, which may be critical to understanding how life on Earth will cope with a changing climate [11, 14].

Climate Change Effects on Extreme Environments

Climate change may be disturbing the fragile balance in these harsh environments. For instance, melting ice caps and glaciers may expose microbial communities previously isolated to new stresses, such as fluctuating temperatures and nutrient cycles, leading to changes in diversity and ecological functioning [14]. On the other hand, deep-sea ecosystems that depend on microbial activity may be disturbed with the rise of ocean temperatures and thus have a broad range of implications for carbon cycling in the oceans [11].

Microbes and Space Exploration

In an interesting twist, scientists also research extremophiles to learn how life could exist on other worlds [34]. As scientists look toward the possibility of colonizing Mars or Jupiter's moons, they look to these tough bacteria to learn how life may persist beyond Earth in high-radiation conditions, temperature changes, and low pressure. This study might also provide crucial insights into the survival of microbes under a warming future [34].

ADAPTATION AND EVOLUTION OF MICROBIAL COMMUNITIES

Microbial communities are very versatile organisms that include bacteria, fungi, and viruses, and have evolved with changes in the environment [4, 35]. They can change quickly in response to environmental changes in temperature, moisture, nutrient availability, and other factors of the environment. This is how they can survive in such diverse and challenging habitats. These communities are crucial to ecosystem functions like nutrient cycling, carbon sequestration, and environmental balance [2, 19]. Understanding the mechanisms through which microbes adapt and evolve helps predict their future responses to climate change and their impacts on the health of humans, agriculture, and the environment [3, 23].

Mechanisms of Microbial Adaptation to Changing Climates

Microbes have several ways to adapt to changing climates. Genetic mutations are one of the primary mechanisms by which microorganisms evolve [16, 24]. Spontaneous changes in their genetic material can result in traits that improve their survival under new conditions, such as the ability to tolerate higher temperatures or increased salinity [12, 24]. In addition to mutations, horizontal gene transfer allows microbes to share genetic material, including genes beneficial for stress resistance [13, 17]. It allows microbes to adapt quickly by promoting the spread of advantageous traits in microbial communities (Fig. **2**). Microbes are also phenotypically plastic, meaning they can alter their physiological characteristics to be better suited to the new environment [13, 15]. For instance, bacteria may modify their metabolism or alter the structure of their cell membranes to be able to withstand extreme conditions better. In addition, microbes often produce protective molecules such as heat shock proteins or spores that safeguard them against stressors such as temperature fluctuations or desiccation [12].

Evolutionary Responses and Potential for Resilience

As environmental conditions change, there is selective pressure on the microbial populations that favors individuals with traits to enhance their survival chances [10, 36]. This process of natural selection leads to the gradual evolution of microbial communities that are better adapted to their environment [36]. This will, over time, manifest as changes in community composition where more resilient species come to dominate. Microbial resilience is the ability of a community to maintain its functions even when environmental disturbances are prevalent [10].

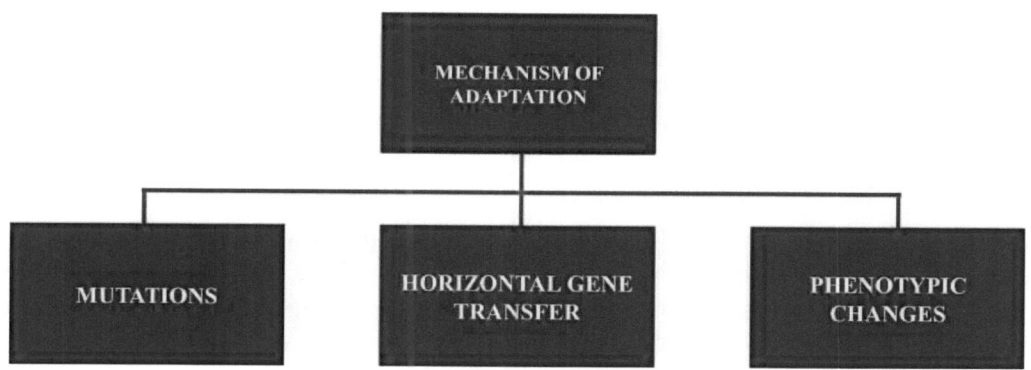

Fig. (2). Mechanisms of microbial adaptation to changing climates.

The evolutionary capacity for resilience is critical as climate change accelerates, allowing microbial communities to continue performing essential roles in ecosystems [8]. Genetic diversity in microbial communities also enhances resilience. A diverse gene pool is likely to increase the possibility that some members of the community will possess traits that are suited to new environmental conditions, ensuring the survival and functionality of the community [36]. For example, soil or aquatic system microorganisms may develop mechanisms for better survival during drought conditions or higher temperatures, therefore stabilizing ecosystems affected by climate change [8, 36]. In conclusion, microbial communities are highly versatile and can adapt to various environmental changes through genetic mutation, gene exchange, and physiological adjustments [20]. These evolution processes allow microbes to survive and maintain ecosystem functionality despite climate disruptions [8, 36].

As climate change continues to influence global environments, the ability of microbial communities to thrive will be vital in ensuring ecological stability and supporting human health and agricultural systems [8, 10]. Understanding these adaptive mechanisms will be important for predicting reactions by microbes to future environmental changes [10].

HUMAN HEALTH IMPLICATIONS

The health of humans is profoundly influenced by microbial communities, both positively and negatively [23, 37]. Microbes play essential roles in human health, but they can also pose significant risks, especially as they evolve and adapt to changing environmental conditions [37]. Two critical concerns for public health are the emergence of new pathogens and the growing challenge of antibiotics [23]. Additionally, microbial activity in the environment, such as their influence on air quality, can have direct impacts on respiratory health [25, 37]. Understanding the dynamics of microbial communities and their health implications is crucial for mitigating these risks [24, 25].

Emerging Pathogens and Disease Spread

As climate change and environmental factors shift, emerging pathogens are becoming a greater threat to human health [25, 37]. Since many microorganisms, such as bacteria, viruses, and fungi, are extremely adaptive, shifts in climate, precipitation patterns, and ecosystem dynamics can facilitate the spread of infectious illnesses [25]. For example, increased rainfall and warmer temperatures can spread vector-borne illnesses into new areas, where ticks and mosquitoes can spread Lyme disease, dengue, and malaria [8, 25]. Furthermore, urbanization, deforestation, and human encroachment into natural ecosystems can put humans into greater contact with wildlife, enabling the spillover of zoonotic diseases [37]. The rapid spread of these pathogens is also aided by globalization, where increased travel and trade enable diseases to spread quickly across borders. Monitoring and understanding microbial evolution are key to predicting and controlling the spread of emerging infectious diseases.

Despite promising forecasts, infectious diseases have continued to be a significant challenge. As early as the 1940s and 1950s, bacteria such as Staphylococcus aureus began developing resistance to penicillin [25, 38]. New strains of influenza emerged in 1957 and 1968 in China and Hong Kong, respectively, spreading rapidly worldwide. In the United States, the 1970s saw a resurgence of sexually transmitted diseases partly because of changes in human sexual behavior and the introduction of antibiotic-resistant gonorrhea strains by soldiers returning from Vietnam [8, 20, 38]. The events of the 1980s marked a turning point: shattering complacency toward infectious diseases with the emergence of AIDS and the re-emergence of tuberculosis, including multidrug-resistant strains. These events underscore the persistent and evolving threat of infectious diseases [8].

Impact on Antibiotic Resistance

One of the most important issues facing human health today is antibiotic resistance. Antibiotic-resistant bacteria have emerged more quickly because of the overuse and abuse of antibiotics in both human treatment and agriculture [8]. Through processes like genetic mutations and horizontal gene transfer, microbes can evolve resistance to commonly used antibiotics, making infections harder to treat [38]. As a result, common infections that were once easily treatable have become life-threatening. Furthermore, the environment plays a critical role in the spread of resistance [8, 38]. For example, antibiotics used in agriculture can enter the environment through runoff and waste, leading to the development of resistant microbial strains in soil and water. Addressing antibiotic resistance requires comprehensive strategies, including better stewardship of antibiotics, development of new treatments, and improved infection control practices [25, 38].

The increasing rate of antibiotic resistance worldwide poses a critical challenge as it makes common antibiotics ineffective against widespread bacterial infections [8, 36]. According to the report in 2022 by the Global Antimicrobial Resistance and Use System (GLASS), resistance rates of key bacterial pathogens are alarmingly high [8, 38]. Median resistance rates were found to be at 42% for third-generation cephalosporin-resistant *E. coli* and at 35% for methicillin-resistant Staphylococcus aureus in 76 countries [20]. Furthermore, in 2020, one in five urinary tract infections caused by *E. coli* were found to be less susceptible to standard antibiotics, including ampicillin, co-trimoxazole, and fluoroquinolones. These trends make it increasingly challenging to treat the most common infections effectively [38].

Microbial Influence on Air Quality

Air quality is also greatly influenced by microbial communities, which may have health consequences. Airborne pollutants, including particulate matter, allergens, and Volatile Organic Compounds (VOCs), are caused by microorganisms, which include bacteria, fungi, and viruses [39]. For example, fungi release spores and VOCs that can affect indoor and outdoor air quality, leading to respiratory issues and allergic reactions [37, 40]. Additionally, microbes can play a role in the degradation of pollutants, such as hydrocarbons, through processes like bioremediation (Fig. **3**). However, microbial activity can also contribute to the formation of secondary pollutants, including ozone and particulate matter, which can worsen air quality and impact public health [24, 37]. Prolonged exposure to poor air quality, influenced in part by microbial activity, can lead to chronic respiratory diseases, cardiovascular problems, and an increased susceptibility to infections. Understanding the role of microbes in air quality is essential for developing strategies to improve environmental health [22, 40].

In conclusion, microbial communities have a profound impact on human health through their roles in the spread of emerging pathogens, the development of antibiotic resistance, and their influence on air quality [8, 38]. As microbes continue to evolve in response to environmental changes, their effects on human health will likely grow in significance. Addressing these challenges requires a multidisciplinary approach, including better surveillance of microbial threats, enhanced public health policies, and environmental management strategies [39].

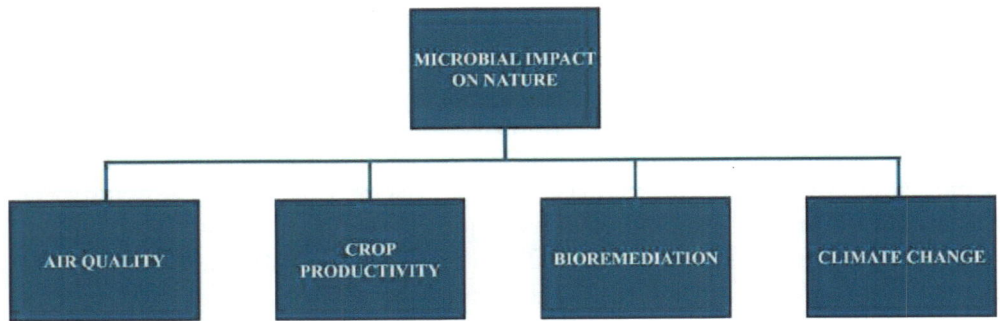

Fig. (3). Microbial impact on nature.

Microorganisms are fundamental to the functioning of ecosystems and play crucial roles in agriculture and environmental management [40]. Their diverse interactions with the soil, crops, and pollutants highlight their importance in both enhancing agricultural productivity and addressing environmental challenges [22, 39]. This article explores three key areas where microorganisms contribute significantly: soil microbes and crop productivity, their role in bioremediation, and their potential in mitigating environmental pollution.

AGRICULTURAL AND ENVIRONMENTAL IMPLICATIONS

Soil Microbes and Crop Productivity

An intricate ecosystem, soil is home to a wide range of microorganisms, such as fungi, bacteria, and archaea [3]. These microbes are essential for crop productivity, plant health, and soil fertility. Soil microorganisms are essential to sustainable agriculture because they can improve soil structure, affect nutrient cycling, and shield plants from disease [24, 32]. Nitrogen fixing is one of the most important ways microorganisms increase crop productivity. By forming symbiotic partnerships with leguminous plants, certain soil bacteria, such as Rhizobium species, transform atmospheric nitrogen into a form that plants can absorb and utilize [20, 23]. Because of this process, fewer chemical nitrogen fertilizers are required, which lowers input prices and lessens the negative environmental effects of excessive fertilizer use [24].

In addition to nitrogen fixation, microbes help in decomposing organic matter, releasing essential nutrients like phosphorus, potassium, and micronutrients into the soil [3]. This microbial activity supports soil health and ensures plants have access to the nutrients they need to grow robustly. Microbes also enhance the soil's ability to retain water and improve soil structure, which can lead to better crop yields, particularly in arid or degraded environments [23]. Microorganisms can also help protect crops from diseases. Beneficial microbes, such as *Bacillus* species and certain fungi, can outcompete or inhibit harmful pathogens, providing natural protection to plants (Fig. **3**). Moreover, the use of microbial inoculants, which introduce beneficial microbes to the soil, has become a common practice to boost soil health and crop productivity [32].

Microbes in Bioremediation: Managing Environmental Pollution

Bioremediation is the process of using microorganisms to clean up contaminated surroundings [41]. By using microorganisms' innate metabolic capacities to break down contaminants, this technology offers a more economical and environmentally benign option to conventional chemical or mechanical pollution management techniques [42].

Microorganisms have the unique capacity to break down a wide range of contaminants, including hydrocarbons, heavy metals, pesticides, and organic solvents [17]. For instance, petroleum hydrocarbons may be broken down by bacteria like Pseudomonas and Bacillus species, which makes bioremediation a practical way to clean up oil spills [32, 42]. Likewise, fungi such as Phanerochaete chrysosporium have been used to break down harmful substances like phenols and dioxins [17].

The advantages of using microorganisms for bioremediation are numerous (Fig. **3**). Microbial processes are often more environmentally friendly than chemical treatments, as they do not produce harmful by-products [17, 41]. They can also be more cost-effective, particularly in large-scale cleanup efforts, and are less disruptive to the local ecosystem. Moreover, bioremediation can be applied *in situ*, meaning that contaminated sites can be treated without the need for excavation or transportation of contaminated materials [40]. Despite its potential, bioremediation faces challenges. The effectiveness of microbial degradation depends on various factors such as temperature, pH, nutrient availability, and the presence of suitable microbial species [3, 5, 39]. In some cases, the introduction of engineered microbes or the addition of nutrients may be necessary to optimize the process. Nonetheless, the use of microbes in environmental pollution management is a rapidly growing field with promising applications for the future [5].

Role of Microorganisms in Mitigating Climate Change

Microorganisms are also playing a role in mitigating climate change (Fig. **3**). They affect carbon and nitrogen cycles. Microorganisms also play a role in climate change mitigation [3, 4]. Their activities can either contribute to or mitigate greenhouse gas emissions, such as carbon dioxide (CO_2), methane (CH_4), and nitrous oxide (N_2O), as they form an important part of the carbon and nitrogen cycles [4, 5].

One of the pivotal functions of any microbial community in ecosystems and soil is carbon sequestration, capturing and holding atmospheric carbon in the soil [5]. Few soils have bacteria and fungi that are capable of carbon fixation through their metabolic activities, which can help mitigate it [4, 5, 7]. For example, an increase in carbon storage in plant biomass together with an increase in soil organic matter is brought about by soil bacteria in the nitrogen cycle as they indirectly promote plant growth and can either promote or alleviate emissions of other greenhouse gases such as carbon dioxide (CO_2), methane (CH_4), and nitrous oxide (N_2O) [4, 22].

Microbes are very important in terms of carbon sequestration, which is the capturing of atmospheric carbon and storing it in the soil [5, 13]. This refers to microbial activity in soil ecosystems. Certain soil bacteria and fungi fix carbon through their metabolic activities, which might lead to a decrease in the amount of CO_2 in the atmosphere [11]. For example, soil bacteria associated with nitrogen cycling can marginally stimulate plant growth and progress toward carbon storage in plant biomass and soil organic matter [4, 26]. Microbes can also help reduce methane emissions, a potent greenhouse gas. Methanotrophic bacteria, which oxidize methane, are important in controlling methane levels in both natural and agricultural environments [3, 39]. In rice paddies, for example, methane emissions are significant, but the presence of methanotrophs can help mitigate this by consuming methane before it escapes into the atmosphere [17, 22].

Furthermore, microorganisms involved in denitrification can reduce nitrous oxide emissions, another greenhouse gas [13, 28]. By converting nitrate into nitrogen gas, these microbes prevent the release of nitrous oxide, which is 300 times more potent than CO_2 in trapping heat in the atmosphere. This process, when managed effectively, can be a critical strategy for mitigating climate change, especially in agricultural systems that rely heavily on fertilizers [13, 39]. In summary, the implications of microorganisms are enormous for agricultural productivity and environmental sustainability, and that is what they are [13]. Hereby, by the role of enhancing soil health, promoting crop growth, remediating polluted environments, and even mitigating the negative impacts of climate change, microorganisms seem to hold some potential as natural allies in responding to some of the significant environmental challenges facing the planet [23, 28]. The potential of such microorganisms may be maximally tapped for creating a more sustainable agricultural environment and environmental systems through persistent research and application [5, 28].

FUTURE RESEARCH DIRECTIONS IN MICROBIAL SCIENCE

Addressing Climate Change and Exploring New Technologies

Microorganisms are key players in the Earth's ecosystems, influencing everything from nutrient cycling to climate regulation [15, 16]. As global challenges such as climate change intensify, understanding microbial responses to these changes and harnessing microbial technologies for climate mitigation is becoming more critical. In this context, future research in microbial science holds enormous potential to offer solutions for both mitigating and adapting to climate change [16, 38]. This article explores key future research directions, focusing on knowledge gaps in microbial responses to climate change, the potential of microbial technologies in climate mitigation, and advances in genomics and microbial ecology [38].

The other name given to Microbial Protein (MP) is Single-Cell Protein (SCP). MP promises to satisfy the future demand for food and feed due to its high productivity and approval for human consumption [17, 38]. MP can be produced in a batch or continuous process within a well-controlled bioreactor, independent of climatic conditions, with minimal usage of natural resources [17]. It can be used as whole dried cells or extracted protein from microbial biomass, containing up to 80% (dry matter) protein and high levels of carbohydrates, minerals, and vitamins [16, 39]. Its amino acid profile, being rich in essential amino acids, makes it highly suitable for use in fish and livestock feed. MP can be produced from photoautotrophic microorganisms, such as microalgae and cyanobacteria, which derive their biomass by photosynthesis. These systems have low nutrient costs as they do not require organic carbon, though they demand substantial land and resources [28, 20, 38]. Alternatively, MP can be produced by chemoautotrophic microorganisms that generate energy by oxidizing inorganic compounds and use CO_2 as a carbon source. For example, hydrogen-oxidizing bacteria can produce MP using hydrogen and CO_2, offering another efficient production method [11, 32].

Knowledge Gaps in Microbial Responses to Climate Change

While microorganisms play a central role in regulating the Earth's climate, many of their responses to climate change remain poorly understood [13, 15]. Climate change is altering temperature patterns, precipitation, soil moisture, and atmospheric CO_2 levels, which are all likely to have profound effects on microbial communities [13]. However, the complexity of these interactions and the wide diversity of microbes make it difficult to predict how these changes will manifest across different ecosystems [3, 16, 28].

There is one significant knowledge gap that needs filling concerning the adjustment of microbial communities due to changes in temperature and nutrient availability. As microbes are quite sensitive to their environments, their abilities to grow or decline will depend on their metabolic flexibility and resilience [24, 43]. Thus, as temperatures increase, certain populations of microbes might increase while others might be unable to survive [15, 44]. Identifying these microbes that are the most vulnerable to these changes, as well as the possible resulting shifts in microbial diversity, is important because it helps project future ecosystem services like soil fertility, carbon sequestration, and the cycles of nutrients such as nitrogen and phosphorus [18, 38, 44].

Another important area of research is understanding how microbial communities will respond to changes in soil moisture and water availability [17]. Many microorganisms, especially those in soil and aquatic environments, are highly sensitive to water stress [3, 44]. Future research should investigate how altered precipitation patterns and drought conditions affect microbial processes, particularly those involved in carbon and nitrogen cycling. Additionally, the impact of climate change on microbial diseases affecting plants and animals must be better understood, as these pathogens could exacerbate the negative effects of climate change on agriculture and biodiversity [4, 20, 45].

Finally, a key area of focus should be the resilience and adaptability of soil microbial communities in the face of changing land-use practices [17, 44]. The conversion of natural landscapes to agriculture, urbanization, and deforestation can disrupt microbial ecosystems [11]. Future research should explore how land-use changes and climate changes together influence soil microbial functions, and how these changes can be mitigated through sustainable land management practices.

Potential for Microbial Technologies in Climate Mitigation

Microbial technologies have immense potential in addressing climate change, particularly in the context of carbon sequestration, greenhouse gas mitigation, and sustainable agriculture [15]. By understanding and harnessing the metabolic capabilities of microorganisms, scientists can develop innovative solutions that contribute to climate change mitigation [13, 17].

Microorganisms such as certain bacteria and fungi have the ability to fix carbon from the atmosphere through natural processes, such as photosynthesis and carbon fixation [4, 5]. These microbes can be utilized to enhance the sequestration of carbon in soils or oceans, providing a natural method of reducing atmospheric CO_2 levels [44, 46]. Additionally, biochar production, where microorganisms convert organic material into stable carbon-rich substances, is an area of growing interest for sequestering carbon in the soil for long periods.

Another area that captures my interest is the possibility of microbes mitigating greenhouse gas emissions, such as methane or nitrous oxide emissions [3, 5]. Methanotrophic bacteria, which consume methane, could potentially be employed in agricultural systems, namely rice fields and livestock rearing operations, to mitigate methanogenesis. Likewise, denitrifying bacteria, which reduce nitrate into nitrogen gas, can control nitrous oxide emissions from fertilizers [20, 45]. Therefore, the incorporation of these microbes into the practice of climate-smart agriculture will contribute quite substantially to the reduction of greenhouse gases affecting global warming.

Sustainable agriculture can also be promoted through microbial technologies by improving soil health and reducing the number of chemical inputs used [45, 47]. Among such microbial applications are microbial inoculants that introduce the right set of beneficial microbes to soils, which enhance soil fertility, improve crop production, and lower the demand for synthetic fertilizers [44]. This will not only lower the carbon footprint associated with agricultural production but will also make farming more secure in the changing climate [23].

Advances in Genomics and Microbial Ecology

Recent advances in genomics and microbial ecology are opening new frontiers in microbial research, providing deeper insights into microbial behavior, community dynamics, and their role in ecosystem functions [11]. The ability to sequence the genomes of microorganisms at an unprecedented scale has enabled scientists to explore microbial diversity and function in much greater detail than ever before [13, 39]. This is particularly important for understanding how microorganisms respond to environmental changes caused by climate change [5, 44, 45].

Next-generation sequencing technologies have revolutionized the study of microbial communities, allowing for the identification of previously unculturable microbes and the discovery of novel enzymes and metabolic pathways [13, 38]. By analyzing the genomic data of microbes from various ecosystems, researchers can uncover how microbial communities are structured and how they interact with their environment. This knowledge is critical for developing strategies to manage microbial ecosystems in ways that optimize their roles in climate change mitigation, such as enhancing carbon sequestration or nutrient cycling [16, 17].

Moreover, with the advancement of synthetic biology and metabolic engineering, there is a possibility to design organisms with specific functionalities [13, 38]. Researchers are working on how microbes can be engineered to do things like degrade environmental pollutants, produce biofuels, or even synthesize valuable chemicals [11, 24]. These kinds of engineered microbes can provide solutions to really challenging environmental problems like pollution and resource depletion. Microbial ecology also concerns understanding how, within ecosystems, the whole microbial community functions and so on, addressing issues where the community needs to come together [25, 40]. This would be extremely pertinent from a climate change perspective: A new introduction among microbial communities might cause a chain effect on functions such as nutrient cycling or disease suppression changes at a broader ecosystem level [44].

Finally, research should now look at how microbes can respond to climate change, develop microbial technologies for climate mitigation, and investigate advances in genomics and microbial ecology [1, 20, 40]. This would contribute meaningfully to understanding and addressing the challenge posed by climate change. In this regard, closing existing knowledge gaps, harnessing microbial capabilities, and leveraging cutting-edge technologies can unlock the full potential of microorganisms to combat climate change and build future sustainability [23, 43]. Just as rapidly emerging trends in science promise, so does microbial science, and this looks set to drive action in global efforts on mitigation and adaptation to the challenges of climate change [15, 47].

CONCLUSION

Microbes and Climate Change: The Hidden Architects of Our Future

As we stand at the crossroads of a rapidly changing climate, it becomes evident that microbes—tiny, often invisible organisms—are far from passive participants in the environmental changes we are witnessing [13, 15]. They play a critical role in determining both the processes driving climate change and the measures we may use to counteract it [5, 13]. The complex interconnections between microorganisms, plants, animals, and ecosystems have an impact on everything from the carbon cycle and greenhouse gas emissions to the resilience of our food systems and health [4]. In this chapter, we looked at how microbes influence climate dynamics and how understanding these microbial processes can provide important insights for mitigating climate change [18, 45].

Synthesis of Microbial Contributions to Climate Dynamics

Microbes are essential to the operation of the Earth's ecology. They are essential to the carbon, nitrogen, sulphur, and phosphorus cycles, which govern the climate system [17, 41]. One of their most essential functions is carbon sequestration. Soil bacteria, for example, degrade organic matter and produce carbon dioxide and methane, but they can also store carbon in the form of organic soil matter [15, 18,

45]. Climate change complicates these microbial processes. Rising temperatures and changed precipitation patterns can change the balance of carbon storage and release, potentially transforming soil from a carbon sink to a carbon source [24, 34, 47].

Similarly, bacteria in the oceans regulate atmospheric carbon levels through photosynthesis. As an example, phytoplankton absorb CO_2 and, through the marine food chain, sequester it deep in the water [11, 13]. Higher CO_2 levels are causing ocean acidification, which may alter microbial ecosystems that act as a buffer against climate change [4, 17]. Microbes are not only responsible for carbon emissions but also for other greenhouse gases. Nitrous oxide (N_2O) is produced in soils largely by the activity of bacteria and has been a very strong greenhouse gas in agricultural situations where nitrogen-containing fertilizer is applied [4, 9]. Thus, understanding the microbiological processes that will result in the generation of nitrous oxide is vital to the reduction of these emissions [3, 21, 48]. Indeed, microorganisms tend to be a problem or a solvation mechanism depending on their operations concerning climate change [5, 9]. Global warming can notably modify the epidemiology of pathogens [6, 26]. Malaria, cholera, and Lyme disease are the finest examples of diseases that have all proven to be tightly coupled to the dynamics of microbes [3, 30]. Changes are not merely theoretical; they spill into real-world implications, whether in terms of food security, public health, or biodiversity.

Integrating Microbial Science into Climate Change Mitigation Strategies

Given their importance in climate dynamics, microorganisms must be included in global climate change mitigation plans [1, 13, 43]. However, microorganisms are frequently disregarded in popular climate models and policy debates. Only by gaining a better grasp of microbial ecology will we be able to fully realize their promise in combating climate change [6, 32, 43].

Microbial science can be applied in several ways to reduce the effects of climate change on:

- Improving Soil Microbial Health to Sequester Carbon

The global carbon cycle depends heavily on soil health. By boosting soil microbial health, we can increase the soil's ability to operate as a carbon sink [8]. This includes fostering techniques that encourage microbial diversity and activity, such as reduced tillage, crop rotation, and organic farming [14, 20]. Inoculating soils with beneficial bacteria, for example, can improve plant growth, nitrogen fixation, and carbon storage capacity. These measures can improve carbon sequestration, reducing the impact of growing CO_2 levels in the atmosphere [20, 34].

- Reducing Greenhouse Gas Emissions *via* Microbial Management

Microbial processes in agriculture provide large greenhouse gas emissions, mainly methane and nitrous oxide. We may directly influence climate change by optimizing farming methods to limit the generation of these gases [18, 23]. This could include utilizing specialized microorganisms to reduce methane emissions from animals or changing fertilizer application to reduce nitrous oxide generation [3, 47, 48]. Researchers are also investigating how certain bacteria might convert greenhouse gases such as methane into less damaging chemicals, which presents another potential technique for cutting emissions [9].

- Using Microbes for Carbon Capture and Conversion

Microbes have the potential to be effective tools in carbon capture technology [10, 37]. Photosynthetic microorganisms, including algae and cyanobacteria, absorb CO_2 from the atmosphere during their life cycle. Advances in biotechnology may allow these creatures to be scaled up for large-scale carbon collection [8, 16, 44]. Microbes can convert CO_2 into biofuels and biodegradable polymers, lowering atmospheric carbon and providing sustainable alternatives to fossil fuels and plastics [18].

- Restoring Ecosystem Balance *via* Microbial Interventions

As climate change continues to destabilize ecosystems, microbial treatments can help restore equilibrium [15, 45]. In damaged environments, such as those affected by deforestation or desertification, the introduction of certain microbial communities can help speed up recovery [13, 38, 42]. For example, certain fungi and bacteria can improve soil fertility and moisture retention, assisting in the restoration of forests and grasslands that serve as key carbon sinks. Similarly, in marine habitats, restoring microbial biodiversity could aid in combating ocean acidification and enhancing marine ecosystem resilience [24, 47].

- Microbial Pathogens and Public Health

The increase in climate change continues to open new areas to microbial diseases that threaten human and animal health. Microbial science integrated with public health policy is fundamental in forecasting and controlling the spread of disease [43, 48]. There is a need to get a better understanding of the vectors and environmental conditions involved in the transmission of climate-sensitive diseases like malaria, dengue fever, and cholera. Hence, microbial research must be included in future public health initiatives to prepare for an altered landscape of infectious diseases [13].

Looking Ahead: Microbes as Partners in The Fight Against Climate Change

When gazed at for long into the future, opportunistic bacteria are an answer to an ever-receding challenge. These organisms are affecting the world's climate [3, 15, 32]. No action of these microbes is without its implications for ecosystems, agriculture, and human health [44]. While they contribute to the problem of climate change, microbes are also important brokers of solutions to the problem: the full acknowledgment of them as major players in addressing and moving climate change into their future microbe knowledge and inclusion in climate policy and mitigation actions, making access to their amazing potential [44, 45, 47].

Finally, the tale of climate change is more than just global warming and increasing sea levels; it's also about adaptation, resilience, and the underlying forces that drive our world. Microbes, in all their diversity and complexity, will be valuable friends in our efforts to negotiate the challenges of a warming planet. It emphasizes the importance of integrating this knowledge into practical initiatives, providing a comprehensive response to the global climate challenge.

REFERENCES

[1] Joshi P, Pande V, Joshi P. Microbial diversity of aquatic ecosystem and its industrial potential. J Bacteriol Mycol 2016; 3(1): 177-9.
[http://dx.doi.org/10.15406/jbmoa.2016.03.00048]

[2] Nelson KS, Baltar F, Lamare MD, Morales SE. Ocean acidification affects microbial community and invertebrate settlement on biofilms. Sci Rep 2020; 10(1): 3274.
[http://dx.doi.org/10.1038/s41598-020-60023-4] [PMID: 32094391]

[3] Widdig M, Heintz-Buschart A, Schleuss P-M, et al. Effects of nitrogen and phosphorus addition on microbial community composition and element cycling in a grassland soil. Soil Biol Biochem 2020; 151: 108041.
[http://dx.doi.org/10.1016/j.soilbio.2020.108041]

[4] Ibáñez A, Garrido-Chamorro S, Barreiro C. Microorganisms and climate change: a not so invisible effect. Microbiol Res (Pavia) 2023; 14(3): 918-47.
[http://dx.doi.org/10.3390/microbiolres14030064]

[5] Classen AT, Sundqvist MK, Henning JA, et al. Direct and indirect effects of climate change on soil microbial and soil microbial-plant interactions: What lies ahead? Ecosphere 2015; 6(8): 1-21.
[http://dx.doi.org/10.1890/ES15-00217.1]

[6] Buragohain P, Nath DJ, Phonglosa A. Role of microbes on carbon sequestration. Int Microbiol Res 2019; 11(1): 1464-8. Available from: https://www.researchgate.net/publication/331564689_ROLE_OF_MICROBES_ON_CARBON_SEQUESTRATION

[7] Delgado-Baquerizo M, Maestre FT, Reich PB, et al. Microbial diversity drives multifunctionality in

[8] Uddin TM, Chakraborty AJ, Khusro A, *et al.* Antibiotic resistance in microbes: History, mechanisms, therapeutic strategies and future prospects. J Infect Public Health 2021; 14(12): 1750-66.
[http://dx.doi.org/10.1016/j.jiph.2021.10.020] [PMID: 34756812]

[9] Lang-Yona N, Flores JM, Haviv R, *et al.* Terrestrial and marine influence on atmospheric bacterial diversity over the north Atlantic and Pacific Oceans. Commun Earth Environ 2022; 3(1): 121.
[http://dx.doi.org/10.1038/s43247-022-00441-6]

[10] Bernhardt JR, O'Connor MI, Sunday JM, Gonzalez A. Life in fluctuating environments. Philos Trans R Soc Lond B Biol Sci 2020; 375(1814): 20190454.
[http://dx.doi.org/10.1098/rstb.2019.0454] [PMID: 33131443]

[11] Kumar A, Alam A, Tripathi D, *et al.* Protein adaptations in extremophiles: An insight into extremophilic connection of mycobacterial proteome. Semin Cell Dev Biol 2018; 84: 147-57.
[http://dx.doi.org/10.1016/j.semcdb.2018.01.003] [PMID: 29331642]

[12] Dastagir MR. Role of Microorganisms in Managing Climate Change Impacts. In: Singh D, Prabha R, eds. Microbial Interventions in Agriculture and Environment. Singapore: Springer; 2019. p. 1–16.
[http://dx.doi.org/10.1007/978-981-32-9084-6_1]

[13] Tiedje JM, Bruns MA, Casadevall A, *et al.* Microbes and climate change: a research prospectus for the future. MBio 2022; 13(3): e00800-22.
[http://dx.doi.org/10.1128/mbio.00800-22] [PMID: 35438534]

[14] Rabari A, Ruparelia JA, Jha CK. Extremophiles: Subsistence of an Extreme Nature Enthusiast. In: Gunjal A, Thombre R, Parray J, eds. Physiology, Genomics, and Biotechnological Applications of Extremophiles. Hershey (PA): IGI Global Scientific Publishing; 2022. p. 1-12.
[http://dx.doi.org/10.4018/978-1-7998-9144-4.ch001]

[15] Parker J. Microbial marvels: could microbes resolve climate change? Biotechniques 2023; 74(6): 279-81.
[http://dx.doi.org/10.2144/btn-2023-0043] [PMID: 37353987]

[16] Matos S, Viardot E, Sovacool BK, Geels FW, Xiong Y. Innovation and climate change: A review and introduction to the special issue. Technovation 2022; 117: 102612.
[http://dx.doi.org/10.1016/j.technovation.2022.102612]

[17] Malkawi HI, Kapiel TYS. Microbial biotechnology: a key tool for addressing climate change and food insecurity. Euro J Bio Biotech 2024; 5(2): 1-15.Available from: https://www.ejbio.org/index.php/ejbio/article/view/503
[http://dx.doi.org/10.24018/ejbio.2024.5.2.503]

[18] Wu KK, Xu PP, Zhao L, Ren NQ, Zhang YF. Microbial conversion of carbon dioxide into premium medium-chain fatty acids: the progress, challenges, and prospects. npj Materials Sustainability 2024; 2(1): 4.
[http://dx.doi.org/10.1038/s44296-024-00008-w]

[19] Ma L, Zhao H, Wu LB, Cheng Z, Liu C. Impact of the microbiome on human, animal, and environmental health from a One Health perspective. Science in One Health 2023; 2: 100037.
[http://dx.doi.org/10.1016/j.soh.2023.100037] [PMID: 39077043]

[20] Wang H, Zhao R, Zhao D, *et al.* Microbial-mediated emissions of greenhouse gas from farmland soils: a review. Processes (Basel) 2022; 10(11): 2361.
[http://dx.doi.org/10.3390/pr10112361]

[21] O'Brien PA, Morrow KM, Willis BL, Bourne DG. Implications of ocean acidification for marine microorganisms from the free-living to the host-associated. Front Mar Sci 2016; 3: 47.
[http://dx.doi.org/10.3389/fmars.2016.00047]

[22] Li J, Benti G, Wang D, Yang Z, Xiao R. Effect of alteration in precipitation amount on soil microbial

community in a semi-arid grassland. Front Microbiol 2022; 13: 842446.
[http://dx.doi.org/10.3389/fmicb.2022.842446] [PMID: 35369529]

[23] Wang B, Yao M, Lv L, Ling Z, Li L. The human microbiota in health and disease. Engineering (Beijing) 2017; 3(1): 71-82.
[http://dx.doi.org/10.1016/J.ENG.2017.01.008]

[24] Bai D, Ye L, Yang Z, Wang G. Impact of climate change on agricultural productivity: a combination of spatial Durbin model and entropy approaches. Int J Clim Chang Strateg Manag 2024; 16(4): 26-48.
[http://dx.doi.org/10.1108/IJCCSM-02-2022-0016]

[25] Philip PM, Polgreen EL. Emerging and re-emerging pathogens and diseases, and health consequences of a changing climate. Infect Dis 2017.
[http://dx.doi.org/10.1016/B978-0-7020-6285-8.00004-6]

[26] Charles, Gupta A, Maurya J, Sohail M. Microbial diversity: exploring microbial diversity of our ecosystem 2022.
https://www.researchgate.net/publication/363739819_Microbial_Diversity_Exploring_Microbial_diversity_of_our_Ecosystem

[27] Graham EB. Microbes as engines of ecosystem function: when does community structure enhance predictions of ecosystem processes? Front Microbiol 2016; 7: 214.
[http://dx.doi.org/10.3389/fmicb.2016.00214] [PMID: 26941732]

[28] Liang C, Amelung W, Lehmann J, Kästner M. Quantitative assessment of microbial necromass contribution to soil organic matter. Glob Change Biol 2019; 25(11): 3578-90.
[http://dx.doi.org/10.1111/gcb.14781] [PMID: 31365780]

[29] Wang Q, Wang C, Yu W, *et al.* Chen. Effects of nitrogen and phosphorus inputs on soil bacterial abundance, diversity, and community composition in Chinese fir plantations. Front Microbiol 2018; 9: 1543.
[http://dx.doi.org/10.3389/fmicb.2018.01543] [PMID: 30072961]

[30] Kharitonov S, Semenov M, Sabrekov A, Kotsyurbenko O, Zhelezova A, Schegolkova N. Microbial communities in methane cycle: modern molecular methods gain insights into their global ecology. Environments (Basel) 2021; 8(2): 16.
[http://dx.doi.org/10.3390/environments8020016]

[31] Liu G, Sun J, Xie P, Guo C, Zhu K, Tian K. Climate warming enhances microbial network complexity by increasing bacterial diversity and fungal interaction strength in litter decomposition. Sci Total Environ 2024; 908: 168444.
[http://dx.doi.org/10.1016/j.scitotenv.2023.168444] [PMID: 37949122]

[32] Das PP, Singh KRB, Nagpure G, *et al.* Plant-soil-microbes: A tripartite interaction for nutrient acquisition and better plant growth for sustainable agricultural practices. Environ Res 2022; 214(Pt 1): 113821.
[http://dx.doi.org/10.1016/j.envres.2022.113821] [PMID: 35810815]

[33] Guo Y, Ryan U, Feng Y, Xiao L. Association of common zoonotic pathogens with concentrated animal feeding operations. Front Microbiol 2022; 12: 810142.
[http://dx.doi.org/10.3389/fmicb.2021.810142] [PMID: 35082774]

[34] Koehle AP, Brumwell SL, Seto EP, Lynch AM, Urbaniak C. Microbial applications for sustainable space exploration beyond low Earth orbit. NPJ Microgravity 2023; 9(1): 47.
[http://dx.doi.org/10.1038/s41526-023-00285-0] [PMID: 37344487]

[35] Tan YS, Zhang RK, Liu ZH, Li BZ, Yuan YJ. Microbial adaptation to enhance stress tolerance. Infect Dis 2017; 1: 40-48
[http://dx.doi.org/10.3389/fmicb.2022.888746] [PMID: 35572687]

[36] Charles, Gupta A, Maurya J, Sohail M. Microbial diversity: exploring microbial diversity of our ecosystem 2022. Available from: https://www.researchgate.netpublication/363739819_

Microbial_Diversity_Exploring_Microbial_diversity_of_our_Ecosystem
[http://dx.doi.org/10.3389/fmars.2022.864797]

[37] Ogunrinola GA, Oyewale JO, Oshamika OO, Olasehinde GI. The human microbiome and its impacts on health. Int J Microbiol 2020; 2020: 1-7.
[http://dx.doi.org/10.1155/2020/8045646] [PMID: 32612660]

[38] Larsson DGJ, Andremont A, Palme R, *et al.* Critical knowledge gaps and research needs related to the environmental dimensions of antibiotic resistance. Environ Int 2023; 117: 132-8.
[http://dx.doi.org/10.1016/j.envint.2018.04.041]

[39] Chawla H, Anand P, Garg K, *et al.* A comprehensive review of microbial contamination in the indoor environment: sources, sampling, health risks, and mitigation strategies. Front Public Health 2023; 11: 1285393.
[http://dx.doi.org/10.3389/fpubh.2023.1285393] [PMID: 38074709]

[40] Yogeswaran K, Azmi L, Bhassu S, Isa HN, Aziz MA. Physical parameters influence the microbial quality of indoor air in research laboratories: A report from Malaysia. Kuwait J Sci 2023; 50(4): 665-73.
[http://dx.doi.org/10.1016/j.kjs.2023.04.009]

[41] Hlihor RM, Cozma P. Microbial bioremediation of environmental pollution. Processes (Basel) 2023; 11(5): 1543.
[http://dx.doi.org/10.3390/pr11051543]

[42] Ayilara MS, Babalola OO. Bioremediation of environmental wastes: the role of microorganisms. Fron Agronomy 2023; 5: 1183691.
[http://dx.doi.org/10.3389/fagro.2023.1183691]

[43] Lennon JT, Abramoff RZ, Allison SD, *et al.* Priorities, opportunities, and challenges for integrating microorganisms into Earth system models for climate change prediction. Enviro Micro 2024; 15(5): e00455-24.
[http://dx.doi.org/10.1128/mbio.00455-24] [PMID: 38526088]

[44] Lemke M, DeSalle R. The next generation of microbial ecology and its importance in environmental sustainability. Microb Ecol 2023; 85(3): 781-95.
[http://dx.doi.org/10.1007/s00248-023-02185-y] [PMID: 36826587]

[45] Cavicchioli R, Ripple WJ, Timmis KN. Scientists' warning to humanity: microorganisms and climate change. Nat Rev Microbiol 2019; 17(9): 569-86.
[http://dx.doi.org/10.1038/s41579-019-0222-5] [PMID: 31213707]

[46] Zhang L, Chen F, Zeng Z, *et al.* Advances in metagenomics and its application in environmental microorganisms. Front Microbiol 2021; 12: 766364.
[http://dx.doi.org/10.3389/fmicb.2021.766364] [PMID: 34975791]

[47] Rawat V, Jasleen K, Sakshi B, Manisha P, Rawat C. Deploying microbes as drivers and indicators in ecological restoration. Restor Ecol 2022; 31(1): 1-16.
[http://dx.doi.org/10.1111/rec.13688]

[48] Elbehiry A, Abalkhail A, Marzouk E, *et al.* Almuzaini, Alfheeaid, Alshahrani MT, Huraysh N, Ibrahem M, Alzaben F. An overview of the public health challenges in diagnosing and controlling human foodborne pathogens. Vaccines (Basel) 2023; 11(4): 725.
[http://dx.doi.org/10.3390/vaccines11040725] [PMID: 37112637]

CHAPTER 8

Advanced Oxidation Process as an Emerging Technology for the Treatment of Pharmaceutical Wastewater

M. Mounica[1], V.V. Vaishnavi[1] and M. Vijay Pradhap Singh[1,*]

[1] *Department of Biotechnology, Vivekanandha College of Engineering for Women (Autonomous), Elayampalayam, Tamil Nadu, India*

Abstract: Water is often regarded as the elixir of life. Approximately one-third of the Earth's freshwater is stored in glaciers and ice caps, while only a limited portion of the remaining freshwater is directly available to sustain the global population. However, growing demand, driven by population growth and urbanization, has led to significant pollution of freshwater resources. The primary route through which pharmaceutical pollutants enter the environment is *via* municipal wastewater treatment plants. Pharmaceutical industries include fermentation and chemical synthesis processes, which discharge pharmaceutical solvents, catalysts, reactants, intermediates, toxic substances, and harmful liquids called effluents into the water stream. Several contaminants in water bodies cause planet-wide environmental impacts and create significant threats to flora and fauna. The conventional wastewater treatment processes often lead to the formation of secondary pollutants, including chemical residues and toxic by-products, in addition to high maintenance costs and issues related to sludge management. Developing a practical, affordable, efficient, and environmentally benign method for treating the pharmaceutical effluents in wastewater is crucial to reducing the release of contaminants into the environment. Advanced Oxidation Processes (AOPs) have attracted considerable attention in recent years for the treatment of industrial and municipal wastewater, owing to their ability to degrade persistent organic pollutants and offer disinfection, decolorization, no formation of secondary pollutants, and deodorization. This chapter discusses the principles and types of Advanced Oxidation Processes, compares the efficiency of different methods, and explores their applications and limitations in achieving a sustainable and cleaner environment.

Keywords: Conventional treatment, Catalysts, Decolorization, Emerging Technology, Effluents, Flora, Fauna, Oxidation, Pharmaceuticals, Pollutants, Wastewater.

---

[*] **Corresponding author M. Vijay Pradhap Singh :** Department of Biotechnology, Vivekanandha College of Engineering for Women (Autonomous), Elayampalayam, Tamil Nadu, India; E-mail vijaypradhapsingh@gmail.com

Rajneesh Kumar, Ram Sharan Singh & Maulin P. Shah. (Eds.)
All rights reserved-© 2026 Bentham Science Publishers

INTRODUCTION

The increasing contamination of water resources is primarily driven by modern lifestyles, industrial technological advancements, and the intensification of agricultural practices, all responding to growing global demands. Among industrial contributors, pharmaceutical companies are notable for the continuous discharge of toxic effluents into aquatic environments. Both organic and inorganic substances, including heavy metals, are present in these effluents. Furthermore, unsustainable agricultural practices contribute to the accumulation of heavy metals such as copper and cadmium in soil and water bodies. The problem is further aggravated by non-segregated waste dumping, including household waste and e-waste, which exacerbates the problem. Moreover, torrential rainfall, surface runoff, and atmospheric deposition contribute to the transfer and accumulation of various contaminants, including organic, biological, inorganic, and heavy metal pollutants, into wastewater and effluent treatment plants (WWTPs and ETPs) [1].

The widespread occurrence of diverse pollutants in aquatic ecosystems significantly intensifies global environmental pollution and poses significant threats to human populations, fauna, and other biotic communities [2]. Among these contaminants, emerging pollutants such as food additives, Pharmaceuticals, and Personal Care Products (PPCPs) are particularly concerning due to their persistence, bioaccumulation potential, and resistance to conventional treatment processes. In conventional wastewater treatment plants, biological processes are often unable to completely remove these effluents, resulting in their continuous release into the environment, even at trace levels. Over time, this contamination threatens the health of both marine and terrestrial organisms while also impacting global water availability and exacerbating the ongoing drinking water crisis in many regions of the world [3, 4].

According to the Food and Agriculture Organization (FAO, 2013), the global demand for potable water is projected to increase by over 40% by 2050. Consequently, the development of innovative, efficient, cost-effective, and environmentally sustainable wastewater purification technologies remains a critical challenge for researchers, as current treatment facilities are unable to completely eliminate the wide range of organic and inorganic contaminants present in industrial and municipal wastewater [5, 6].

The second most critical environmental challenge facing humanity today is water contamination, broadly defined as the degradation of water quality caused by the discharge of untreated toxic substances from industrial and other anthropogenic sources [7]. The marine environment is severely affected, and these pollutants infiltrate the soil, thereby contaminating local water supplies [8]. The factors responsible for this pollution can be physical, chemical, or biological in nature. It can be categorized as biodegradable pollutants, which have short-term impacts, and persistent toxins, such as heavy metals, plastics, and chemicals like DDT. When DDT breaks down, it forms DDE (dichlorodiphenyl dichloroethylene), leading to long-lasting toxic effects. These persistent toxins are known as stock pollutants. They have low absorption rates, resulting in accumulation in water resources and increased toxicity over time. In contrast, fund pollutants have higher absorption potential, and their concentration can be reduced through natural dilution, minimizing their harmful effects unless their levels exceed the environment's absorption capacity. Physical factors, including elevated temperature and radiation, should influence water quality, while bio pollutants consist of microorganisms and dangerous pathogens. Among the various causes of water pollution, the discharge of untreated industrial effluents is a significant contributor. When the pollution originates from a single identifiable source, it is referred to as point source pollution. Conversely, non-point source pollution arises from multiple, diffuse sources, producing a mixture of pollutants that often act synergistically, complicating treatment processes [9].

Pharmaceutical contaminants pose serious risks to human health and the environment worldwide. These compounds are now recognized as emerging pollutants, with residues commonly detected in river discharges, wastewater treatment plants (WWTPs), groundwater, industrial effluents, and even drinking water. Pharmaceuticals comprise a wide range of synthetic and naturally derived chemical substances designed to treat diseases and sustain human health. They are typically potent at low concentrations, readily absorbed, and remain within the body until their therapeutic action is complete [10]. However, due to rapid industrialization and population growth, nearly one-third of global freshwater resources have been impacted by the release of pharmaceutical contaminants into aquatic systems [11]. The expansion of the pharmaceutical industry has therefore raised major concerns regarding its role in increasing environmental pollution [12]. Conventional wastewater treatment methods, which rely on physicochemical and biological mechanisms, are only partially effective in removing these microbial contaminants in municipal wastewater. Many pharmaceutical contaminants cannot be metabolized by microbes and may adversely affect microbial activity. Historically, environmental protection strategies have primarily focused on detecting and monitoring pollution, indirectly addressing its impacts on ecosystems. With advances in science and technology development, along with a deeper understanding of the chemistry of pharmaceutical pollutants, new technologies such as Advanced Oxidation Processes (AOPs) have been developed to treat and degrade these contaminants in wastewater [13]. Among these refractory compounds, pharmaceutical waste represents one of the most persistent and challenging pollutants to degrade using conventional treatment methods.

These include various categories of pharmaceuticals such as antibiotics, blood lipid regulators, hormones, β-blockers, serotonin reuptake inhibitors, antihistamines, antiepileptics, non-steroidal anti-inflammatory drugs (NSAIDs), psychotropic medications, anti-ulcer agents, and anti-asthma drugs [14]. In addition to their toxic effects, certain pharmaceutical compounds, particularly antibiotics, can cause irreversible changes in microorganisms and their genetic material, remaining resilient in the environment even at minimal concentrations. Furthermore, the presence of Endocrine-Disrupting Substances (EDSs) in the aquatic environment has emerged as a significant issue, as these compounds are known to interfere with the human endocrine system [15]. Active pharmaceutical ingredients (APIs) are the chemically and biologically active components of drugs that exert therapeutic effects within the body. Because of their persistence and widespread use, the extensive manufacturing, utilization, and improper disposal of pharmaceutical compounds pose serious threats to humans, aquatic ecosystems, and other biological forms of life due to their uncontrolled release into the environment. Moreover, the presence of these compounds has been detected at trace levels from nanograms to micrograms in wastewater samples over the past decade. Active Pharmaceutical Ingredients (APIs) are synthesized through multiple chemical steps, often beginning with and involving several intermediate compounds to convert raw materials into the final API [16]. The pharmaceutical industry produces thousands of APIs for a wide range of therapeutic applications, including cancer therapies, pain management, and antidepressants. The most common route through which pharmaceuticals enter the environment is *via* human excretion, as only a fraction of administered drugs are metabolized by the body, while the remainder is released into sewage and wastewater systems. Another major source is the improper disposal of expired or unused pharmaceuticals into the wastewater. WWTPs have been identified as a significant source of these pharmaceuticals, collecting discharge from hospitals, veterinary clinics, households, industries, and pharmacies. The availability and quality of freshwater resources have been severely affected by overpopulation and poor water management, leading to the failure of conventional wastewater treatment, resulting in a critical situation. Moreover, the increasing human population has amplified the demand for safe drinking water, highlighting the urgent need for efficient wastewater treatment technologies. Among the most advanced and effective methods, Advanced Oxidation Processes (AOPs) have gained recognition for their ability to generate powerful reactive oxygen species (ROS) and free radicals capable of degrading a wide variety of persistent organic contaminants, including micropollutants and total organic carbon [17]. Industries worldwide are increasingly acknowledging the importance of wastewater treatment and the adoption of sustainable technologies. Effective wastewater management enhances environmental quality, safeguards public health, and promotes long-term industrial sustainability [18]. Despite these advancements, several challenges remain in achieving effluent wastewater treatment. The primary issues arise from a limited understanding of pollutant origins and decontamination technologies. The emergence of new and complex contaminants, particularly from the pharmaceutical sector, often renders traditional treatment methods insufficient to achieve the desired purification outcomes. Improper treatment of pharmaceutical wastewater can result in severe health risks and environmental damage [19]. This review, therefore, focuses on examining the extent of pharmaceutical pollution and evaluating the efficiency of Advanced Oxidation Processes (AOPs) in eliminating pharmaceutical contaminants from various aquatic systems.

PHARMACEUTICAL EFFLUENTS

In the environment, pharmaceuticals, therapeutics, cosmetics, and personal care products are typically at low pollutant concentrations. However, the molecular behavior of these pollutants, their pathogen resilience, and interaction effects remain unresolved, creating scientific ambiguity. Pharmaceutical drugs and residues infiltrate the ecosystem and sewage systems from various dispersed sources. The major origins of these pollutants include drug manufacturing units, waste disposal sites, healthcare facilities, refuse dumps, and even cemeteries. One of the most extensively studied pathways for these pollutants entering the environment is through wastewater treatment plants. Metabolic waste from humans, either unaltered or only partially metabolized, often contains active therapeutic compounds that are transformed into ionized molecules before entering sewage systems. Within these systems, the compounds can undergo decomposition and may be released as their original Active Pharmaceutical Ingredients (APIs) into treatment sites. Oversight of API release from pharmaceutical industries is not standardized, and pharmaceutical effluents may contain volatile industrial solvents, facilitating agents, additives, substrate, intermediates, feedstock, and APIs, complicating the treatment process. The presence of harmful or resistant materials in wastewater can reduce Chemical Oxygen Demand (COD) concentration, indicating impaired treatment efficiency. Fig. (1) illustrates the various categories of pharmaceuticals, including active pharmaceutical ingredients and personal care products (PPCPs). Once utilized, most PPCPs enter the environment through several pathways. Unprocessed industrial runoff and effluents from healthcare services containing resistant compounds and personal care residues are often partially degraded, and they can be released directly into various water bodies. Common contributors include perfumes, preservatives, sunscreen agents, mosquito and pest repellents, disinfectants are among the common contributors [20]. PPCPs can disrupt biological treatment processes in sewage treatment plants and pose risks to both human and ecological health. Pharmaceuticals are designed to modulate cellular signalling and enzymatic activity, thereby influencing biological functions. Some drugs, such as paracetamol, are metabolized into inactive forms before environmental release. However, others are excreted in their active form, potentially altering the physicochemical properties of aquatic ecosystems upon entry. This review highlights the importance of understanding the environmental fate of pharmaceuticals and highlights the need for advanced techniques such as Advanced Oxidation Processes (AOPs).

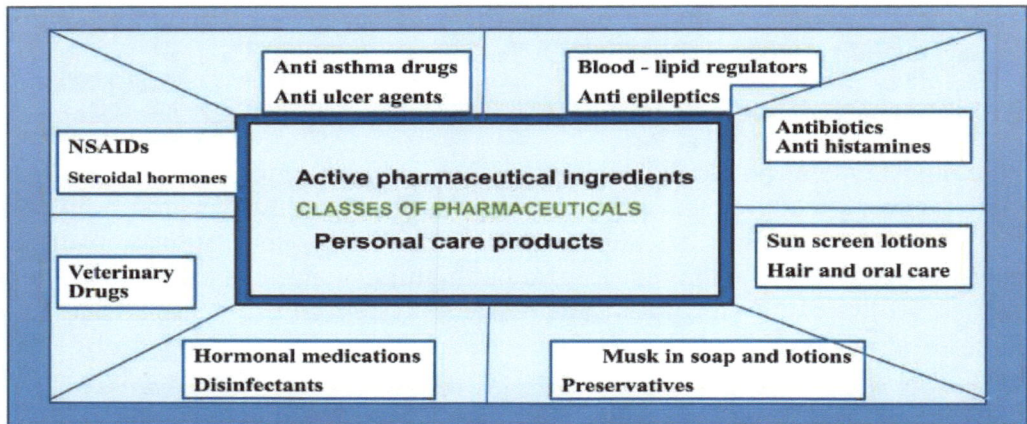

Fig. (1). Classes of pharmaceuticals.

Different Sources of Pharmaceutical Products That Cause Pollution To The Environment

Medicinal products play a vital role in improving human health and are collectively referred to as medications or pharmaceuticals. These compounds are used to promote well-being and treat a wide range of diseases and health disorders. However, the increasing occurrence of pharmaceutical compounds in the environment has become a growing global concern. Recent studies have identified approximately 631 pharmaceutical agents or their metabolites in environmental samples from over 71 countries, underscoring the widespread nature of this issue. Pharmaceutical products enter the environment through multiple pathways, including wastewater treatment facilities, landfills, hospitals, cemeteries, and pharmaceutical manufacturing plants, primarily due to inadequate waste management practices [21, 22]. Fig. (**2**) illustrates the major routes of pharmaceutical products' entry into the environment, which include human and veterinary drugs, manure discharge, household wastewater, disposal of unused medicines, effluents from wastewater treatment plants, and medical waste such as discarded or expired drugs, bandages, cotton swabs, and biosolids. The combination of these diverse contaminants significantly complicates the removal of pharmaceuticals from aquatic environments. If pharmaceutical contamination in aquatic systems is not effectively managed, the long-term ecological and health consequences may become increasingly severe [23]. Among the various pathways, the two primary routes through which pharmaceuticals enter the environment are human excretion and the improper disposal of over-the-counter or prescription drugs. Although anticancer drugs are typically released in low concentrations, they possess mutagenic and carcinogenic properties, posing significant risks to aquatic organisms and potentially disrupting aquatic ecosystems.

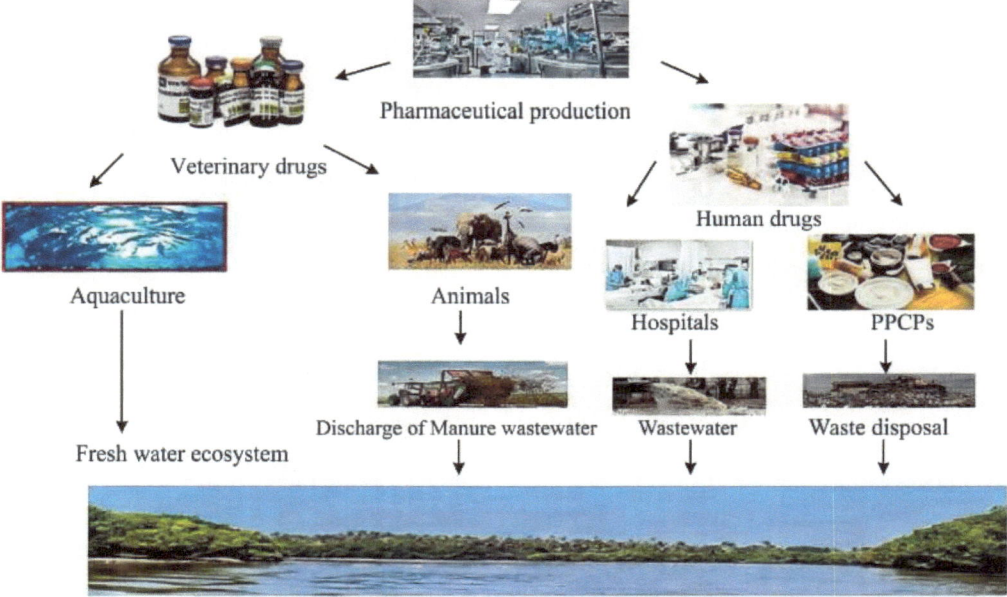

Fig. (2). Sources of pharmaceuticals into the environment.

ADVANCED OXIDATION PROCESS

Theory of Advanced Oxidation Process

Advanced Oxidation Processes (AOPs) are generally defined as treatment methods that utilize highly reactive radicals to oxidatively degrade pollutants. These are in situ processes, including hydroxy radicals, singlet oxygen, and radicals derived from persulfate, among others. Fig. (3) illustrates the classification of Advanced Oxidation Processes into two main types: homogeneous and heterogeneous processes.

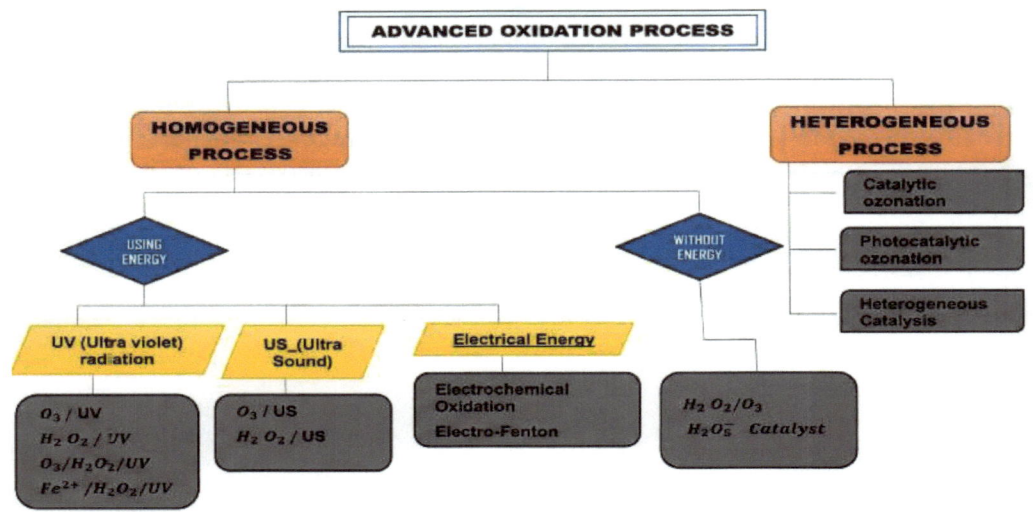

Fig. (3). Classification of advanced oxidation processes.

It is effective in eliminating pollutants that exhibit low biodegradability or high chemical stability [24]. These techniques rely on the generation of hydroxyl free radicals (HO*), which are highly powerful oxidizing agents capable of breaking down compounds that cannot be degraded by conventional oxidants [25]. Table **1** presents the redox potentials of several common oxidants relative to the normal hydrogen electrode. AOPs are considered a promising approach for removing such contaminants from aqueous solutions. The application of AOPs has proven to be an efficient and reliable method for eliminating trace levels of Emerging Contaminants (ECs) from both water and wastewater [26]. The fundamental principle of AOPs lies in the generation of highly oxidative radicals that rapidly degrade or mineralize a wide variety of organic contaminants in wastewater. Hydroxyl radicals can be generated through multiple mechanisms within AOP systems, allowing the processes to be tailored to specific treatment requirements. These technologies possess the capability to partially or completely oxidize complex compounds into smaller, less harmful by-products that can be subsequently removed through conventional biological treatment processes. In AOPs, hydroxyl free radicals are produced using different methods such as ozonation, heterogeneous photocatalysis processes, or homogeneous photocatalysis processes. Ozone, with an oxidation potential of ~2.07V at low pH, has been extensively utilized in water treatment, because it can react directly or indirectly with numerous organic compounds, leading to effective oxidation. In heterogeneous photocatalytic processes, photons absorbed at the surface of the photocatalyst excites electrons from the valence band to the conduction band, generating positive holes in the valence band. These holes facilitate the formation of hydroxyl radicals that participate in pollutant degradation. Titanium dioxide (TiO_2) is commonly used as a photocatalyst due to its high chemical stability, low toxicity, and cost-effectiveness, and exceptional ability to absorb solar radiation. Homogeneous photocatalytic proceeses process usually occurs in a single phase in a liquid medium or the gaseous medium where oxidizing agents react uniformly throughout the medium. These reactions typically involve oxidants such as molecular oxygen, ozone, hydrogen peroxide, or other electron acceptors. Depending on the source of activation energy, oxidation processes may be thermal, catalytic, or photo-induced [27].

Table 1. Redox potential of oxidants [24].

Redox potential of oxidants	Species E_0 (V, t = 25°C)
Hydroxy radical	2.80
Fluorine	3.03
Ozone (O_3)	2.07
Chlorine (Cl_2)	1.36
Bromine (Br_2)	1.09
Iodine (I_2)	0.54
Atomic oxygen (O)	2.42

Strong oxidizing agents, such as the reactive oxygen species (ROS) or free radicals, play a crucial role in initiating AOPs for the degradation of pollutants into nontoxic molecules. Free radicals are atoms and molecules that contain one or more unpaired electrons, such as the alkoxyl radical (RO), per hydroxyl radical (HO_2•), oxygen free radical O_2 or other reactive oxygen species (HO). Among these, the hydroxyl radical has received the greatest attention due to its exceptional reactivity and effectiveness in pollutant degradation. The hydroxyl radical reacts with organic compounds (RH) to form an organic radical (R°) and water, as shown in Equation 1. The resulting organic radical (R•) can further react with molecular oxygen (O_2) to produce a peroxyl radical (RO_2•), as illustrated in Equation 2. Alternatively, the hydroxyl radical (HO·) can also participate in substitution reactions, forming RX• and hydroxide ions (HO^-), as depicted in Equation 3. Hydroxyl radicals are characterized by their strong oxidizing abilities, high reactivity, and non-selective nature [28]. They readily attack a wide range of organic compounds, making them a central component in AOPs for the effective breakdown of persistent and complex pollutants.

$$HO^{\cdot} + RH \rightarrow R^{\circ} + H_2O \quad (1)$$

$$R^{\circ} + O_2 \rightarrow RO_2^{\circ} \quad (2)$$

$$HO^{\cdot} + RH \rightarrow RX^{\circ} + HO^{-} \quad (3)$$

Status and Emergence of the AOP Process

During the 1970s and 1980s, research primarily focused on using ozone (O_3) and ultraviolet (UV) radiation for the degradation of organic pollutants in water. The concept of AOPs gained significant momentum with the introduction of the Fenton reaction ($H_2H_2 + Fe^{2+}$ and photo-catalytic oxidation (TiO^2/UV). Over time, AOPs have evolved to play a major role in nanotechnology, electrochemical oxidation, and hybrid AOP systems, which have enhanced both scalability and treatment efficiency. Furthermore, the Fenton process has been recognized as an effective method for treating industrial wastewater. While ozone and ultraviolet irradiation have a long history in wastewater treatment, the advent of advanced AOP technologies has introduced processes that demonstrate higher energy efficiency and are tailored for specific applications such as organic pollutant degradation, environmental remediation, and ballast water management.

Homogeneous Process

Homogeneous processes can be classified based on the type of energy utilization-those with external energy sources, such as ultraviolet radiation, ultrasonication, and electrical energy, and without energy input. The term "homogeneous advanced oxidation process" (HO-AOP) refers to a water treatment technique in which organic pollutants are degraded within the bulk liquid phase through the uniform generation of Reactive Oxygen Species (ROS)—primarily hydroxyl radicals (HO•)—via chemical reactions that occur without the need for a solid catalyst surface.

(O_3/UV) Ozone couples with UV

Ozone is a powerful oxidizing agent capable of degrading most organic and inorganic materials in a water body across a broad pH range. Ozonation can proceed through two main reaction pathways: direct and indirect oxidation. In the direct electrophilic reaction, ozone interacts directly with organic molecules. In the indirect reaction, ozone decomposes in the presence of UV light and water (H_2O) to generate hydroxyl radicals (HO•), as shown in Equation (4). These radicals further enhance the oxidation process, enhancing the degradation efficiency of organic pollutants [29].

$$O_3 + h\nu + H_2O \rightarrow 2\,OH + O_2 \tag{4}$$

Ozone (O_3) possesses a reduction potential of 2.07 V. The UV/O_3 process enhances the degradation rate and broadens the range of organic compounds broken down by ozone oxidation, as shown in Fig. (4).

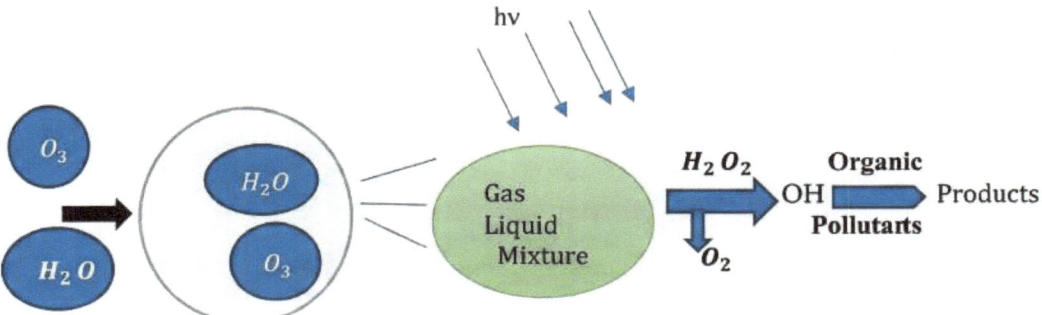

Fig. (4). Degradation of pollutants by UV/O_3 method.

When exposed to ultraviolet radiation, ozone generates highly reactive oxidizing radicals, making the process more efficient—a phenomenon commonly referred to as catalytic ozonation. The enhanced mass transfer in the efficiency of catalytic ozonation facilitates the effective removal of pollutants [30]. Moreover, UV radiation stimulates ozone decomposition, leading to the formation of additional radicals and creating a stronger oxidative environment [31].

(H_2O_2/UV) (H_2O_2) coupled with UV

The primary advantages of this technique include the high solubility of hydrogen peroxide (H_2O_2) and its ease of incorporation into the source water, even at relatively high concentrations. Additionally, the H_2O_2/UV processes can generate a greater quantity of hydroxyl radicals per unit of energy consumed compared to the UV/O_3 process, as shown in (Equation 5) [32].

$$H_2O_2 + h\nu \rightarrow 2HO \tag{5}$$

Water and CO_2 are the byproducts of the complete mineralization of organic compounds in the H_2O_2/UV system. Direct photolysis is a specialized process in which light reacts with molecules (in the presence of water), causing them to break down into smaller fragments through the mechanism illustrated below

$$R + h\nu \rightarrow \text{Intermediates} \qquad (6)$$

$$\text{Intermediates} + h\nu \rightarrow CO_2 + H_2O + R- \qquad (7)$$

Equations 6 and 7 describe the formation of intermediates when the hydroxy radical interacts with the light, followed by the subsequent transformation of these intermediates into carbon dioxide and water.

(H_2O_2/O_3) H_2O_2 couples with O_3

The H_2O_2/O_3 system, also known as the peroxone advanced oxidation process (AOP), utilizes catalysis to degrade refractory pollutants present in wastewater. This process operates through a radical chain mechanism, initiated by the hydroperoxide anion (HO^{2-}), which promotes the decomposition of ozone (Eq. 8).

$$H_2O_2 \rightarrow HO_2^- + H^+ \qquad (8)$$

O_3/Fe^{2+}) process

The mechanism primarily depends on two pathways:

i. the generation of reactive oxygen species through the decomposition of ozone, facilitated by charged metal species
ii. the formation of a complex between the catalyst and the organic molecule, followed by the oxidation of this complex to enhance pollutant degradation.

ULTRA SOUND IRRADIATION

Ultrasound irradiation, also known as sonolysis, can be combined with UV light, hydrogen peroxide (H_2O_2) and ozone (O_3). In sonolysis, three key mechanisms are involved: the formation of cavitation bubbles, the interaction between the bubbles and the surrounding liquid, and the intermediate region between the bulk solution and the liquid surrounding the bubble, where chemical or pyroelectric reactions may occur. The ultrasonic frequency plays a crucial role in the formation and behavior of these cavitation bubbles. Generally, lower frequencies produce a reduced concentration of active bubbles due to the high vapor content within the collapsing bubbles. Conversely, high frequencies cause increased bubble collisions, which lead to the generation of fewer reactive species and a consequent decrease in overall efficiency [33].

FENTON PROCESS AND PHOTO FENTON PROCESS

The Fenton reaction is a chemical process discovered in 1894 by Henry John

Horstman Fenton. The Fenton reagent is a mixture of hydrogen peroxide and iron (II) salt. Fenton demonstrated that hydrogen peroxide can oxidize a wide range of organic molecules in the presence of Fe (II) as a catalyst, producing hydroxyl radicals (OH) from H_2O_2. The fundamental mechanism involves the decomposition of hydrogen peroxide to generate free radicals. This reaction proceeds through a series of seven steps in an acidic medium, influenced by dark conditions, and it involves further oxidation. Fe^{2+} is regenerated, and the resulting precipitates readily form as ferric oxyhydroxides, which remain stable for potential future applications. The photo-Fenton reaction is a combined process in which Fenton's reagent is used alongside ultraviolet (UV) irradiation. The Fenton process is considered a simple and user-friendly method, suitable for the efficient removal of biological compounds and contaminants. A notable application of the Fenton process is the degradation of antibiotics, which highlights its specificity and effectiveness for improving treatment processes. The Fenton treatment offers several advantages, including the use of H_2O_2/UV as a catalyst to enhance reaction efficiency without adversely affecting reactant or product, by altering the pH in acidic or basic conditions, a basic mechanism has been produced. Fig. (**5**) illustrates the mechanism of the Fenton reaction under acidic conditions.

$$Fe^{2+} + H_2O_2 \rightarrow Fe^{3+} + HO\cdot + OH^-$$

$$Fe^{3+} + H_2O_2 \rightarrow Fe^{2+} + HOO\cdot + H^-$$

$$2\,H_2O_2 \rightarrow HO\cdot + HOO\cdot + H_2O$$

Fig. (5). Mechanism of Fenton.

Another advantage of the Fenton process is the use of UV irradiation, which facilitates the recycling of the ferric ions (Fe^{3+}) into ferrous ions (Fe^{2+}). This recycling enhances the overall efficiency and accelerates the chemical reaction [34]. Numerous catalysts, including carbon-based materials, transition metal complexes, and Zero-Valent Iron (ZVI), have been investigated to improve the efficiency of Fenton-like processes. By promoting the decomposition of H_2O_2 and facilitating the electron transfer Fe^{2+} and H_2O_2, as originally demonstrated by Henry J. H. Fenton, these catalysts enhance both the production and stability of hydroxyl (•OH) radicals [35]. Table **2** illustrates the Fenton process, highlighting the pharmaceutical classes, reaction products, treatment efficiency within the framework of Advanced Oxidation Processes (AOPs), the type of matrix, and the methods used for the removal of pharmaceutical effluents. The degradation efficiency largely depends on persistence, concentration, and level of the pharmaceutical pollutants present in the wastewater, which can be quantitatively assessed.

Table 2. Pharmaceutical products and corresponding treatment processes with reported efficiencies.

Pharmaceutical Class	Pharmaceutical Product	Treatment Efficiency	Matrix	Refs.
Antibiotics	Penicillin	Photo fenton like 66%	Antibiotic formulation effluent	[36]
Anti phlogistic	Diclofenac	Degradation following 60 mins	Demineralised water	[36]
Antibiotic	Amoxicillin	Total mineralization in <60 mins	Distilled water	[36]
Analgesic	Paracetamol	98% degrade after 5 mins	WWTP effluent	[36]
Anti-bacterial	Metronidazole	76% received after 5 minutes of degradation	Deionized water	[36]
Anti depressants	Fluoxetine	89% degradation after 5 mins	WWTP effluent	[36]

Electrochemical Oxidation

Electrochemical methods are primarily based on electron transfer mechanisms, in which Electrochemical Oxidation (EO) plays a crucial role in environmental sustainability. Unlike other Advanced Oxidation Processes (AOPs), EO does not require additional chemical catalysts. The main objective of electrochemical oxidation is to destroy pollutants through electron-mediated reactions using a clean reagent.

During the EO process, two mechanisms are involved in the oxidation of organic pollutants: direct and indirect mechanisms. In the direct mechanism, pollutants are adsorbed directly onto the electrode surface, and the efficiency depends on the type of electrode material used. In the indirect mechanism, electro-generated reactive species are produced, which subsequently degrade the pollutants and organic compounds [37]. Anodes are classified as active or inactive. Active anodes exhibit a low oxygen evolution potential, whereas inactive anodes have a high oxygen evolution potential. The two primary reactions in electrochemical oxidation are anodic oxidation and cathodic reduction. At the anode, water molecules are oxidized (Eq. 9) to generate hydroxyl radicals (•OH), which are highly reactive oxidizers. At the cathode, dissolved oxygen or other species are reduced, producing reactive species such as hydrogen peroxide (H_2O_2) that further facilitate the oxidation process.

$$\text{Mechanism at the anode:} \quad MO_X + H_2O \rightarrow MO_X(\cdot OH) + H^+ + e^- \quad (9)$$

The efficiency of anodic oxidation depends on several factors, including the concentration of organic pollutants, pH, mass transfer between the anode and cathode, applied current intensity, and the nature of the electrode material [38].

Electro Fenton

Electro Fenton is an advanced oxidation process in which hydroxy radicals are generated in combination with an electrochemical process. Its primary function is the degradation of toxic pollutants through highly reactive oxidative processes [39]. Among electrochemical methods, electro-Fenton is notable for consuming less energy per mole of contaminants removed, of contaminants removed, which is attributed to the lower required potential. The process involves the in-situ generation of hydrogen peroxide (H_2O_2) and its subsequent reaction with ferrous ions (Fe^{2+}).

The process can be described in three steps:

1. At the cathode (typically carbon felt), oxygen is reduced to hydrogen peroxide (Equation 10).
2. Ferrous ions catalyze the decomposition of hydrogen peroxide, producing hydroxyl radicals (Equation 11).
3. Ferric ions are regenerated, completing the cycle (Equation 12).

$$O_2 + 2H^+ + 2e^- \rightarrow H_2O_2 \quad (10)$$

$$Fe^{2+} + H_2O_2 \rightarrow Fe^{3+} + OH^- + \cdot OH \quad (11)$$

$$Fe^{3+} + e^- \rightarrow Fe^{2+} \quad (12)$$

HETEROGENEOUS PROCESS

Heterogeneous processes can be classified into catalytic ozonation, photocatalytic ozonation, and heterogeneous catalysis. Globally, confined within AOPs such as Fenton-like and photothermal catalysis has been successfully applied for the removal of antibiotics and pharmaceutical pollutants [40]. Heterogeneous chemical oxidation reactions are widespread in nature and play a crucial role in numerous chemical reaction systems related to catalysis, environmental remediation, and energy conversion [41]. AOPs, including photocatalysis and photo-Fenton processes, have been proposed as tertiary treatments in urban sewage systems due to their efficiency in purifying streams containing persistent contaminants [42].

Heterogeneous Photocatalysis

Photocatalysis is a process in which a chemical reaction is initiated by photon absorption, where the energy band of the material plays a critical role in accelerating the reaction. When a semiconductor is exposed to light with energy equal to or greater than its band gap, a valence band electron is excited to the conduction band, creating a photogenerated hole (h^+) in the valence band. Heterogeneous photocatalysis refers to a system in which the catalyst is in a different phase and the pollutants are in a fluid phase. Over time, various modifications of this technique have led to the development of new Advanced Oxidation Process (AOP) technologies for energy and environmental applications. For example, when a semiconductor catalyst such as TiO_2 absorbs photons with energy equal to or greater than its band gap, electrons are excited from the valence band to the conduction band, thereby accelerating chemical reactions (Equation 13). These photogenerated charge carriers react with water or hydroxide ions to produce hydroxyl radicals (Equations 14–15), while the electrons reduce oxygen species, leading to further radical formation (Equations 16–20). Catalysts play a key role in AOPs, enhancing the generation of reactive oxidizing species. Recent developments include novel heterogeneous catalysts, such as metal-organic frameworks (MOFs), which exhibit high surface area, tunable porosity, and unique catalytic properties, making them effective in promoting the oxidation of pollutants [43].

$$TiO_2 + h\nu \rightarrow TiO_2\ (e^-CB/\ h^+VB) \quad (13)$$

$$TiO_2 + h\nu \rightarrow TiO_2\ (e^-CB/\ h^+VB) \quad (14)$$

$$TiO_2(h^+VB) + H_2O \rightarrow TiO_2 + H^+ + \cdot OH \quad (15)$$

$$TiO_2(h^+VB) + HO^- \rightarrow TiO_2 + \cdot OH \quad (16)$$

$$O_2^- + H^+ \rightarrow HO_2^\cdot \quad (17)$$

$$TiO_2\ (e^-CB) + H_2O_2 \rightarrow \cdot OH + OH^- \quad (18)$$

$$H_2O_2 + O_2 \rightarrow \cdot OH + OH^- + O_2 \quad (19)$$

$$H_2O_2 + h\nu \rightarrow 2\cdot OH \quad (20)$$

However, compared to homogeneous photocatalysts, heterogeneous photocatalysis offers several advantages, including easy catalyst separation and recycling. Despite the promising results reported by many researchers at the laboratory scale, scaling up these systems presents significant operational challenges that require extensive technical expertise. A key aspect of photocatalysis is the development of suitable catalysts capable of overcoming the limitations of titanium dioxide (TiO_2), such as electron-hole recombination, rapid backward reactions, and low photocatalytic activity under visible light.

Sonochemical Process

This process involves the generation of an oxidizing agent using ultrasound as the energy source. It can occur as a purely physical phenomenon or in combination with catalysts or other chemical reagents. The primary energy for the process is provided by acoustic cavitation, which is created when bubbles collapse under ultrasonic sound waves in the frequency range of 20–1000 kHz. The collapse of these bubbles releases a tremendous amount of thermal energy, leading to the splitting of water molecules into highly reactive hydroxyl radicals (·OH). As one of the most powerful oxidizing species, the hydroxyl radical subsequently participates in the degradation of contaminants.

INTEGRATED AOP

Advanced oxidation has evolved to include hybrid and biological methods for getting a more efficient process for better removal of pollutants. Although AOPs and adsorption techniques have been applied individually to eliminate pharmaceuticals from water bodies, these approaches have not completely removed the contaminants. Therefore, to develop an effective method for eliminating pharmaceutical residues, even at low concentrations, future research should focus on integrating oxidation with adsorption processes. Additionally, studies on biological treatment should explore optimal conditions and microbial mechanisms that enable simultaneous removal of pharmaceutical compounds, as well as the influence of different environmental factors on treatment efficiency. Investigating innovative constructed wetland technologies is also crucial to understanding the synergistic removal of pharmaceuticals, antibiotic resistance genes, and conventional contaminants. Table 3 presents the advantages and disadvantages of various methods of AOP.

Table 3. Advantages and disadvantages of various methods of AOP.

Aop	Advantages	Disadvantages	Ref.
UV / UV/H_2O_2	Significant usage of UV sterilization Low light scattering	Elevated operational costs Large quantities	[44]
Fenton – based	Acidophilic Multifunctionality of Fe ions to react with pollutants	Slow reaction rate Preparation cost of the Fenton reagent	[44]
Ozone-based	Stable derivatives ffective in pretreatment	Drawbacks in aqueous reactions Self-decomposition	[45]

(Table 3) cont.....

Aop	Advantages	Disadvantages	Ref.
Photocatalysis	Reliability Cost-effective Eco-friendliness	The selection of the proper catalyst is difficult Radiation HV cost is high	[45]
Sonolysis	The reagent is lacking Versatility of the parameter influence	Uncontrollable byproducts Target low concentration	[45]

FACTORS AFFECTING AOP

Higher pollutant concentrations often require longer treatment times and increased doses of reagents. The pH of the solution influences both the generation and stability of hydroxyl radicals. Certain substances, such as bicarbonates, chlorides, and organic matter, can react with hydroxyl radicals, thereby reducing AOP effectiveness. The presence of competing species must also be considered to optimize process performance. While higher temperatures generally increase reaction rates, they may also accelerate the decomposition of oxidants before they can act on pollutants. The choice of catalyst should take into account factors such as cost, availability, and the characteristics of the wastewater. Processes like UV/H_2O_2 require sufficient energy to generate free radicals efficiently. High turbidity can reduce light penetration in UV-based AOPs, decreasing their efficiency; therefore, pre-treatment to remove solids may be necessary. The feasibility of AOPs depends on operational and maintenance costs, and large-scale applications require optimized reagent use and energy-efficient designs. By carefully optimizing these factors, AOPs can effectively treat wastewater, degrade persistent organic pollutants, and improve overall water quality [45].

CONCLUSION

This review has comprehensively demonstrated the effectiveness of Advanced Oxidation Processes (AOPs) as a viable solution for the treatment of pharmaceutical wastewater. To overcome the limitations of conventional treatment methods, AOPs offer a superior alternative. Evidence from multiple studies highlights the significant enhancement in contaminant removal and overall improvement in water quality achieved through AOPs. Their ability to degrade a broad spectrum of pharmaceutical pollutants, including recalcitrant compounds, underscores their potential to mitigate environmental concerns associated with these emerging contaminants. AOPs also address the shortcomings of traditional wastewater treatment processes, which often struggle to remove pharmaceutical residues effectively. With the growing global concern over pharmaceutical pollution, adopting AOPs as a treatment strategy can play a pivotal role in protecting aquatic ecosystems and ensuring a safer environment for future generations. Therefore, it is essential to further investigate and optimize AOPs for large-scale implementation, paving the way for a more sustainable and environmentally responsible management of pharmaceutical wastewater. Since Pharmaceuticals and Personal Care Products (PPCPs) cannot be efficiently removed by conventional methods, advanced oxidation techniques such as electrochemical oxidation, high-energy radiation, and ultrasonication are necessary to eliminate these contaminants, including painkillers, antiseptics, and anti-infectives, from sewage effluents. During their manufacture, use, and disposal, Active Pharmaceutical Ingredients (APIs) are released into the environment, posing risks to both ecosystems and human health. While single AOP methods can provide efficient wastewater treatment, their combination with biological processes or hybrid AOP techniques can enhance overall performance. For instance, antibiotics can be effectively degraded using Fenton-based processes and UV treatment to remove contaminants from effluent streams. In combined AOP-biological processes, key considerations include minimizing operational costs, reducing byproduct formation, and optimizing photocatalytic activity. The photocatalytic performance of heterogeneous catalysts depends on factors such as charge carrier separation, redox properties, and interfacial charge transfer. Future research should also focus on how factors such as molecular polarizability, dispersion, spatial interference, and the photonic properties of pharmaceutical compounds influence their degradation by AOPs. Additionally, mass spectrometry and biological evaluation of byproducts generated during photocatalysis should be incorporated to fully understand the transformation of pharmaceutical compounds. Overall, this review provides important insights for future research and innovation in wastewater treatment, offering a roadmap for the development of long-term solutions to pharmaceutical pollution.

REFERENCES

[1] Babu Ponnusami A, Sinha S, Ashokan H, *et al*. Advanced oxidation process (AOP) combined biological process for wastewater treatment: A review on advancements, feasibility of combined techniques. Environ Res. 2023;237:116944. Available from: https://www.sciencedirect.com/science/article/pii/S0013935123017486

[2] Saravanan A, Deivayanai VC, Kumar PS, *et al*. A detailed review on advanced oxidation process in treatment of wastewater: Mechanism, challenges and future outlook. Chemosphere 2022; 308(Pt 3): 136524. Available from: https://www.sciencedirect.com/science/article/pii/S004565352203017X
[http://dx.doi.org/10.1016/j.chemosphere.2022.136524] [PMID: 36165838]

[3] Fdez-Sanromán A, Pazos M, Rosales E, Sanromán MÁ. Prospects on integrated electrokinetic systems for decontamination of soil polluted with organic contaminants. Curr Opin Electrochem 2021; 27: 100692.
[http://dx.doi.org/10.1016/j.coelec.2021.100692]

[4] Osuoha JO, Anyanwu BO, Ejileugha C. Pharmaceuticals and personal care products as emerging contaminants: Need for combined treatment strategy. J Hazard Mater Adv. 2023;9: 100206.
[http://dx.doi.org/10.1016/j.hazadv.2022.100206]

[5] Casado J. Towards industrial implementation of Electro-Fenton and derived technologies for wastewater treatment: A review. J Environ Chem Eng 2019; 7(1): 102823.

[http://dx.doi.org/10.1016/j.jece.2018.102823]

[6] Lama G, Meijide J, Sanromán A, Pazos M. Heterogeneous Advanced Oxidation Processes: Current Approaches for Wastewater Treatment. Catalysts 2022; 12(3): 344.
[http://dx.doi.org/10.3390/catal12030344]

[7] Yang Y, Ricoveri A, Demeestere K, Van Hulle S. Advanced treatment of landfill leachate through combined Anammox-based biotreatment, O_3/H_2O_2 oxidation, and activated carbon adsorption: technical performance, surrogate-based control strategy, and operational cost analysis. J Hazard Mater 2022; 430: 128481.
[http://dx.doi.org/10.1016/j.jhazmat.2022.128481] [PMID: 35176699]

[8] Liu L, Chen Z, Zhang J, et al. Treatment of industrial dye wastewater and pharmaceutical residue wastewater by advanced oxidation processes and its combination with nanocatalysts: A review. J Water Process Eng 2021; 42(22): 102122.
[http://dx.doi.org/10.1016/j.jwpe.2021.102122]

[9] Li J, Zhou C, Xu H, et al. Ambient air pollution is associated with HDL (High-Density Lipoprotein) dysfunction in healthy adults. Arterioscler Thromb Vasc Biol 2019; 39(3): 513-22.
[http://dx.doi.org/10.1161/ATVBAHA.118.311749] [PMID: 30700134]

[10] Mezzelani M, Gorbi S, Regoli F. Pharmaceuticals in the aquatic environments: Evidence of emerged threat and future challenges for marine organisms. Mar Environ Res 2018; 140(April): 41-60.
[http://dx.doi.org/10.1016/j.marenvres.2018.05.001] [PMID: 29859717]

[11] Vijay Pradhap Singh M, Ravi Shankar K. Next-generation hybrid technologies for the treatment of pharmaceutical industry effluents. J Environ Manage 2024; 353: 120197.
[PMID: 38301475]

[12] Phoon BL, Ong CC, Mohamed Saheed MS, et al. Conventional and emerging technologies for removal of antibiotics from wastewater. J Hazard Mater 2020; 400: 122961.
[http://dx.doi.org/10.1016/j.jhazmat.2020.122961] [PMID: 32947727]

[13] Deegan AM, Shaik B, Nolan K, et al. Treatment options for wastewater effluents from pharmaceutical companies. Int J Environ Sci Technol 2011; 8(3): 649-66.
[http://dx.doi.org/10.1007/BF03326250]

[14] Fouad K, Bassyouni M, Alalm MG, Saleh MY. The Treatment of Wastewater Containing Pharmaceuticals. J Environ Treat Tech 2021; 9(2): 499-504.
[http://dx.doi.org/10.47277/JETT/9(2)504]

[15] Klavarioti M, Mantzavinos D, Kassinos D. Removal of residual pharmaceuticals from aqueous systems by advanced oxidation processes. Environ Int 2009; 35(2): 402-17.
[http://dx.doi.org/10.1016/j.envint.2008.07.009] [PMID: 18760478]

[16] Kumar V, Bansal V, Madhavan A, et al. Active pharmaceutical ingredient (API) chemicals: a critical review of current biotechnological approaches. Bioengineered 2022; 13(2): 4309-27.
[http://dx.doi.org/10.1080/21655979.2022.2031412] [PMID: 35135435]

[17] Mishra N, Reddy R, Kuila A, Rani A, Nawaz A, Pichiah S. A Review on Advanced Oxidation Processes for Effective Water Treatment. Curr World Environ 2017; 12(3): 469-89.
[http://dx.doi.org/10.12944/CWE.12.3.02]

[18] Obaideen K, Shehata N, Sayed ET, Abdelkareem MA, Mahmoud MS, Olabi AG. The role of wastewater treatment in achieving sustainable development goals (SDGs) and sustainability guideline. Energy Nexus. 2022; 7: 100112.
[http://dx.doi.org/10.1016/j.nexus.2022.100112]

[19] Nishat A, Yusuf M, Qadir A, et al. Wastewater treatment: A short assessment on available techniques. Alex Eng J 2023; 76: 505-16.
[http://dx.doi.org/10.1016/j.aej.2023.06.054]

[20] Nakada N, Shinohara H, Murata A, et al. Removal of selected pharmaceuticals and personal care

products (PPCPs) and endocrine-disrupting chemicals (EDCs) during sand filtration and ozonation at a municipal sewage treatment plant. Water Res 2007; 41(19): 4373-82.
[http://dx.doi.org/10.1016/j.watres.2007.06.038] [PMID: 17632207]

[21] Ariffin M, Zakili TST. Household pharmaceutical waste disposal in Selangor, Malaysia—policy, public perception, and current practices. Environ Manage 2019; 64(4): 509-19.
[http://dx.doi.org/10.1007/s00267-019-01199-y] [PMID: 31399770]

[22] Karungamye P, Rugaika A, Mtei K, Machunda R. The pharmaceutical disposal practices and environmental contamination: a review in East African countries. HydroResearch 2022; 5: 99–107.
[http://dx.doi.org/10.1016/j.hydres.2022.11.001]

[23] Mansouri F, Chouchene K, Roche N, Ksibi M. Removal of pharmaceuticals from water by adsorption and advanced oxidation processes: State of the art and trends. Appl Sci (Basel) 2021; 11(14): 6659.
[http://dx.doi.org/10.3390/app11146659]

[24] Pandis PK, Kalogirou C, Kanellou E, et al. Key points of advanced oxidation processes (AOPs) for wastewater, organic pollutants and pharmaceutical waste treatment: a mini review. ChemEngineering 2022 Jan 12; 6(1): 8.
[http://dx.doi.org/10.3390/chemengineering6010008]

[25] Adeoye JB, Tan YH, Lau SY, et al. Advanced oxidation and biological integrated processes for pharmaceutical wastewater treatment: A review. J Environ Manage 2024; 353(January): 120170.
[http://dx.doi.org/10.1016/j.jenvman.2024.120170] [PMID: 38308991]

[26] Lupu GI, Orbeci C, Bobirică L, Bobirică C, Pascu LF. Key Principles of Advanced Oxidation Processes: A Systematic Analysis of Current and Future Perspectives of the Removal of Antibiotics from Wastewater. Catalysts 2023; 13(9): 1280.
[http://dx.doi.org/10.3390/catal13091280]

[27] Almomani F, Bhosale R, Kumar A, Khraisheh M. Potential use of solar photocatalytic oxidation in removing emerging pharmaceuticals from wastewater: A pilot plant study. Sol Energy 2018; 172(172): 128-40.
[http://dx.doi.org/10.1016/j.solener.2018.07.041]

[28] Kanakaraju D, Glass BD, Oelgemöller M. Advanced oxidation process-mediated removal of pharmaceuticals from water: A review. J Environ Manage 2018; 219: 189-207.
[http://dx.doi.org/10.1016/j.jenvman.2018.04.103] [PMID: 29747102]

[29] Ma D, Yi H, Lai C, et al. Critical review of advanced oxidation processes in organic wastewater treatment. Chemosphere 2021; 275: 130104.
[http://dx.doi.org/10.1016/j.chemosphere.2021.130104] [PMID: 33984911]

[30] Roslan NN, Lau HLH, Suhaimi NAA, et al. Recent Advances in Advanced Oxidation Processes for Degrading Pharmaceuticals in Wastewater—A Review. Catalysts 2024; 14(3): 189.
[http://dx.doi.org/10.3390/catal14030189]

[31] Mahmoodi M, Pishbin E. Ozone-based advanced oxidation processes in water treatment: recent advances, challenges, and perspective. Environ Sci Pollut Res Int 2025; 32(7): 3531-70.
[http://dx.doi.org/10.1007/s11356-024-35835-w] [PMID: 39827442]

[32] Mayyahi A Al, Al-asadi HAA. Advanced Oxidation Processes (AOPs) for Wastewater Treatment and Reuse : A Brief Review. 2018;2(3):18–30.

[33] Antiñolo Bermúdez L, Martín Pascual J, Muñio Martínez MM, Poyatos Capilla JM. Effectiveness of advanced oxidation processes in wastewater treatment: state of the art. Water. 2021 Aug 1; 13(15): 2094.
[http://dx.doi.org/10.3390/w13152094]

[34] Mahdi MH, Mohammed TJ, Al-Najar JA. Advanced Oxidation Processes (AOPs) for treatment of antibiotics in wastewater: A review. IOP Conf Ser Earth Environ Sci 2021; 779(1): 012109.
[http://dx.doi.org/10.1088/1755-1315/779/1/012109]

[35] Wang J, Tang J. Fe-based Fenton-like catalysts for water treatment: preparation, characterization and modification. Chemosphere 2021 Aug; 276: 130177.
[http://dx.doi.org/10.1016/j.chemosphere.2021.130177]

[36] Adamek E, Masternak E, Sapińska D, Baran W. Degradation of the Selected Antibiotic in an Aqueous Solution by the Fenton Process: Kinetics, Products and Ecotoxicity. Int J Mol Sci 2022; 23(24): 15676.
[http://dx.doi.org/10.3390/ijms232415676] [PMID: 36555316]

[37] Bashir Y, Raj R, Ghangrekar MM, Nema AK, Das S. Critical assessment of advanced oxidation processes and bio-electrochemical integrated systems for removing emerging contaminants from wastewater. RSC Sustainability 2023; 1(8): 1912-31.
[http://dx.doi.org/10.1039/D3SU00112A]

[38] Kumari P, Kumar A. ADVANCED OXIDATION PROCESS: A remediation technique for organic and non-biodegradable pollutant. Results Surf Inter 2023; 11: 100122.
[http://dx.doi.org/10.1016/j.rsurfi.2023.100122]

[39] Ganzenko O, Trellu C, Papirio S, *et al.* Bioelectro-Fenton: evaluation of a combined biological—advanced oxidation treatment for pharmaceutical wastewater. Environ Sci Pollut Res Int 2018; 25(21): 20283-92.
[http://dx.doi.org/10.1007/s11356-017-8450-6] [PMID: 28144861]

[40] Niu J, Yuan R, Chen H, Zhou B, Luo S. Heterogeneous catalytic ozonation for the removal of antibiotics in water: a review. Environ Res. 2024 Dec 1; 262(Pt 2): 119889.
[http://dx.doi.org/10.1016/j.envres.2024.119889]

[41] Jiang HJ, Underwood TC, Bell JG, Ranjan S, Sasselov D, Whitesides GM. Mimicking Lighting-Induced Electrochemistry on the Early Earth. Proc Natl Acad Sci USA 2017; 120: 2017.
[http://dx.doi.org/10.1073/pnas.2400819121]

[42] Garrido-Cardenas JA, Esteban-García B, Agüera A, Sánchez-Pérez JA, Manzano-Agugliaro F. Wastewater treatment by advanced oxidation process and their worldwide research trends. Int J Environ Res Public Health 2019; 17(1): 170.
[http://dx.doi.org/10.3390/ijerph17010170] [PMID: 31881722]

[43] Zhang X, Wang J, Dong XX, Lv YK. Functionalized metal-organic frameworks for photocatalytic degradation of organic pollutants in environment. Chemosphere. 2020 Mar; 242: 125144.
[http://dx.doi.org/10.1016/j.chemosphere.2019.125144]

[44] Camargo-Perea AL, Rubio-Clemente A, Peñuela GA. Use of ultrasound as an advanced oxidation process for the degradation of emerging pollutants in water. Water 2020; 12(4): 1068.
[http://dx.doi.org/10.3390/w12041068]

[45] Baniamerian H, Ghofrani Isfahani P, Tsapekos P, *et al.* Application of nano-structured materials in anaerobic digestion: current status and perspectives. Chemosphere. 2019 Dec; 237: 124521.
[http://dx.doi.org/10.1016/j.chemosphere.2019.04.193]

CHAPTER 9

Digital Technologies Used in Environmental Management

G. Vijaya Laxmi[1,*], **V. Sai Nikhitha**[1], **K. Sree Saahitthi Reddy**[1], **B. Indira**[2,*], **Vanitha Guda**[3], **Keshetti Sreekala**[4], **Ramesh Ponnala**[5], **Sanjeeb Kumar Mandal**[1] and **Bishwambhar Mishra**[1]

[1] *Department of Biotechnology, Chaitanya Bharathi Institute of Technology, Hyderabad, Telangana, India*

[2] *Department of Master of Computer Applications, Chaitanya Bharathi Institute of Technology, Hyderabad, Telangana, India*

[3] *Department of Computer Science and Engineering, Chaitanya Bharathi Institute of Technology, Hyderabad, Telangana, India*

[4] *Department of Computer Science and Engineering, Mahatma Gandhi Institute of Technology, Hyderabad, Telangana, India*

[5] *Department of Computer Science and Engineering, Faculty of Science and Technology (ICFAITech), ICFAI Foundation for Higher Education, Hyderabad, Telangana, India*

Abstract: The incorporation of digital technologies into environmental management is critical for meeting modern challenges. Geographic Information Systems (GIS) and remote sensing provide detailed mapping of land usage and natural resources, whereas the Internet of Things (IoT) allows for real-time monitoring of air quality, water, and pollution. Big Data analytics and Artificial Intelligence (AI) improve decision-making by forecasting climate trends and maximizing conservation efforts. Blockchain enables transparency in sustainability programs like carbon credit certification and responsible supply chain tracking. Digital twins replicate environmental conditions to optimize resource consumption in smart cities and ecosystems. In the energy sector, digital platforms incorporate renewable sources such as solar, wind, and hydropower with smart networks to improve energy distribution. Precision agricultural instruments, such as drones, sensors, and self-driving tractors, improve crop yields by assessing soil health and water levels. Similarly, IoT-enabled irrigation systems save water by responding to changing weather conditions in real time. Advanced recycling technologies, aided by machine learning algorithms, improve garbage sorting, while robotic devices increase productivity in waste management facilities. These advancements extend to wildlife protection, where AI-powered cameras and drones

---

[*] **Corresponding authors G. Vijaya Laxmi and B. Indira:** Department of Biotechnology, Chaitanya Bharathi Institute of Technology, Hyderabad, Telangana, India; and ; Department of Master of Computer Applications, Chaitanya Bharathi Institute of Technology, Hyderabad, Telangana, India;
E-mails: drgvlaxmi_ biotech@cbit.ac.in, bindira_mca@cbit.ac.in

monitor endangered species and prevent poaching. These digital developments encourage climate action, promote a circular economy, and help to achieve global sustainability goals. The United Nations Environment Program (UNEP) actively encourages projects such as the Coalition for Digital Environmental Sustainability, which aim to eliminate data gaps and accelerate progress towards the Sustainable Development Goals.

Keywords: Artificial intelligence (AI), Blockchain, Internet of things, Renewable energy, Sustainable Development Goals (SDGs).

INTRODUCTION

What are Digital Technologies in Environmental Management?

Over the past few years, the need to address environmental problems has become more and more apparent. Just a few of the most urgent environmental problems include climate change, mainstreaming biodiversity and ecosystems, water and air pollution, and depletion of resources. Innovative technologies appear to be key assets in changing the way we think and act for the protection and sustainability of our environment. These technologies are helping not only to resolve and prevent the environmental challenges but also to determine what the world will look like in years to come [1, 2]. Information technologies in environmental management include a range of advanced technologies and systems used to inform process, control, and protect the environment based on IT, data analysis, automation, and AI (Fig. 1). Table 1 summarizes the information technology and its applications in biotechnology.

Table 1. Information technologies and their applications.

Technologies	Applications
Geospatial technologies (GIS, remote sensing)	Used for mapping and monitoring environmental changes.
IoT (Internet of Things)	Sensors that collect real-time data on air quality, water levels, temperature, *etc.*
Big Data and AI	Used for predictive analytics, pattern recognition, and data-driven environmental decision-making
Blockchain	Applied for tracking environmental credits (*e.g.*, carbon trading), ensuring transparency in environmental transactions.
Cloud Computing	Storing and processing large volumes of environmental data.

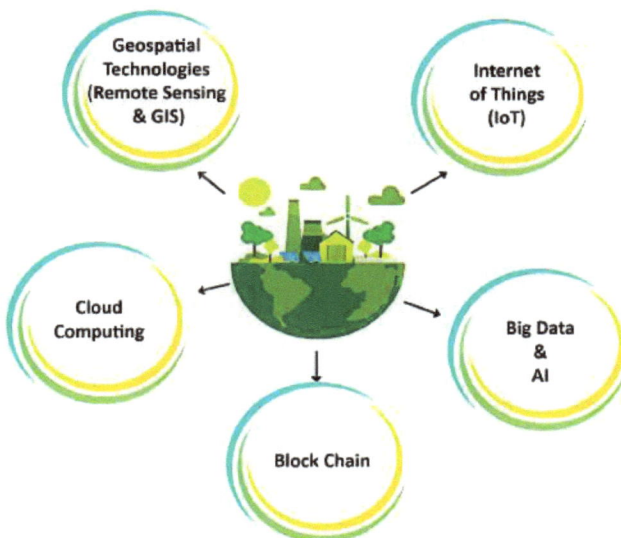

Fig. (1). Technologies used in environmental management.

Evolution of Digital Technologies in Environmental Conservation

The history of environmental conservation shows that as the technologies improved, there was an increased awareness of environmentally related issues. Current technologies were first developed in the early years of the 1990s during the first phase of the environmental conservation era, with more emphasis on data collection, automation, and predictive analytics being recorded post-2000. Within the realms of environmental conservation, GIS technology began to gain attention in the 1990s as the technology made it possible to collect and analyze very large volumes of environmental data. GIS was applied in mapping broad land-use distributions, assessing the space and rate of deforestation, as well as mapping out biodiversity hotspots, which enabled them to see how the environment undergoes changes over huge areas of land. This use of GIS for the first time enabled more robust decisions to be made regarding environmental aspects such as forest management, urban development, and biodiversity conservation. Along with GIS, imagery from satellites has also become an invaluable resource for monitoring and evaluating large-scale ecological phenomena, which is called remote sensing technology. Remote sensing has contributed to assessing the dynamics of vegetation, glacier relocation, and deforestation changes, thereby enhancing one's view of the Earth's ecosystem [3].

By the 2000s, advances in technology increased the scope of satellite-based monitoring. NASA's Landsat and other satellites, such as TERRA-AQUA, met the need for more undiluted and frequent tracking of environmental alterations, thus leading to vast investigations on issues such as urban expansion, wetland shrinking, coral reef bleaching, and illegal fishing, which were rampant during that time. It was during this period that the IoT gained traction. The most marked developments during the 2010s were, however, the introduction of big data analytics and AI into environmental management [4]. With the deluge of data from satellites, sensors, and others, conservationists were able to develop complex models to predict environmental changes. AI is in use now to detect deforestation trends, assess the vulnerability of a biome to climate change, and hypothesize conservation strategies. Machine learning, AI, and blockchain are today fully integrated into environmental protection activities [5]. Functions such as the observation of endangered species, fire risk assessment, and the management of protected areas are now carried out with minimal or no human effort. Machine learning models are now used to predict long-term environmental variables like sea level rise and climate change, and over time, the models get better [6].

Importance of Digital Solutions for Sustainable Environmental Management

Digital solutions are essential for ensuring the sustainable management of the environment as they present new ways to perform complex environmental tasks more efficiently. The potential of these technologies can be appreciated in terms of their roles in key areas such as data analytics, monitoring and transparency, efficiency, public involvement, *etc.* Together, these efforts contribute to promoting sustainability and safeguarding ecosystems in alignment with global initiatives like the United Nations Sustainable Development Goals (SDGs).

Data Driven Decision Making

With the advent of digital technologies, decisions relating to environmental management can now be made based on an analysis of the available information. Big data, GIS, and IoT (Internet of Things) technologies have made it possible to gather a large amount of real-time data over expansive areas from biosensors monitoring air conditions, water quality, soil nutrients, and the population of biological organisms. Such a steady supply of information enables governments, institutions, and non-governmental organizations pertaining to the environment to make better decisions [4].

Monitoring and Transparency

With the advent of digital technologies, better monitoring and transparency in activities, including continuous "greening" of forests, wetlands, and other fragile ecosystems through satellite-based remote sensing and drone technology, are upscaled. Such systems have the capability to detect unlawful activities, such as deforestation and poaching, as well as overfishing, in real time to enable quick responses from the authorities [7]. Secondly, blockchain technology can enable openness in carbon trading and other environmental activities that call for accountability and traceability. This level of transparency further induces corporations and governments to operate in even greener manners since their policies are openly viewed by other international organizations as well as the rest of society [5].

Eco-Friendliness Efficiency

Among the consequences of automation and digital technologies—such as AI, machine learning, and robotics—is increased efficiency in operations through reduced human intervention and costs. For instance, AI allows for increased efficiency in the speed of environmental impact assessments through scanning of big datasets for potential hazards or opportunities [4].

Support sustainability goals

The achievement of global sustainability goals, such as the United Nations Sustainable Development Goals (SDGs), is very central in the implementation of digital technologies. Digital solutions are, therefore, looking to ensure long-term environmental sustainability in improving resource management, wasting fewer resources, boosting energy efficiency, and promoting biodiversity conservation. The use of climate models by AI powers predictions of climate change effects, thus enabling policymakers to design adaptation and mitigation methods. Similarly, digital air and water quality monitoring techniques can enable governments to confirm sustainable progress towards the targets for clean water and air [8].

Scope of Digital Technologies in Various Environmental Sectors

Digital technologies have various applications in several environmental areas and, hence, positively charge management and conservation efforts. Utilization of water management incorporates the use of IoT sensors and their associated cloud-based platforms to monitor water quality, detect contaminants, and enhance the utilization of water resources. Websites such as Water.org provide information on sustainable water management strategies. Urban air quality monitors incorporate the use of IoT-based air pollution sensors to detect pollutants in real-time, whereby AirVisual gives interactive map systems.

In the energy sector, smart grids employ AI and automation in this best practice to use energy optimally through the utilization of renewable energy sources while reducing carbon footprint. IRENA and other websites focus on the global adoption of renewable energy. In animal and forest conservation, AI uses drones to track endangered species, monitor deforestation, and prevent unlawful poaching. Global Forest Watch uses satellite data to monitor forests in real time. Furthermore, Blockchain technology is applied in waste management with the aim of increasing transparency in the operations of trash disposal and recycling systems, with companies like Waste Management at the forefront of sustainable methods of waste treatment. Some of these precision farming technologies are changing farming practices at the ground level regarding water use efficiency and better crop productivity without harming the environment [9].

TYPES OF DIGITAL TECHNOLOGIES USED IN ENVIRONMENTAL MANAGEMENT

Geographic Information System (GIS)

GIS is one incredibly powerful and vital tool in helping protect our planet, its inhabitants, and its resources. Relying upon GIS in conservation and protection offers a critical edge when trying to help protect, preserve, conserve, and restore the environment. As people continue to multiply and as the climate alters, environmental agencies and conservationists do their best with help from GIS technology to monitor and help manage natural resources and the environment. GIS effectively monitors the analysis of environmental data and helps in proper planning because it enables the viewing and tracking of current data, as well as some historical data. It also includes the overlay of many factors such as slopes, vegetation, threatened species, watersheds, and other aspects to calculate several environmental parameters, as well as the impact analysis [10].

GIS is also helpful in soil suitability analysis, which provides critical information through soil mapping for better land use and the prevention of environmental degradation. Farmers use GIS in order to optimize crop yields by assessing the levels of nutrients in soils. Remote sensing, one of the primary techniques involved in GIS, gathers biological data by detecting reflected energy using satellites, drones, and aircraft, making this a vital resource for environmental and agricultural management.

Remote Sensing

Remote sensing in geospatial technology collects electromagnetic radiation reflected from Earth's atmospheric, terrestrial, and aquatic ecosystems, making it possible to study the physical characteristics without physical contact. The tools for monitoring the radiation, mostly reflected sunlight, through passive remote sensing are charge-coupled devices, radiometers, film photography, and infrared sensors. Remote sensing employs satellite imaging, aerial photography, and radar to understand the environment through electromagnetic wave measurements. Earth observation satellites, commonly applied for monitoring the environment, meteorology, and surveying, are earth-imaging satellites and GNSS radio occultation systems [11]. Active remote sensing techniques differ based on the emission in terms of light or waves and the measurements taken:

Radar: An active technique that emits radio frequency pulses; the distance is calculated based on the scattered energy and the time taken to travel.

Lidar: Uses light pulses to determine distances and positions by calculating the time light travels, with applications like elevation measurement *via* laser altimeters.

Scatterometer: Measures backscattered radiation to analyze surface properties.

Internet of Things (IoT) and Environmental Sensors

IoT environment monitoring is widely spread in all spheres: from agriculture and forestry to urban planning, generating energy, and distributing it. Crop monitoring, health of soil, water quality, and weather conditions are monitored in IoT-based systems within agricultural industries [2].

Systems in the energy sector use IoT to monitor air and weather conditions, emissions. Consequently, such systems assist public bodies, companies, and environmental agencies in monitoring and taking measures against the negative impacts of pollution. In urban planning, the IoT-based systems used in monitoring the traffic congestion or air pollution levels in smart cities may be implemented for application in decision-making processes as to how future urban development can reduce environmental impacts (Table 2).

Table 2. Applications of IoT in environmental monitoring.

Application	Description	IoT Usage
Air-Quality Monitoring	Tracks air pollutants to protect health.	Monitors air quality and detects issues.
Water-Quality Monitoring	Ensures safe water by tracking contaminants.	Detects water pollution and leaks.
Energy Monitoring	Manages energy usage and distribution.	Tracks consumption and identifies issues.
Commercial Farming	Improves agriculture using sensors.	Monitors soil, crops, and livestock.
Toxic Gas Detection	Detects harmful gases for safety.	Sends alerts to prevent hazards.
Environmental Sensors	Measures conditions like air and soil.	Measures conditions like air and soil.

Artificial Intelligence and Machine Learning

AI and ML are quickly developing areas that significantly impact environmental science and human health. With AI and ML, it is possible to gather and analyze large volumes of data; therefore, there will be enhanced understanding of complex environmental systems and their potential future changes with greater accuracy.

The effects of climate change and population aging did not significantly contribute to reductions in cardiovascular mortality. A time-series analysis of daily PM2.5 concentrations, temperature, humidity, cardiovascular mortality, and population demographics in Guangzhou from 2013 to 2021 reveals that improvements in air quality led to a decrease in cardiovascular mortality. However, this effect was influenced by factors such as temperature and the age of the population. The author also emphasized the role of Machine Learning (ML) approaches in processing large raw metabolomics datasets obtained through techniques like mass spectrometry and nuclear magnetic resonance. A comprehensive review of ML applications for screening is currently in progress [12].

AI to ML Relationship:

AI is the higher goal to create intelligent machines that can mimic human intelligence. Machine learning is an area of AI that majorly explains how to make machines learn from data and improve their performance. In other words, machine learning is the method through which AI can be achieved. Whereas AI encompasses rule-based systems, expert systems, and logic-based reasoning, ML relies on a data-driven approach to enable systems to change and learn without any explicit programming. AI aims at the replication of human-like intelligence; however, ML presents one of the most important ways to make this possible by providing the ability to learn from experience.

Deep Learning: A Subfield of Machine Learning

Deep learning is an application of ML with more than one layer of artificial neural networks. They are specifically designed to model simple and multiple patterns in large data sets. As such, deep learning has been very successful in applications like:

- Computer Vision: Image recognition, object detection, analysis for medical imaging
- Speech Recognition: Smart assistants like Siri, Alexa, and Google Assistant

- Natural Language Processing: Continuing work on translation, summarization, and sentiment analysis.

Big Data Analytics

Big Data has revolutionized our lives in countless ways. Leading companies in Big Data Analytics have created advanced algorithms that can accurately predict consumer behavior. In an era where traditional media like cinema and newspapers no longer dominate, data plays a pivotal role in shaping our preferences, from the food we eat to the movies we watch, and even the products we choose to buy.

Prominent global organizations like the UN, EPA, and Microsoft are harnessing the potential of Big Data Analytics to tackle climate change. By employing advanced predictive algorithms, they are devising impactful strategies aimed at addressing this human-driven crisis and safeguarding the planet's future [13].

The government is able to diagnose environmental problems in real-time situations through a high-tech sensor. The Big Data tools give this data more precision and thereby result in more effective enforcement of stricter environmental regulations. As an example, smoke presence sensors can ensure that the thermal power plant emits smoke as specified by measuring the same in the latter. Big Data analytics, in addition, provides the stakeholders with the most imperative observations into their problems and routes to solve them. Analyzing Big Data gives stakeholders valuable insights into challenges they are facing. Research using Big Data analytics is able to spot patterns and linkages that one would fail to notice otherwise. By making use of a vast network of sensors, scientists collect data about several pollutants, and this enables the examination of how levels of pollution are in relation to industries, geographic locations, and other relevant aspects.

Ongoing Environmental Projects Utilizing Big Data Analytics

Recently, the United States EPA joined the National Consortium for Data Science (NCDS), which leads the way in big data across the world. Some of the activities in the EPA include creating the stream catchment dataset for an entire map of the water bodies and streams throughout the United States.

The EPA wants to know the current state of vital water resources. A new major effort in the battle against climate change is the use of Big Data and Artificial Intelligence by Microsoft on a large scale. This is the first of its kind initiative wherein the company plans to simulate life on Earth by creating a model of the entire biosphere, which will enable the accurate measurement of the effects of major events such as deforestation and environmental pollution.

The United Nations is extremely eager to explore the potential of big data for countering climate change. Through Global Forest Watch, it conveys very important information on the global forest. This also helps countries like Guyana, Canada, Russia, or Congo come up with policies intended to save these valuable ecological treasures.

Key Factors of Big Data in Environmental Management

Data Gathering and Integration: Data are derived from a variety of environmental sources, including sensors (IoT devices), satellites, weather stations, social media, and mobile apps. The data sources may include everything from the current time, which includes real-time data, such as levels of air or water, to historical data, like records of climate for decades. Big Data Analytics is the integration and processing of data obtained from multiple sources to generate an overall view of environmental conditions.

Predictive analytics on climate change and natural disasters: Big data analytics employs past data and machine learning algorithms, which predict environmental trends and natural disasters.

Application of big data analytics in climate modeling: These help researchers simulate different climatic scenarios of the future. This enables them to project long-term effects of climate change, such as sea-level rise, precipitation shift, and temperature changes. Predictive analytics helps in disaster management by predicting natural calamities, including earthquakes, floods, droughts, and hurricanes; thus, intervention is timely, and preparedness is improved.

Pollution Monitoring and Mitigation: Big data analytics is required to monitor contamination of air, water, and soil. Sensors and monitoring devices are constantly collecting data on the concentration of pollutants in various places. The mashup between big data and IoT devices allows one to monitor real-time sources of pollution, hence assisting agencies concerned with the environment to identify people who are polluters or those who are breaking the law.

Big data analytics is one of the key components of sustainable urban settings in smart cities. It evaluates the output of traffic sensors, trash management systems, and air quality monitors to enhance energy efficiency, reduce emissions, and minimize waste [13].

Drones and UAVs

UAVs (Unmanned Aerial Vehicles) are controlled aerial systems designed to carry out multiple tasks without human intervention. These vehicles are operated using electronic devices like microprocessors and sensors, enabling autonomous operation in environments where human presence would be hazardous. Although UAVs can perform independently, they still require a human operator for remote control. A number of studies have examined different aspects of UAV performance. For instance, some research has focused on enhancing drone battery life through mobile charging features. To minimize the UAV's weight, a wirelessly powered receiver has been used, and several wireless charging methods have been proposed to extend flight time, a charging system powered by Wireless Power Transfer (WPT) to improve UAV efficiency has been developed.

UAVs, or what most people commonly call drones, are aircraft that do not have a human pilot on board but instead are piloted through a remote or autonomous basis on pre-programmed flight plans and sensors to collect information. This has made UAVs of great importance in many industries, one of which is environmental management, because they can rapidly scan large areas, penetrate difficult terrains, and collect high-resolution data (Table 3).

Table 3. Components of UAVs.

UAV Component	Description	Key components
Airframe	The main structure of the UAV houses subsystems like cameras, sensors, and communication equipment.	Fixed-wing UAVs: Resemble airplanes, ideal for long-range, high-altitude flights over large areas (*e.g.*, agricultural fields). Rotary-wing UAVs (Multicopters): Includes quadcopters; can hover for detailed inspection.
Propulsion System	Provides the thrust needed to maintain stable flight, powered by electric motors or combustion engines.	Electric motors are quiet and eco-friendly, suitable for small UAVs; larger UAVs may use internal combustion engines for longer flights.
Sensors and Payloads	Sensors and Payloads	Cameras: Optical, infrared, and thermal for imaging. Multispectral and Hyperspectral Sensors: Analyze vegetation health and water quality. LiDAR: Creates 3D terrain models. Gas and Water Sensors.
Communication System	Transmits data between UAV and ground control, including telemetry, flight status, and real-time sensor data.	Uses radio frequencies or satellite communications for long-range flights, ensuring consistent data transfer between UAV and operators.

APPLICATIONS OF DIGITAL TECHNOLOGIES IN ENVIRONMENTAL MANAGEMENT

Climate Monitoring and Weather Forecasting

Digital technology has transformed climate monitoring and weather forecasting by enabling the collection, analysis, and modeling of vast atmospheric datasets. Satellite remote sensing, ML, AI, and High-Performance Computing (HPC) have significantly improved the accuracy and reliability of climate and weather predictions. Satellite remote sensing monitors patterns of weather, global temperatures, cloud types, and other essential factors. Real-time and historical climate data are obtained from satellites like NOAA's GOES and NASA's Aqua and Terra.

Example: The ECMWF uses AI and supercomputing to enhance the precision of weather forecasting models [14].

Biodiversity Conservation

Digital technologies help in biodiversity monitoring and conservation since one is able to trace the population of different species, map ecosystems, and better understand the effects of human activities on animals.

Remote Sensing with UAVs: Drones that are equipped with cameras and thermal sensors map wildlife habitats and track poaching. Satellite image captures large-scale environmental information useful in mapping ecosystems and detecting deforestation or land-use change [15].

Bioacoustic Monitoring: AI algorithms detect and monitor species in remote ecosystems by analyzing animal sounds and bird calls captured using acoustic sensors [15].

Genomics and AI: Digital genomics allows for the study of DNA from species, which contributes to conservation efforts. AI helps to analyze genetic diversity, identify endangered species, and predict how climate change may impact biodiversity [15].

Pollution Prevention and Waste Management

Digital technologies also help in controlling pollution by enhancing the capability to observe air, water, and soil quality, plus optimizing waste management processes to minimize environmental effects. IoT devices, equipped with sensors for air and water quality monitoring, can provide real-time information regarding the aforementioned pollutants, including CO_2, PM2.5, and toxic substances in the water bodies. These sensors transmit data to the main systems for analysis.

AI-powered robots and image recognition technology separate the waste at recycling factories and increase effectiveness in separating plastics, metals, among others.

Water Resource Management

The management of water resources is necessary for the sustainable usage and protection of freshwater resources. Through digital tools, water resources can be monitored, managed, and predicted more efficiently in terms of availability and usage.

IoT with Remote Sensing: IoT sensors monitor water quality in lakes, rivers, and reservoirs through pH, dissolved oxygen, and pollutant level detection. Satellite imagery and UAVs track water bodies and changes in volume.

AI and Data Analytics: AI algorithms use historical and real-time data to forecast water shortages, floods, and pollutants. Machine learning algorithms improve water distribution in cities, minimizing waste [2].

Disaster Risk Reduction and Mitigation

Digital technologies enhance our ability to predict, prepare for, and respond better to disasters such as floods, earthquakes, and wildfires. As the Internet of Things sensors report on seismic activity, water levels, and other meteorological variables in a real-time basis, early warning systems about tsunamis, hurricanes, and floods are powered.

AI and Big Data Analytics: AI applications analyze past disaster data and current conditions to predict the likelihood and intensity of natural disasters. Machine learning identifies trends and increases the accuracy of predictions.

Drone Surveillance: Immediately after disasters, drones survey damage, identify missing persons, and monitor inaccessible areas for ground rescue teams [2].

Sustainable Farming and Land Management Planning

Sustainable agriculture with land use planning is vital to optimize land use with reduced environmental impacts. Digital technologies support precision farming, crop monitoring, and land management.

Precision agriculture utilizes UAVs, IoT sensors, and satellite imaging for the monitoring of soil conditions, crop health, and water usage. Data analytics supports effective irrigation, pesticide, and fertilizer use, reducing waste and increasing yields.

AI & Machine Learning: AI can be used in optimizing planting schedules, predicting crop yields, and assessing the effect of climate change on agriculture. Further, it might analyze soil sensor data so that it can properly utilize resources like water and fertilizer.

GIS and Remote Sensing: Through GIS and satellite imagery, mapping and analysis of land use patterns are done, identification of areas prone to erosion and deforestation, and planning for sustainable development projects on the ground [16].

Renewable Energy Monitoring and Optimization

Digital technologies play a critical role in increasing the efficiency of renewable energy sources, such as solar, wind, and hydroelectric power. Optimizing energy production on-site while integrating into power systems is done with advanced data analytics and artificial intelligence.

Smart Grids and IoT: IoT-enabled smart grids allow monitoring of energy production and Consumption, thereby enabling better management between supply and demand. Through Real-time data, intermittent sources of renewable energy, like solar and wind power, are monitored.

AI and Predictive Analytics: AI-based systems predict consumption, optimize storage, and determine peak renewable energy generation hours through forecasts and historical data on weather conditions [4].

BENEFITS OF DIGITAL TECHNOLOGIES IN ENVIRONMENTAL MANAGEMENT

Internet of Things (IoT)

IoT has transformed the nature of environmental data collection and analytics. IoT represents a network of interconnected devices, sensors, and systems that can collect, transmit, and process real-time information. Other than that, IoT has been applied for environmental monitoring to monitor air quality.

Water Quality, Temperature, and Biodiversity

Air quality monitoring stations can be equipped with IoT sensors to continuously measure pollutants such as particulate matter, nitrogen dioxide, and ozone. The real-time data collected can trigger pollution alerts and provide the public with up-to-date air quality information. Similarly, IoT-enabled water quality sensors monitor rivers, lakes, and oceans, helping to detect contamination and safeguard aquatic ecosystems [17].

Renewable Energy Technologies

Moving from fossil fuels to renewable sources is an important step in controlling greenhouse gas emissions and combating climate change. The major technologies that are altering the energy sector include solar panels, wind turbines, and hydropower. The surface of the sun catches the sunlight through solar panels and directs it as a clean source of energy. Wind turbines capture the power from the wind. Hydropower uses moving water for electric power production [18].

These renewable energy sources decrease carbon output and reduce fossil fuel usage while there is ample availability. Lastly, energy storage, for example, by high-capacity batteries, makes it possible to store excess generation during peak usage and use it during periods when demand arises.

Green Building

The construction and use of buildings significantly affect the environment. Sustainable architecture seeks to reduce these impacts by utilizing eco-friendly materials and designing energy-efficient buildings. Some of the technologies used in building designs include smart windows that adapt to sunlight, energy-efficient insulation, and rooftop gardens that help lower energy consumption.

Electric Vehicles (EVs)

Harmful gas emissions are one of the major hindrances in the transportation sector. In a relatively short period, electric vehicles have emerged as a very favorable alternative to fossil fuel-based cars. These vehicles are powered by electricity stored in batteries, and when feasible, they can also be powered by electricity generated from renewable sources such as solar and wind power.

No tailpipe emissions; therefore, reducing air pollution in city centers lowers the carbon dioxide level. Improvements in the technology of batteries are providing more value to the customer concerning driving distance. Charging lasts for a shorter period, hence there is a fast move toward green transportation, enhanced by government as well as car manufacturers' incentive forms worldwide.

Recycling Technologies

Effective waste management is crucial for reducing environmental impact. Automated sorting systems are a key innovation in this area, as they efficiently separate recyclable materials from general waste. This not only enhances recycling rates but also helps decrease the amount of waste that is sent to landfills.

Because they reduce the need for landfills by converting non-recyclable garbage into heat or power, technologies like waste-to-energy facilities are crucial to waste management. Additionally, methane emissions, a powerful greenhouse gas, are decreased by diverting organic wastes from landfills through composting and organic waste management systems.

AI and ML

Machine learning and artificial intelligence become vital tools to preserve the environment. They analyze huge data volumes, trend detection capabilities, and future prediction capabilities through which they give a boost to protection and resource exploitation. AI, along with ML, is integrated into many imperative environmental protection application domains.

In order to help communities and governments remain ready and successfully respond to possible dangers, Artificial Intelligence (AI) may analyse past environmental data to discover trends, such as climate changes and patterns of natural disasters. In order to support conservation efforts, machine learning algorithms are also employed to interpret sensor data and satellite imagery. This allows for the tracking of deforestation, the monitoring of ecosystem changes, and the identification of habitats for endangered species.

Carbon Capture and Storage (CCS)

The goal of CCS technology is to safely store CO_2 emissions underground by capturing them from power stations and industrial operations. Particularly in industries like cement and heavy industry, the materials and procedures used in CCS technologies are therefore always evolving, making it simpler and less expensive to absorb and store CO_2. This technique reduces CO_2 emissions from major industrial sources, which is a key component of the plan to slow down climate change [19].

Better data collection with accuracy

The use of IoT devices has significantly enhanced the efficacy and efficiency of environmental monitoring in terms of accuracy. Traditional systems usually depend on random sampling, and at times, critical changes go unnoticed. IoT sensors collect continuous, highly accurate data that leaves little room for error. Data that is measured more precisely leads to more informed decision-making when talking about governmental regulations, corporate compliance, or community awareness. For example, while decent lab sensors applied in monitoring water quality provide quantified measurements, they could improve the efficiency of pollution control activities [20].

Real-time environmental monitoring

The IoT solution has highly improved the way we track and manage pollution. Sensors placed across all cities and industrial hubs continually collect data on different types of pollutants within the atmosphere and water. Real-time tracking systems managed to improve London's air quality in 2023 by 20%. According to the Breathe London initiative, immediate alerts and actionable data have resulted in quicker reactions and effective mitigation strategies. This capacity is not only supportive of regulators in enforcing environmental laws but also energizes local communities toward proactive actions [21].

Tremendous Advantages of IoT Technologies in Wildlife and Ecosystem Monitoring: IoT technologies can offer tremendous advantages in monitoring wildlife and ecosystems. For instance, in Kenya, an IoT sensor is tracking endangered rhinoceroses, giving data on the movements and health of the rhinos. Such information allows conservationists to act in a timely manner, leading to a 15% reduction in poaching occurrences in the tracked regions. Sensors of forest ecosystems collect continuous data about the moisture content and temperature in the soil, thereby allowing more adequate practices that are in line with responsible forestry approaches.

Predictive modeling and forecasting

Digital technologies applied to environmental management have various very important benefits, mainly in areas such as predictive modeling and forecasting. Among the most critical advantages are:

Where digital technologies improve data gathering and further analysis: 'The availability of digital technologies has made it easier to collect vast amounts of data from many sources such as satellite images, sensors, or IoTs'. All this data can be analyzed for trends in environmental changes, such as fluctuations in climate, loss of biodiversity, and pollution levels.

Improved Predictive Accuracy: Predictive modeling uses algorithms and machine learning techniques to develop models that mimic nature, simulating the environment, and predicting future scenarios. It can thus be used in making more accurate predictions about the climate impact, resource availability, and changes in ecosystems for better decision-making.

Real-Time Monitoring of Environmental Conditions: With digital tools, environmental managers can monitor real-time environmental conditions. This allows the timely response to emerging issues, such as the impact of natural disasters or pollution events, to reduce their effects.

Scenario Planning: Digital technologies allow for building alternate scenarios, based on parameters and assumptions. This allows for the exploration of possible outcomes and the planning of management decisions on resources, conservation, and urban planning.

Hence, stakeholder engagement can be enhanced by digital platforms for governmental agencies, NGOs, and the public sector through better coordination and communication. Also, the influence of the cause on stakeholders can be communicated and understood through visualization tools and interactive dashboards.

Resource Optimization: Predictive modeling identifies spaces where interventions will have the most impact. Again, resource optimization is a huge advantage in terms of managing natural resources sustainably because it could contribute to saving more water, energy, and land.

Risk assessment and management: Digital technologies support the assessment of environmental risks faced by a variety of projects or policies. Based on the prediction of possible environmental impacts, managers can formulate strategies to reduce the detrimental impacts and build resilience.

Integrate Different Data Sources: Digital technologies can integrate most social, economic, and ecological data. This holistic perspective could lead to even more effective solutions to environmental problems.

Cost-effectiveness: Automation of processes used to collect and analyze data would minimize costs of operation and improve efficiency in environmental management practices.

Improved decision-making and policy formulation

Digital technologies significantly enhance decision-making and policy formulation in environmental management through various benefits (Table 4).

Table 4. Benefits of IoT systems in environmental policy and decision-making.

Benefit	Impact on Policy and Decision-Making
Data-Driven Insights	Informs policymakers of trends, risks, and impacts, allowing for evidence-based and informed policies to address environmental challenges.
Real-Time Monitoring and Adaptability	Supports adaptive policy measures that evolve with real-time environmental conditions, ensuring timely responses to emergent environmental issues.
Scenario Modeling and Predictive Analysis	Enables policymakers to forecast impacts and select the most effective policies, enhancing proactive environmental management and strategy selection.
Enhanced Public Engagement and Transparency	Promotes transparency and public understanding, encouraging community involvement in policy discussions and fostering greater support and compliance with environmental regulations.

Cost-effectiveness in Environmental Conservation Efforts

Less Costly Monitoring and Data Collection: Robotic data collection techniques, such as drones, satellite imagery, and IoT sensors, can monitor large areas using little human labour. This results in saving money related to traditional methods and allows for more frequent and detailed data collection, providing high-quality insights at a fraction of the cost usually allocated to them.

Targeted Resource Allocation: Digital technologies allow for accurate mapping and analysis, thus enabling conservationists to concentrate efforts in areas with the highest potential value for ecological measures or degradation risk. Identifying critical zones and predicting outcomes allows resources to be allocated more effectively and to minimize the incidence of frivolous expenditure on less effective interventions.

Efficient Predictive Modelling and Risk Management: Advanced predictive modelling tools predict environmental risks like deforestation, pollution, or climate-related impacts. Proactive management of actual risks means that emergency responses will not be needed, and for the most part, prevention measures are cheaper than the cure.

Lowered Costs in Stakeholder Collaboration: Digital platforms enable stakeholders to communicate and collaborate in real time, reducing the extent of physical meetings and travel. This reduces logistics costs while speeding up decision-making to implement conservation measures even faster [22].

LIMITATIONS AND CHALLENGES OF DIGITAL TECHNOLOGIES IN ENVIRONMENTAL MANAGEMENT

Data Privacy and Security Issues

Environmental data typically contains sensitive information about ecosystems, resources, and biodiversity. Sensitive ecological data is mismanaged, protected areas are harmed, and natural resources are exploited due to unauthorized access to this data. Cyberattacks on environmental management systems, particularly those connected with public or cloud-based platforms, can be accomplished. Hackers may interfere with monitoring systems, jeopardize the integrity of data, or release private information, with potentially serious effects on the environment and the economy [23]. For example, environmental data is regularly released into public view and to other agencies for accountability and transparency purposes. However, there could be an entanglement of balance between the protection of data and open access. Careful management and robust security protocols are required to make sure that only essential pieces of information are released and that other sensitive data is well protected. Many digital tools within environmental management have embedded user data, such as personal information from citizen scientists or app users. This data may thus be exploited or used for purposes other than environmental management, violating privacy rights and eroding public trust if it is not adequately protected. Building secure, resilient digital infrastructure often requires substantial investment. For smaller organizations or initiatives, these costs can be prohibitively expensive, leaving data less protected.

High Cost of Implementing Digital Technologies

Implementing digital technologies for environmental management involves a lot of financial challenges. The initial capital costs, such as buying equipment, drones, sensors, satellite imagery services, and data processing software, can be quite prohibitive, even for organizations operating in developing regions, as these organizations might find such tools unaffordable. More than the upfront capital cost, the running of digital infrastructure incurs running expenses. Maintenance, upgrades, and repairs of high-tech equipment frequently require supplementary financing as well as technical resources, often straining already limited budgets for environmental projects [24].

Further, processing the enormous volumes of environmental data collected from digital sources requires considerable expenses for data management and storage. Hence, data management and analysis across long terms can be expensive because of the recurring costs associated with high-capacity storage solutions, such as cloud services and large data storage systems that increase with data size. Because environmental solutions need to be all-inclusive, the sector often relies on integrating many technologies, including GIS systems, remote sensing, and IoT networks. The complexity and the cost of ensuring compatibility and integration across these different systems, however, increase the financial burden of environmental management's move towards a digital platform.

Digital Gap and Access to Technology in Developing Areas

There are a lot of challenges to be overcome when putting up digital environmental management systems in poor countries. The rollout of new technology is restricted by a lack of access to infrastructure, including reliable internet, electricity, and data storage facilities. In rural or isolated areas, where conservation efforts are frequently most crucial, this is especially true. Governments and organisations with limited resources have additional difficulties due to the high costs of purchasing, setting up, and maintaining digital technologies like satellites, drones, and Internet of Things sensors. This cost could restrict access to the new tools needed for efficient environmental monitoring and data collection if outside funding or partnerships are not available.

Lack of technical skills also prevents digital solutions from being widely adopted because developing nations frequently lack specialised knowledge in fields like data science, software engineering, and environmental informatics. Without these abilities, it is challenging to comprehend, manage, and fully utilise data-driven environmental solutions. When resources are limited, data inaccuracies and gaps frequently result in major restrictions that hinder sound decision-making and result in less precise forecasts [25].

The digital gap occurs because of social and economic barriers such as poverty and limited educational opportunities, which tend to restrict access to and the use of digital technologies for environmental protection. Local involvement and acceptance, in particular, are critical to ensuring that technology-based projects are implemented effectively; otherwise, communities may have difficulty garnering such support if faced with challenges such as these. Thirdly, for many regions, foreign finance and technology are relied upon for economic and infrastructure reasons, which creates sustainability concerns if and when external funding or support is reduced or phased out.

Limited understanding and skill gaps in using advanced technologies

The efficient application of cutting-edge digital tools like machine learning models, GIS, and IoT devices can be severely impacted by a lack of technical skills among environmental managers and stakeholders. The quality and efficacy of environmental management decision-making may suffer from underutilisation of available technology if the required skills are not there. The closure of such skill gaps calls for continuous training and capacity building, yet requires a huge financial and temporal investment. It proves to be a very challenging task for the companies that have less funding to find the money and time to train employees on new digital technology [23].

It requires expert understanding to grasp complicated information generated by new technologies, and incorrect interpretation or oversimplification of the data may lead to poor conservation strategies or mistaken environmental management decisions. These skills deficits lead many organisations to rely on outside consultants or third-party vendors to create and implement digital technology. The hazards associated with this dependence on outside expertise include the possibility of inefficiency and a misalignment with the organization's objectives if internal staff members are unable to comprehend or utilise the technology effectively.

Technological Aging and the Necessity for Continuous Updates

The high costs of frequent technology replacements and upgrades place a heavy load on environmental management agencies, which may have little money. Regular upgrades are necessary to stay current with evolving instruments, such as complex sensors, software, and data processing systems, in order to maintain efficacy. This can increase operational expenses and limit access to the most recent advancements. These frequent changes may also cause compatibility problems. Data continuity is necessary for trend analysis and long-term monitoring. This continuity may be broken by new technology or software that does not function well with existing systems. Compatibility problems could result in data gaps or necessitate additional resources to integrate old data into a new system [26].

Every technological advancement usually results in a rise in the need for training since environmental managers and analysts need to adjust to new instruments and procedures. A learning curve like this might slow down operations and take time and resources away from important tasks. E-waste, a by-product of digital technology that frequently ages overnight, is another major worry since it can cause environmental disruptions. For businesses that value conservation, the disposal of outdated sensors, drones, and other equipment can contaminate the environment, which is a paradox.

FUTURE DIRECTIONS IN DIGITAL TECHNOLOGIES FOR ENVIRONMENTAL MANAGEMENT

AI and IoT-Based Intelligent Environmental Systems

Systems that can automatically monitor, assess, and adapt to environmental changes are made possible by the marriage of AI and IoT. Automatic response and real-time environmental monitoring, where IoT sensors continuously collect data on soil health, water levels, and air quality, are among the most promising fields. It can be integrated with artificial intelligence to promptly analyze the data, issue alerts to the concerned authorities about pollution surges, and even take autonomous action, such as controlling water supply in sensitive drought-prone areas.

AI-IoT integration also makes precision environmental interventions possible; for instance, managers can focus on targeted areas, such as early indicators of tree disease in forests, so that a targeted solution with minimal environmental impact can be achieved.

IoT sensors integrated with AI image recognition help improve biodiversity monitoring by automatically counting numbers and recording behaviors of species without having to use invasive methods, thus supporting conservation efforts. Beyond tracking, AI's ability to learn from past data could enhance adaptive environmental management by allowing systems to change resource use in real time, such as smart grids that balance water or urban energy use based on conditions [27].

Future integration and interoperability of these data will make a comprehensive understanding of environmental aspects possible. AI can aggregate data from various sources, such as weather and air quality, in order to produce better-informed decisions. More comprehensively, applications empowered by IoT can contribute to citizen science and community participation, where users can contribute data, such as photographs of wildlife. These can be aggregated and analyzed by AI, encouraging the public to participate in conservation and spread knowledge.

The Development of Affordable and Scalable Digital Solutions

The development of low-cost and scalable solutions will spur the adoption of digital technology in environmental management to strengthen the impact and reach of cutting-edge instruments in the future. More NGOs and local governments will be able to access environmental management tools as a result of increased access to technology, thus enhancing grassroots conservation initiatives and fostering more participatory approaches in decision-making.

Because open-source software eliminates expensive license fees, it will encourage collaboration and knowledge transfer among environmental professionals, allowing them to tailor tools to suit localized needs. Increased widespread environmental monitoring will be enabled by the increasing adoption of low-cost sensors and IoT devices, especially in rural areas. In this way, biodiversity, climate factors, and air and water quality metrics will then become affordable to provide useful real-time information.

Cloud computing and data analytics will also make scalable solutions for data processing and storage possible, and businesses will be able to analyze environmental data in an effective manner without substantial infrastructure investments. Innovation collaborations involving academic institutions, environmental organizations, and technology companies can accelerate the uptake of efficient technologies by generating customized, cost-effective solutions for all types of environmental problems.

Strengthening Public-Private Partnerships in Digital Environmental Initiatives

Fortifying Public-Private Partnerships (PPPs) in digital environmental initiatives could significantly enhance the impact and efficiency of environmental management programs. One possible means is collaborative data sharing. These collaborations can improve data accessibility and transparency by setting up platforms where the public and private sectors can exchange environmental data, such as pollution levels and climate effect assessments. Consequently, people will make better decisions and have greater faith in environmental management initiatives [28].

Examining new funding models is another critical issue. PPPs can collaborate in a study of funding alternatives for digital environmental projects, such as social bonds and impact investing, by sharing risks and pooling resources to develop and make new environmental sustainability projects and technologies more available to a broader segment of the public.

It can also foster innovation in frontier technologies like blockchain, AI, and IoT for environmental monitoring and management through public-private collaborative research and development projects. Innovative solutions that can better cope with urgent environmental issues could be developed by tapping into the R&D capabilities of both sectors. Environmental protection and monitoring can also be promoted at the local community level through community engagement initiatives supported by PPP. People can make reports on shifts in biodiversity or changes in levels of pollution with the help of digital tools such as applications available in smart phones, which enhances the feeling of accountability and ownership while providing assistance for the collection of massive amounts of data.

Digital Engagement Through Public Participation: Future Directions

Digital engagement for environmental management will increasingly feature interactive platforms and innovative tools that foster public participation, education, and engagement. Interactive Web Platforms-Interactive web platforms play a crucial role, providing accessible websites and applications for real-time data sharing. With interactive maps and live monitoring systems, citizens can monitor and understand environmental data such as air quality, water levels, or biodiversity, engaging with the environment much more. In addition, publicity through social media campaigns can be strongly emphasized to share local conservation efforts, success stories, and sustainability practices. These campaigns reach a much wider population and motivate grassroots action and collective responsibility towards environmental protection.

Crowdsourcing data and citizen science projects also play an important role. It allows the public to contribute their observations through mobile apps. Reporting wildlife sightings and pollution incidents helps citizens contribute rich collaborative datasets that support traditional monitoring methods and empower communities in making decisions for environmental management. More importantly, gamification in digital platforms makes environmental education more interesting and contextual as well. These challenges, competition, and rewards will be motivating towards sustainable actions that encourage individuals and communities to be active for conservation.

Mobile applications focusing on sustainability can enable people to take informed, eco-friendly decisions. Features such as hints on reducing carbon footprints, reporting environmental hazards, or connecting users with local initiatives will support sustainability at the personal and community level. Moreover, feedback mechanisms on digital platforms should be included so that the general public can contribute their input on proposed environmental projects and have their perspectives shape the decisions and policies [29].

Open Data and Cooperative Environmental Monitoring

The other promising potential area for the application of digital technology in environmental management is open data and cooperative environmental monitoring. Probably the most important advantage of open data projects is that environmental data becomes more accessible, allowing it to be freely used among various stakeholders, including the public, NGOs, politicians, and researchers. Such openness not only promotes trust but also helps in making better decisions through the transparency it gives people access to the finest information available. Data gathering is also one of the most important components that will form growing environmental monitoring initiatives, which involves crowdsourced data collecting [29].

This approach expands the horizon of collecting data by using citizen power of science, in which volunteers collect data through Internet platforms and mobile applications, and it provides a range of perspectives in solving issues related to the environment. As networks of gathering data are stretched, the knowledge gained is richer and comprehensive. Open data promotes interdisciplinary collaboration, so that experts from other fields, such as ecology, economics, urban planning, or public health, can easily collaborate. This interaction of diverse expertise contributes to more holistic and innovative solutions, considering interconnections between environmental issues. Open data helps in easy real-time monitoring and reporting while ensuring access to the latest information by stakeholders; hence, it becomes possible to respond rapidly to emergencies or changes in the environment.

This can accelerate the development of new technologies or approaches to environmental issues and our ability to understand them. Moreover, open data promotes standardization and interoperability by promoting the use of common protocols for data gathering and sharing. This makes cross-platform integration of data relatively smooth and allows for comparisons between datasets that are more effective and powerful [30].

CONCLUSION

Digital technology has transformed environmental management into a more accurate, inclusive, and responsive decision-making process. For the first time in history, it is possible to monitor, analyze, and forecast environmental conditions using data analytics, IoT sensors, artificial intelligence, and digital communication platforms. Policymakers may make well-informed decisions quickly and flexibly thanks to real-time data and predictive insights, which promotes a change from reactive to proactive management. By employing data-driven insights to guide their decisions, policymakers may ensure that policies reflect the current state of the environment.

Comprehensive environmental assessments can be accomplished by using huge datasets on variables such as deforestation, water pollution, and air quality. These evaluations can be used to evaluate the long-term effects of regulatory measures, identify issues that require immediate attention, and track progress. This tactic is further enhanced by real-time monitoring, which enables quick responses to changes in pollution levels or temperature.

This foundation offers a policy that is flexible enough to be adjusted as the environmental situation changes. Examining the likely effects of various policy choices is made feasible by scenario modelling and predictive analysis using AI and simulation approaches.

By simulating the consequences of conservation measures, limiting emissions, or allocating resources, policymakers can choose policies that maximise environmental benefit. By visualising long-term outcomes, policymakers may give sustainable solutions high priority, lessen the risk of environmental harm, and improve resource conservation.

Environmental data is now more easily accessible and freely available thanks to digital platforms, which have also significantly improved public participation and transparency. Dashboards, interactive maps, and other visual aids facilitate public participation in policy-related conversations in addition to assisting communities in understanding environmental concerns. This transparency is closely related to trust and collective responsibility, both of which are prerequisites for widespread support and adherence to sustainable practices.

The efficiency, flexibility, and inclusivity of environmental management can be improved with the help of digital technologies. They help legislators create data-driven, flexible, and socially responsible programs to address current environmental problems and anticipate future needs. Governments and organisations can use these cutting-edge technologies to safeguard the environment, advance sustainable development, and raise the level of living for current and future generations.

REFERENCES

[1] Feroz AK, Zo H, Chiravuri A. Digital transformation and environmental sustainability: A review and research agenda. Sustainability (Basel) 2021; 13(3): 1530.
[http://dx.doi.org/10.3390/su13031530]

[2] Ahmad A, Alshurideh M, Al Kurdi B, Aburayya A, Hamadneh S. Digital transformation metrics: a conceptual view. J Manag Inform Dec Sci 2021; 24(7): 1-8. Available from: https://nchr.elsevierpure.com/en/publications/digital-transformation-metrics-a-conceptual-view/

[3] Goyal MK, Sharma A, Surampalli RY. Remote sensing and GIS applications in sustainability. In: Surampalli RY, Zhang TC, Goyal MK, Brar SK, Tyagi RD, eds. Sustainability: fundamentals and applications. Hoboken (NJ): John Wiley & Sons; 2020. p. 605–26.
[http://dx.doi.org/10.1002/9781119434016.ch28]

[4] Ye Z, Yang J, Zhong N, Tu X, Jia J, Wang J. Tackling environmental challenges in pollution controls using artificial intelligence: A review. Sci Total Environ 2020; 699: 134279.

[http://dx.doi.org/10.1016/j.scitotenv.2019.134279] [PMID: 33736193]

[5] Anuradha Yadav, Shivani Shivani, Vikramaditya Sangwan, Demkiv A, Demkiv A. Blockchain technology for ecological and environmental applications. Ecol Quest 2024; 35(4): 1-20.
[http://dx.doi.org/10.12775/EQ.2024.050]

[6] Nieves V, Radin C, Camps-Valls G. Predicting regional coastal sea level changes with machine learning. Sci Rep 2021; 11(1): 7650.
[http://dx.doi.org/10.1038/s41598-021-87460-z] [PMID: 33828225]

[7] Arvidsson R, Tillman AM, Sandén BA, et al. Environmental assessment of emerging technologies: recommendations for prospective LCA. J Ind Ecol 2018; 22(6): 1286-94.
[http://dx.doi.org/10.1111/jiec.12690]

[8] Amsel N, Ibrahim Z, Malik A, Tomlinson B. Toward sustainable software engineering (nier track). In: Proceedings of the 33rd International Conference on Software Engineering. Piscataway (NJ): IEEE; 2011. p. 976-9.

[9] Dwivedi YK, Hughes L, Kar AK, et al. Climate change and COP26: Are digital technologies and information management part of the problem or the solution? An editorial reflection and call to action. Int J Inf Manage 2022; 63: 102456.
[http://dx.doi.org/10.1016/j.ijinfomgt.2021.102456]

[10] Zhu X. GIS for environmental applications: a practical approach. 1st ed. London: Routledge; 2016.
[http://dx.doi.org/10.4324/9780203383124]

[11] Melesse AM, Weng Q, Thenkabail PS, Senay GB. Remote sensing sensors and applications in environmental resources mapping and modelling. Sensors (Basel) 2007; 7(12): 3209-41.
[http://dx.doi.org/10.3390/s7123209] [PMID: 28903290]

[12] Wu C, Raghavendra R, Gupta U, et al. Sustainable AI: environmental implications, challenges and opportunities. arXiv [Preprint]. 2021 Nov 1 [cited 2025 Dec 19]: arXiv:2111.00364.
[http://dx.doi.org/10.48550/arXiv.2111.00364]

[13] Lucivero F. Big data, big waste? A reflection on the environmental sustainability of big data initiatives. Sci Eng Ethics 2020; 26(2): 1009-30.
[http://dx.doi.org/10.1007/s11948-019-00171-7] [PMID: 31893331]

[14] Chen X, Kurdve M, Johansson B, Despeisse M. Enabling the twin transitions: Digital technologies support environmental sustainability through lean principles. Sustain Prod Consum 2023; 38: 13-27.
[http://dx.doi.org/10.1016/j.spc.2023.03.020]

[15] Arts K, van der Wal R, Adams WM. Digital technology and the conservation of nature. Ambio 2015; 44(S4) (Suppl. 4): 661-73.
[http://dx.doi.org/10.1007/s13280-015-0705-1] [PMID: 26508352]

[16] Finger R. Digital innovations for sustainable and resilient agricultural systems. Eur Rev Agric Econ 2023; 50(4): 1277-309.
[http://dx.doi.org/10.1093/erae/jbad021]

[17] Katie B. Internet of Things (IoT) for Environmental Monitoring. Intl J Comp Eng 2024; 6(3): 29-42.
[http://dx.doi.org/10.47941/ijce.2139]

[18] Spellman FR. Mathematics manual for water and wastewater treatment plant operators: basic mathematics for water and wastewater operators. CRC Press 2014 May 7.
[http://dx.doi.org/10.1201/b17023]

[19] Bashmakov IA, Nilsson LJ, Acquaye A, et al. Industry. In: Shukla PR, Skea J, Slade R, et al., eds. Climate change 2022: mitigation of climate change contribution of Working Group III to the Sixth Assessment Report of the Intergovernmental Panel on Climate Change. Cambridge (UK): Cambridge University Press; 2023. p. 1161–256.
[http://dx.doi.org/10.1017/9781009157926.013]

[20] Mörth O, Eder M, Holzegger L, Ramsauer C. IoT-based monitoring of environmental conditions to improve the production performance. Procedia Manuf 2020; 45: 283-8.
[http://dx.doi.org/10.1016/j.promfg.2020.04.018]

[21] Ramadan MNA, Ali MAH, Khoo SY, Alkhedher M, Alherbawi M. Real-time IoT-powered AI system for monitoring and forecasting of air pollution in industrial environment. Ecotoxicol Environ Saf 2024; 283: 116856.
[http://dx.doi.org/10.1016/j.ecoenv.2024.116856] [PMID: 39151373]

[22] Zakari A, Khan I, Tan D, Alvarado R, Dagar V. Energy efficiency and sustainable development goals (SDGs). Energy 2022; 239: 122365.
[http://dx.doi.org/10.1016/j.energy.2021.122365]

[23] Kritika. The role of transformational leadership in addressing complex challenges in the cybersecurity landscape. Int J Sci Res Comp Sci Eng. 2025 Aug; 13(4): 28–40.

[24] Cybersecurity CI. Framework for improving critical infrastructure cybersecurity 2018; (7): https://nvlpubs.nist. Gov/nistpubs/CSWP/NIST

[25] World Bank Group World development report 2016: Digital dividends. World Bank Publications. 2016 Jan 14.

[26] National Research Council, Global Affairs, Technology for Sustainability Program, Committee on Incorporating Sustainability in the US Environmental Protection Agency Sustainability and the US EPA. National Academies Press. 2011 Oct 8.

[27] Popescu SM, Mansoor S, Wani OA, *et al.* Artificial intelligence and IoT driven technologies for environmental pollution monitoring and management. Front Environ Sci 2024; 12: 1336088.
[http://dx.doi.org/10.3389/fenvs.2024.1336088]

[28] van Driel M, Biermann F, Kim RE, Vijge MJ. The UN regional commissions as orchestrators for the sustainable development goals. Global Governance: A Review of Multilateralism and International Organizations 2023 Dec 21; 29(4): 561-90.
[http://dx.doi.org/10.1163/19426720-02904006]

[29] Dubey R, Gunasekaran A, Childe SJ, *et al.* Can big data and predictive analytics improve social and environmental sustainability? Technol Forecast Soc Change 2019; 144: 534-45.
[http://dx.doi.org/10.1016/j.techfore.2017.06.020]

[30] Ehrich LN. Cooperative vehicle environmental monitoring. In: T.B Curtain. ed. Springer Handbook of Ocean Engineering. Cham: Springer International Publishing; 2016. p. 441-58.

CHAPTER 10

Sustainable Development Goals of Health and Environment and Current Status of India with Measurement Strategies for Future

Rajal Dave[1] **and Abhijeet Joshi**[1,*]

[1] *Department of Microbiology, Faculty of Science, Atmiya University, Rajkot, Gujarat, India*

Abstract: Sustainable Development Goals (SDGs) represent a comprehensive strategy aimed at transforming society, public health, and various sectors, such as the economy, technology, healthcare, and business, by enhancing resources and investments. Established by the United Nations in 2015, the SDGs aim to create a better and more sustainable life by addressing each dimension through distinct approaches encapsulated in 17 different goals. Progress towards any of these goals through strategic implementation is ultimately reflected in Goal 3, which focuses on improving public health. The Sustainable Development Goals (SDGs) are applicable to many countries, with 17 goals forming a unique and broad range that are interrelated. This chapter summarizes the current report on India, highlighting every key feature of the goals related to health and the environment and discussing their interconnections. Furthermore, the chapter emphasizes Goal 3, which focuses on good health and well-being, by detailing stakeholder initiatives and related indicators. We also discuss the challenges and measurement strategies that help align with the targets for achieving good health and well-being. Additionally, this chapter provides insights into various SDGs and the current situation in India.

Keywords : Current status of india, Good health and wellbeing, Measurement strategies, Sustainable environment, Sustainable development goals (SDGs).

INTRODUCTION

The first idea of the Sustainable Development Goals (SDGs) originated from the Brundtland Commission's 1987 report on the UN system. The Brundtland Commission is also known as the World Commission on Environment and Development [1 - 5]. The concept of SDGs was highlighted by the United Nations Conference on Environment and Development (UNCED), also known as the Earth Summit, which was held in 1992 [4, 5]. The decision to re-establish Sus-

---

[*] **Corresponding author Abhijeet Joshi:** Department of Microbiology, Faculty of Science, Atmiya University, Rajkot, Gujarat, India; E-mail abhijeet.joshi@atmiyauni.ac.in.

Rajneesh Kumar, Ram Sharan Singh & Maulin P. Shah. (Eds.)
All rights reserved-© 2026 Bentham Science Publishers

tainable Development Goals (SDGs), adopted on September 25, 2015, has opened up many opportunities and hopes for a healthy future with the principle of "leaving no one behind" to protect the integrity of the planet by 2030 [1 - 6].

SDGs have created well-defined policies for every government sector and have been implemented in several countries. The structure of the Sustainable Development Goals (SDGs) is a unique phase of public benefit at the global level. This is the first time that defined policies have been universally launched across all sectors, regardless of the income of the countries. This aspiring and visionary change has made all countries responsible for their achievements [2, 3].

Understanding the interconnections between the planet, environment, and human life is critical, as some environmental challenges are associated with human growth and survival. The basic Earth environment processes include seasonal climate change, ozone depletion, aerosol-related air pollution, freshwater misuse, biochemical imbalances, ocean acidification, land transformation, and biodiversity loss.

Around 2015, the majority of these processes began to deteriorate, ultimately leading to the loss of biodiversity integrity. Activities such as the incorporation of nitrogen and phosphorus into crops, deforestation, and changes in land-use systems have hampered ecosystems. Through rigorous analysis, the government has designed 17 unique goals that cover 169 different targets within the 2030 baseline agenda [1]. After designing the SDG indicators, health was found to be an environmental factor associated with most sectors. The basic principles of all SDGs are interlinked, including social, economic, and environmental characteristics [2, 4, 5].

The scope of SDG application extends to every developed and developing country with the intention of building a better world. The current trend of each SDG indicator reveals that some have invigorated the development of policies (Fig. **1**, Table **1**), whereas others still face challenges between strategy and implementation [2, 3, 5]. Specifically, regarding the environment and health, allied SDG indicators require more development and vigorous effort.

Globally, especially in India, there is a strong inference that we must identify and accept ecosystem disruption, which inherently reduces the potential benefits that ecosystems provide, including increases in disease, changes in crop production, declining water levels, increased waste, and disruption of the nutrient cycle. The world is interconnected with the environment and ecology. Any change or damage may ultimately affect health, such as the migration of families due to food insecurity.

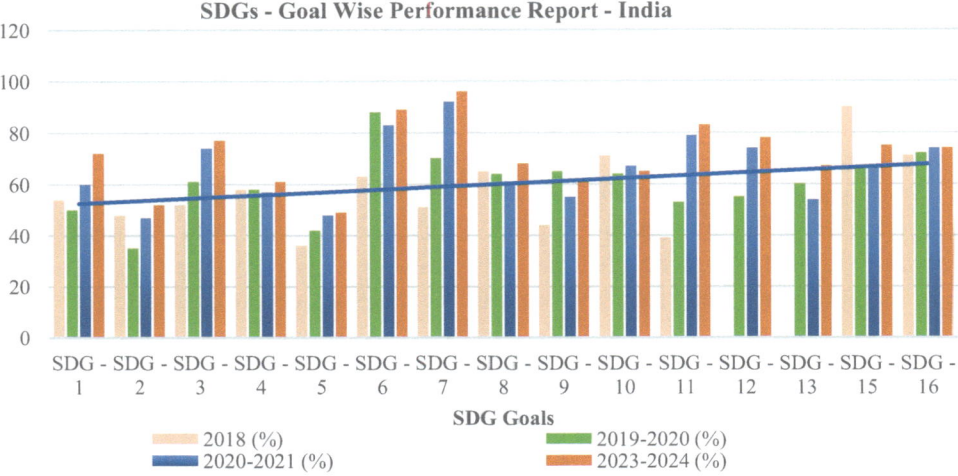

Fig. (1). SDGs – Goalwise performance report – India.

Table 1. Framework and current status of SDGs, which are interlinked for health and environment [2, 3, 29].

	SDG 3: Good Health and Well-being			
Indicator	Description	Result	Target	Achieved States
Maternal Mortality Rate	Women aged 15–49 who die due to pregnancy-related causes within 42 days of termination, per 100,000 live births.	102.7	3.4	Delhi, Gujarat
Neonatal Mortality Rate	Neonates who die within the first 28 days of life, per 1,000 live births.	18.13	1.1	—
Under-Five Mortality Rate	Probability that a newborn dies before age 5, per 1,000 live births.	29.07	2.6	—
Tuberculosis Incidence	New and relapsed TB cases per 100,000 population, including HIV-positive individuals.	199	0	—
New HIV Infections	Newly infected individuals per 1,000 uninfected population.	0.05	0	—

(Table 1) cont.....

NCD Mortality (Age 30–70)	Deaths due to CVD, cancer, diabetes, or respiratory disease between the ages of 30 and 70.	21.85	9.3	—
Air Pollution–Related Mortality	Deaths due to household and ambient air pollution.	139	0	—
Road Traffic Death Rate	Deaths due to road traffic injuries per 100,000 population.	14.6	3.2	—
Life Expectancy at Birth	The average years a newborn is expected to live.	67.24	83	—
Adolescent Fertility Rate	Births per 1,000 females aged 15–19.	11.3	—	—
Skilled Birth Attendance	Births attended by trained healthcare personnel.	89.4	100	—
Infant Immunization Coverage	Infants receiving measles and DTP vaccines.	95	100	—
Universal Health Coverage Index	Coverage of essential health services.	63	100	—
Subjective Well-being	Self-reported happiness score (0–10 scale).	4.68	7.6	—
	SDG 4: Quality Education			
Pre-primary Participation Rate	Estimated as the net enrollment rate in organized learning programs prior to primary education.	91.39 (Target Achieved)	100	—
Literacy Rate	Estimated as the percentage of individuals with basic reading and writing skills in daily life.	96.54 (Target Achieved)	100	Delhi, Kerala
	SDG 6: Clean Water and Sanitation			
Access to Basic Drinking Water	Estimated as the percentage of the population using improved water sources with collection time under 30 minutes.	93.30 (Challenging)	100	—

(Table 1) cont.....

Access to Basic Sanitation	Estimated as the percentage of the population using improved sanitation facilities not shared with other households.	78.39 (Challenging)	100	Andaman & Nicobar Islands
Access to Piped Water	The urban population with piped water is protected from contamination.	65.56	100	—
7 AFFORDABLE AND CLEAN ENERGY	SDG 7: Affordable and Clean Energy			
Access to Electricity	Estimated as the percentage of the population with access to electricity.	99.57 (Target Achieved)	100	Multiple states listed
Clean Cooking Fuel Access	Estimated as the percentage of the population using clean cooking fuels (excluding kerosene).	71.10 (Challenging)	100	—
CO_2 Emissions per Electricity Output	Estimated as the carbon intensity of electricity production, measured in megatonnes of CO_2 per total output.	1.61	0	—
Renewable Energy Share	Estimated as the percentage of renewable energy in total final energy consumption.	17.70 (Challenging)	55	—
11 SUSTAINABLE CITIES AND COMMUNITIES	SDG 11: Sustainable Cities and Communities			
Access to Public Transport	Population with access to transport within 500m of key locations.	69.85	100	—
Urban Slum Population	Urban population living in slum conditions.	49.01	0	Chandigarh, Maharashtra
13 CLIMATE ACTION / 15 LIFE ON LAND	SDG 13 and 15: Climate Action and Life on Land			

(Table 1) cont.....

PM2.5 Concentration	Annual mean PM2.5 in urban areas.	50	6.3	—
Deforestation Rate	Permanent forest removal for development.	0.04	0	Tripura, Jharkhand, Daman and Diu
Biodiversity Protection (Marine)	Protected marine areas for biodiversity.	4.23	100	West Bengal
Biodiversity Protection (Terrestrial)	Protected land areas for biodiversity.	6.32	100	Tripura, Jharkhand
Red List Index	Survival rate of endangered species.	0.67	1	—

Additionally, there are many challenges to human health and the environment, including increases in water, air, and noise pollution and exposure to toxic chemicals. According to the World Health Organization (WHO), approximately 1.4 million deaths occur annually due to environmental factors. These included heart disease (26%), cancer (17%), and stroke (25%). Most deaths were attributed to indoor and outdoor air pollution, which also contributes to non-communicable diseases, accounting for 20% of the total deaths [7].

Hence, five to ten years of SDG progress data reflect that all indicators are directly or indirectly linked to health and the environment of the country. All SDGs are closely related to both. Therefore, this chapter discusses the current report in terms of policies related to health and the environment, measurement strategies, and action plans. This highlights the areas where vigorous strategic planning is required to achieve these goals. Moreover, this chapter delves into the current status and the measurement strategies for each indicator. The final section concludes with the role of the health sector and communities in leading actions to accomplish targets on time.

SUSTAINABLE DEVELOPMENT GOALS (SDGS) – STATUS OF INDIA

NITI Aayog serves as India's ideal functional model. This leads to policy design, promotes competition and innovation among states, and stimulates cooperative federalism. India has made significant progress in achieving targets such as poverty reduction, economic growth, decent work, climate action, and life on land (Fig. **2**) [8].

Fig. (2). Interlinking of SDG 3 with other SDGs [27].

India has adopted a metadata index score-based system that compiles raw data from selected indicators, identifies errors, and converts them into normalized scores. The metadata value represents the average score across all goals (Table **1**) [7 - 11].

SUSTAINABLE DEVELOPMENT GOALS RELATED TO HEALTH

Good health is fundamental to achieving all SDGs. It supports freedom from poverty and hunger, access to education and employment, gender equality, and safe environments. To meet the 2030 Agenda, the United Nations Millennium Project has targeted the highest causes of disease and death, including child and maternal mortality and communicable diseases [1, 12 - 14].

Fig. (**2**) illustrates how SDG 3 (Good Health and Well-Being) is interlinked with other goals [28]. While the Millennium Development Goals (MDGs) focus on four pillars—child and maternal mortality (SDG 3), social equity (SDGs 4, 5, 6, 16, 17), environmental health (SDGs 2, 6, 7, 11–15), and collective welfare (SDGs 1, 8, 9)—the SDGs integrate and expand upon these to create a more holistic framework [14].

Health is defined as a state of complete physical, mental, and social well-being and not merely the absence of disease [13 - 15].

In contrast to SDGs, Millennium Development Goals (MDGs) focus on four pillar sectors: child and maternal mortality rate (SDGs 3) through social equality (SDGs 4, 5, 6, 16, 17), providing a healthy environment (SDGs -2,6,7,11,12,13,14,15), and collective welfare (SDGs – 1,8,9) [13]. Therefore, the MDGs and SDGs are integrated to achieve a better life. Health refers to overall physical, social, and mental well-being [13 - 15].

Risk Factors Associated with Health and Well-being

Nutritional Risk

Nutrition is a critical determinant of good health and well-being. The nutritional risk factors are listed in Table **2**. It is closely tied to income, which depends on access to quality education. Affordability influences nutritional status, and malnutrition can often lead to anemia. India continues to face challenges in meeting its nutritional targets, particularly in reducing anemia in women. More than 53% of Indian women aged 15–49 years are affected, and data on low birth weight are lacking.

Table 2. Risk factors responsible for illness & death [28].

Nutritional Risk Factor	Anaemia among women	Women aged 15-49 years. (SDG – 2.2.3)
-	Trans fatty acids (TFA)	Best practices policies are 2 gm of industrially produced food per 100 gm of total fat in food.
Behavioural risk factor	Tobacco Use and Tobacco Control	As per the report of 2022, the male-to-female ratio is higher than that of women. Affects the SDG 3.a.1.
	Alcohol consumption	SDGs indicator 3.5.2. As per the 2020 report, global consumption is estimated at 4.9 litres.
	Metabolic Risk Factors	Hypertension, overweight, and obesity
	Environmental Risk Factors	WASH, Safety from chemicals, clean (Pure) Air, a Healthy environment, Good agricultural practices
		Protection of Natural Areas and Water Sources
		Air pollution, Related SDG 7.1.2
	Risk to women's and girls' health	Adolescent pregnancy
		Violence against women

India has made progress in exclusive breastfeeding of infants aged 0–5 months old. Good nutrition is essential for maternal and infant health and forms the foundation of a strong immune system. It reduces the risk of noncommunicable diseases (NCDs), anemia, and other health complications. Globally, the burden of malnutrition, overweight, and obesity is increasing. Women with moderate-to-severe anemia often experience fatigue, cognitive impairment, and reduced physical capacities.

According to the latest WHO report, the target of reducing anemia in women aged 15–49 by 50% is unlikely to be achieved [7]. Additionally, high intake of trans fatty acids (TFA) is linked to heart attacks and coronary heart disease. The WHO recommends limiting industrially produced TFA to no more than 2 g per 100 g of total fat in food products [28].

Behaviour Risk Factor

Some factors are responsible for modifiable behaviors, such as tobacco, alcohol, and physical immobility. The behavioral risk factors are presented in Table **2**. All of these diseases are associated with non-communicable diseases. The Global Action Plan for the Prevention and Control of NCDs targets for 2013-2020, includes the global prevalence of tobacco use. In 2022, the prevalence of tobacco consumption among individuals aged 15 years and above was estimated at a rate of 20.9%, whereas in 2010, the rate was 26.5%. If the same trend is pursued, there is a possibility of a reduction of more than 25% by the end of 2025. Tobacco consumption differs remarkably between sexes. The prevalence rate in women was lower than that in men in 2022. MPOWER monitors tobacco use and prevention policies, protects people from tobacco smoke, offers help in quitting tobacco use, warns about the dangers of tobacco, enforces bans on tobacco advertising and sponsorship, and raises taxes on tobacco.

It is a technical tool of the WHO for measuring the reduction in tobacco consumption, as per the WHO Work Plan Convention on Tobacco Control. An ideal example is the Netherlands (Kingdom of the), which has adapted the WHO work plan for tobacco control since 2008. The estimated rate of tobacco consumption (age standardized at 15 years and above) has reduced to 34.5% in 2000, 27.7% in 2010, and 21.3% in 2022. This would highlight 28% of the overall reduction from 2010 to 2025 and be very close to the target of 30% reduction determined by the Global Action Plan for the Prevention and Control of Non-Communicable Diseases (NCDs). Alcohol consumption was described under SDG 3.5.2, which is the total record and without record APC per litre per year, which was decided for tourists and persons aged 15 years and older. Globally, the total APC was 5.7 litres in 2010, while 4.5% declined to 5.5 litres in 2019, and is expected to reach a 10% reduction by the end of 2025 and more than 20% reduction by 2030 [1, 16]. Compared to all other regions of the country, Southeast Asia has shown a continuous increase in the consumption of these products. The metabolic risk factors responsible for NCDs are being overweight, obesity, hypertension, high blood glucose levels, and high levels of fat (TFA) in the blood [7]. The prevalence of hypertension was standardized between 30-79-year-olds who were taking hypertension medications [7]. It has been observed that in the last few decades, the prevalence rate has decreased in European countries. Since 2015, the number of people receiving basic and proper hygiene services has increased to more than one billion, and more improvements have been observed in rural areas between 2015 and 2022 [7].

Mortality from Air Pollution

Environments possess both biotic factors, including living organisms, and abiotic factors, such as environmental layers. According to the United States Environmental Agency, major pollutants are Particulate Matter (PM), Ground-Level Ozone (GLO), lead, carbon monoxide (CO), nitrous oxide (NO), and polycyclic hydrocarbons (PAH). To improve indoor and outdoor air quality, advanced technologies such as nano/micro particle filtration, nano aerosol filtration units, and heating, ventilation, and air conditioning (HVAC) systems have been developed for the control of indoor air pollution [17 - 19].

Measurement Strategies for Infectious Diseases and Non-communicable Diseases

SDG target 3 aims to end tuberculosis, HIV, malaria, Neglected Tropical diseases (NTDs), and hepatitis epidemics. Along with all diseases, SDG targets polio and controls antimicrobial resistance. In 2022, approximately 39.00 million people globally were living with HIV, of which 37.50 million were more than 15 years old and 1.5% were under the age of 15 years. In 2022, 1.3 million new cases of HIV out of 100000 population. This is a reduction compared to the 2.8 million in 2000. In India, the rate of new case development was reduced to 31%, suggesting that we are in the stage of improvement to achieve the target (Table **3**).

Apart from that, there are many other points to consider for better achievement,

Universal Health Coverage (UHC) is key to achieving SDGs and strengthening the health sector by scaling up coverage services with more targets for Primary Health Care (PHC) [4, 20].

Table 3. Diseases that are covered under SDGs and strategies for control [27].

Diseases	Strategies for improvement
HIV/AIDS	1. Increase the HIV testing, including antenatal care testing, to avoid transmission from mother to child. 2. More accessibility for Antiretroviral therapy (ART). 3. Promotion of Pre-exposure prophylaxis programs 4. Increase engagement and partnerships 5. Increase the public awareness programs
Tuberculosis (TB)	1. Promotes the early detection and diagnosis programs 2. Increases rapid testing centres for fast treatment 3. Treatment support with assurance of Directly Observed Treatment Short Course (DOTS) 4. Promotes a health coverage scheme for BCG immunization 5. Improves awareness, sanitation, and hygiene.
Malaria	1. Control of vectors by Long Lasting Insecticidal Nets. 2. Larval control & Management 3. Increase the early detection testing 4. Increases the awareness and surveillance. 5. Monitoring data for drug resistance to malaria, possibility to research on the vector change of their dynamics for survival.
Neglected Tropical Disease (NTDs)	1. Promote mass drug administration of albendazole and diethylcarbamazine to control Lymphatic Filariasis (LF) 2. Insecticides contained net to control Kala Azar. 3. Multiple drug therapy for Leprosy 4. Development of new drugs and vaccines
Hepatitis	1. Mandatory action for immunization against hepatitis for newborns. 2. Compulsory testing of blood to diagnose hepatitis B &hepatitis C in health care workers as a screening test, and the same for blood donors. 3. Promote the public awareness program related to the use of only sterilized needles. 4. Increase drugs and rapid diagnosis testing.
Polio	1. Mass immunization campaign, surveillance, and monitoring of data.
Antimicrobial Resistance	1. Good hand hygiene facility, Infection prevention and control program, use of antibiotics rationally, adopt the One Health approach, avoid common use of antibiotics between animal, human, and agriculture systems. Proper diagnosis and research on the development of new potential antimicrobial therapeutic agents [30].

Increase manpower, recruitment, training programs, and Development [12].

Improvements in health techniques, access to medicines, development of drug delivery systems through nanotechnology, delivery of drugs through nanoparticles, and conjugation minimize the side effects of drugs, especially in cancer (Fig. 3) [12, 21].

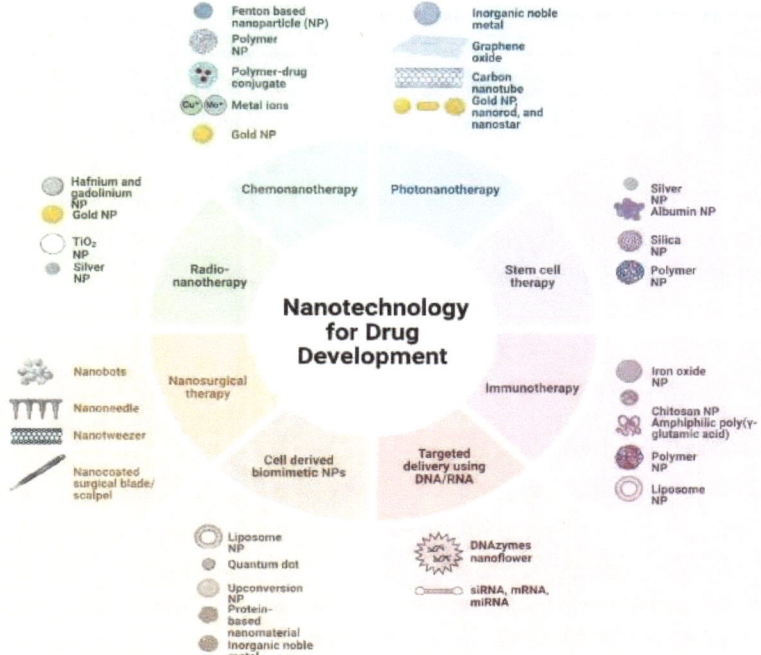

Fig. (3). Application of nanotechnology for drug and vaccine development.

Implementing health regulations to encourage resilience [12].

Arranging health finances by increasing resource mobilization and financial security [12].

SUSTAINABLE DEVELOPMENT GOALS FOR SUSTAINABLE ENVIRONMENT

A sustainable environment addresses the interdisciplinary challenges of nature. It secures the health, well-being, and integrity of ecosystems, including biodiversity, wetlands, forests, and oceans. Major issues include water scarcity, mismanagement of waste generated from households, industries, agriculture, poverty, and climate change. Any imbalance that directly impacts health; therefore, all sectors are interrelated. A sustainable environment provides a framework for addressing all issues in a coordinated approach by fostering a work plan among stakeholders of government, business organizations, institutes, and individuals. Preservation of ecosystem integrity through the restoration of sustainable practices. There is also an interlinking between SDGs and the environment (Table **4**).

Table 4. Framework and measurement strategies of SDGs that are interlinked for the Environment.

SDGs	Framework	Measurement Strategies
SDGs 6 – Clean Water & Sanitation	This focuses on the security of the supply of clean water for every household.	1. Development of the infrastructure of water, 2. Promotion for sustainable water management techniques, improvement of sanitation, 3. Increase awareness about hygiene and the development of water-efficient technology, such as nanotechnology for wastewater treatment, and the use of biological and green technology for solid waste management.
SDG 7 – Affordable and Clean Energy	It focuses on securing the provision of affordable and sustainable energy for all. It targets renewable energy sources, potential techniques that can reduce greenhouse gases, and are compatible with climate actions.	1. High-scale solar energy production farm 2. Biomass energy 3. Hydroelectric power 4. Increase the policies and support from regulations – Portfolio for renewable standard, Tax credit for clean energy users. 5. Develop technology – Green Hydrogen production 6. Geothermal energy
SDG 11 - Sustainable cities and communities	It focuses on cities' development in terms of waste management, infrastructure, Urbanization, and control of air pollution. Promotion of green energy access.	1. Promotion of green infrastructure, green roof, develop rain gardens, green walls. 2. Plan the compact cities, develop major nodes, *i.e.*, shops, hotels, *etc.*, near railway stations that can reduce traffic, pollution, and energy consumption.
Goal 12 – Responsible utilization and production of goods	It focuses on sustainable development in production, consumption, and waste reduction.	1. Reduce the food waste 2. More use of disposable items like paper cups reduces plastic utilization 3. Reduce the use of paper and promote online tickets and passes 4. Increase the donation thing that can circulate in society.
Goal 13 – Climate Issues	It focuses on urgent actions from climate disasters, reducing the emission of greenhouse gases. Promotion of precautionary measures related to climate disasters	1. Climate change adaptation 2. Promotion of renewable energy sources 3. Design the infrastructure with better insulation 4. Storage of carbon 5. Sustainable agriculture practices

(Table 4) cont.....

SDGs	Framework	Measurement Strategies
Goal 14 – Life Below Water Goal 15 – Life on Land	It focuses on ecosystem conservation by reducing marine pollution, deforestation, overfishing, and loss of habitat.	1. Promotion of beach clean-up activity 2. No use of plastic at the beach 3. Promotion of sustainable practice of fishing 4. Conservation of marine coastal areas. 5. Development of water harvesting technology 6. Species conservation
Goals 17 – Partnership for the Goals	It focuses on strengthening the means of implementation and revitalizing the global partnership for sustainable development, emphasizing collaboration between governments, private sector, and civil society to mobilize finance, technology, capacity-building, trade, and data for achieving all 17 Sustainable Development Goals (SDGs) by 2030.	1. Identify the good partners 2. Make a strong relationship with business or deal partners, NGOs, and individual institutes, and promote public-private partnerships. 3. Promote the knowledge-sharing programs. 4. Strong Fund building 5. Promote the SDGs-related partnerships

Strategies for Sustainable Environment

To develop a sustainable environment, the major areas that should be covered are agricultural systems, bioremediation of environments, and the possibility of enhancing microorganism-mediated solid waste management technology and wastewater, specifically industrial waste effluent treatment using nanotechnology.

Practice of Organic farming. Variables such as location, stakeholders, crop type, machinery, storage, pesticides, fertilizers, agrochemicals, cultivation process, transportation, and economic viability.

The practice of genetic engineering, plant omics for high yield, and climate-resilient crop production [22, 23].

Water management techniques include the following: To adapt green technology, which minimizes the use of chemicals and high-energy-consuming instruments to treat effluents, and to develop eco-friendly technologies, such as microbes mediated by a combination of nanotechnology.

Bioremediation processes for pollutants and contamination are used to remove them from the environment.

Promote the Genome Editing technology, Organic Farming for high yields, and climate-resilient crop production for a Sustainable Environment

Good agricultural practices can improve health, reduce environmental pollution, decrease poverty, and protect the ecosystem. Sustainable agricultural practices and improvements in overall SDGs related to health and the environment.

Compared to conventional farming, the adaptation of organic farming techniques can increase nutritional value, improve overall yield and soil quality, maintain and increase biodiversity, reduce water pollution, and increase productivity. Organic farming can increase the chances of employment, thus improving the rate of SDGs related to poverty (Fig. 4). Sudden climate change can affect plant growth and reduce crop production, thereby leading to increased poverty and reduced financial security. The application of plant genome engineering by editing target genes during seed germination can increase the nutritional value of crops and mitigate climate-related actions (Fig. 5). Recent genome techniques combined with artificial intelligence can overcome many environmental and biological limitations, such as salinity and heat, decrease the rate of pathogens, and increase the rate of nitrifying and phosphorus-solubilizing bacteria in soil through fertilizer [17, 18].

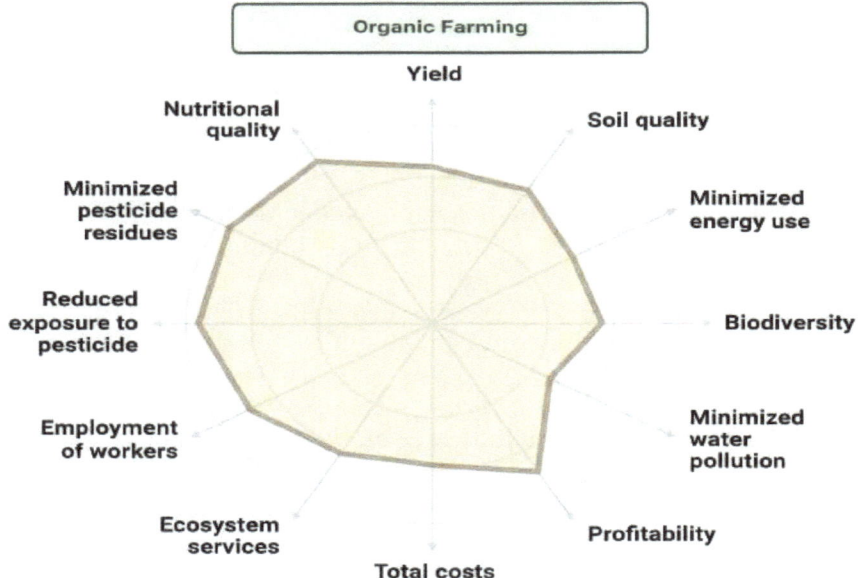

Fig. (4). Advantages of organic farming [9].

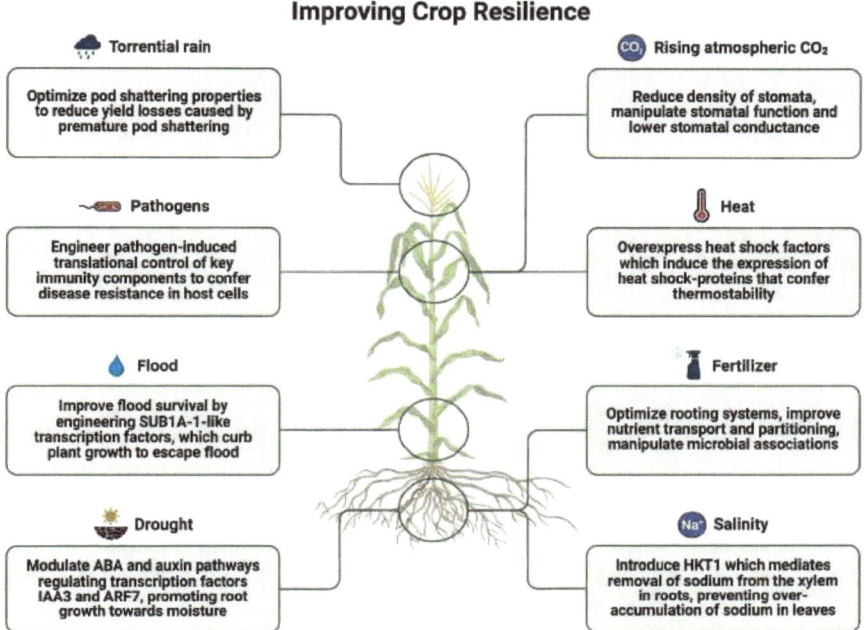

Fig. (5). Genome editing technology for climate-resilient crop production [10].

Promote the Application of Biopesticides Compared to the use of Conventional Chemical Pesticides

Compared to conventional chemical pesticides, the application of biological pesticides can maintain soil diversity and ecosystems. Biopesticides can also promote agricultural systems towards a sustainable, green, and eco-friendly approach that ultimately leads to a sustainable environment (Fig. **6**).

Bioremediation Technology for the Removal of Pollutants from the Environment

Rapid industrialization and urbanization have led to a rapid increase in waste generated by human activities. It is necessary to remove toxins from the environment through sustainable and eco-friendly techniques such as "bioremediation." Word remediation refers to the treatment of things and making them safe and free from danger. This process uses a biological approach by applying microorganisms, such as bacteria, fungi, and actinomycetes, for various bioremediation processes. Bioremediation is an advanced technology that involves degradation, detoxification, immobilization, and eradication of pollutants. All of these techniques work on the basic principle of toxic waste removal and make it safer through biostimulation, bioventing, bio-attenuation, and bio-piles. This process uses microorganisms to eliminate toxic pollutants, such as heavy metals, plastics, chemical-based fertilizers, herbicides, insecticides, hydrocarbons, and greenhouse gases, from waste.

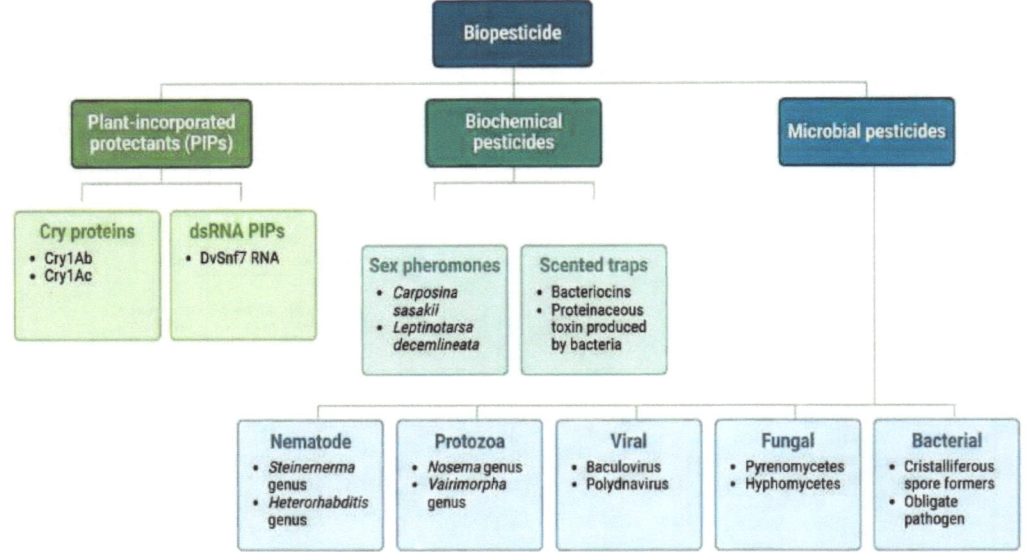

Fig. (6). Types of biopesticides.

The products of bioremediation can be combined with advanced technologies, such as chromatography and mass spectrophotometry, to identify intermediate metabolic products, characterize metabolites, and identify potent microorganisms with the highest degradation capacity (Fig. 7) [24 - 29].

Achievements of India Towards Environmental Sustainability

India jointly founded the Solar Alliance collaboration with France to promote the use of renewable energy. Swachh Bharat Abhiyan targeted solid waste management and sanitation [24]. To preserve natural resources, the government initiated the Jal Shakti Abhiyan, which targets the conservation of water, groundwater recharge, and rainwater harvesting [24].

MoHUA - Ministry of Housing and Urban Affairs has launched the project AMRUT-Atal Mission for Rejuvenation and Urban Transformation for the management of sewage and septage [24].

To promote organic farming and soil health management, the ovt. Krishi Vigyan Kendra (KVK) and Pramparagat Krishi Vikas Yogna (PKVY) [24].

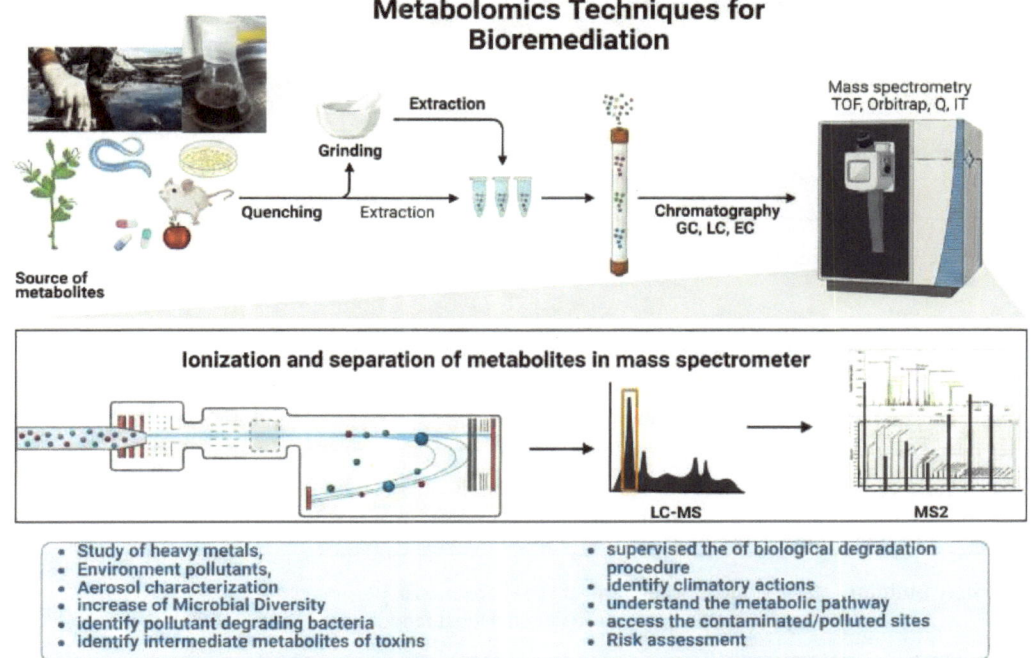

Fig. (7). Metabolomics technology for bioremediation.

To clean and rejuvenate the Ganga, the Government launched a program called Namami Gange [24].

For the provision of Function Household Tap connections (FHTC), the Government has reformed the National Rural Drinking Water Programme (NRDWP) into the Jal Jeevan Mission (JJM) and Har Ghar Nal Se Jal (HGNSJ) by 2024. This mission aims to arrange hygiene services to provide safe drinking water to every household [24].

Challenges

However, to achieve the target agenda of 2030, there are many challenges, such as a lack of clear concepts, a lack of data extraction technology, and more data, especially from the health sector, which require analysis of the current status for each indicator (Table **5**). Therefore, it is necessary to involve private hospitals, institutes, and colleges to obtain more patient records.

Table 5. Challenges related to environment and health-related SDGs [24 - 30].

Challenges related to the Environment	Challenges related to health
Political level – Continuous changing of political commitment and environment, Lack of priority for planning & policies for implementation.	Monitoring and evaluation of the status based on data Issue with reliable data, Data monitoring issues Lack of technical knowledge for data collection and management of infrastructure development. Issues with periodic data Lack of private sector data Lack of coordination between national and sub-national targets. Lack of survey based data collection. Lack of technical capacity.
Institute level - Lack of coordination between national and sub-national agencies.	
Financial Level - High dependence on external resources	
Multisector level - Lack of empowerment in local government	
Role of stakeholders – Lack of resources, their role, and limited participation of stakeholders.	
Capacity building – lack of monitoring, limited capacity for cost, designing of policies, management, and monitoring of statistical reports	

Changes in data collection strategies through stratification were related to age, sex, wealth, education, family background, and geography. These strategies can resolve several issues in future research.

Measuring Strategies

Since the establishment of the SDGs with the 2030 Agenda, global-level policies as indicators have been designed with a strong foundation and clear targets that authorize the sectoral health area to participate actively in achieving results [1]. Members of the WHO from European regions have already committed to advancing actions for the improvement of health and the environment at national, international, and subnational levels. The 6th Ministerial Conference on Health and Environment was held between 13th and 15th June 2017 in Ostrava, Czech Republic [24]. The Declaration of Ostrava identified six primary areas of concern: Water& Air pollution, hygiene and sanitation, identification of waste and areas of contamination, safety from chemicals, change of climate and urban region environmental actions, and development of sustainable health technology and systems (Table 6). The summary points of the Ostrava declaration for possible actions for advanced improvement on SDGs related to health and environment are [25]:

Table 6. Global and regional level framework of policies [26 - 30].

Framework of policies	Goal
Ostrava declaration – 2017	Has committed
WHO – Thirteenth general programme of work (2019-2024)	Has committed
United Nations plan for the General Assembly meeting for the control and preventive strategies of non-communicable diseases (2018)	It identifies
WHO plan for Global strategy for Health-Environment and climate change (2019)	This

Improvement on Indoor and Outdoor Air

Ensuring uniform, equal, and sustainable access to safe drinking water, hygiene, and sanitation across every sector, the reuse of wastewater, and the promotion and management of integrated water resources are crucial.

Minimization of adverse reactions of chemicals to the environment and human health by replacing toxic chemicals with safe alternatives, reducing exposure to toxic chemicals, especially during human development, strengthening risk assessment policies, increasing research related to disease, and applying precautions where necessary.

Waste and contaminated site management with respect to human health, the adaptation of advanced technology for illegal and uncontrolled waste, and sound management generated from waste disposal direct the transition of the economy.

Strengthen the capacity for health risks related to the climate crisis and adaptation to the Paris Agreement.

Support cities and regions to transform into safer, healthier, and sustainable using a smart approach, implement effective strategies across government sectors, and arrange conferences and meetings to share experiences and approaches in accordance with the New Urban Agenda.

Building sustainable health systems by adapting potential technology to minimize the sound of clinical products and chemicals disposal in line with the agenda of reducing pollution of waste and wastewater, with the preconception of the sanitation mission

Global strategies have been designed to monitor the progress and current status of each indicator, apart from those of Ostrava and the WHO, to ensure consistency with the SDG structure. Other policies include European Health policies for health and well-being [26] and UN conventional policies related to environmental protocols, agreements, environmental factors related to air, sanitation, water, and waste, safety of chemicals, work injuries, and climate crises.

DISCUSSION

After observing the 2022 SDG report, we can say that India is rapidly moving towards transformation. India has launched several initiatives to develop new sources of renewable energy, particularly wind and solar systems. Until now, the majority of targets have been achieved, and we can say that around 2030, the Global Biodiversity Framework's four goals with 23 targets will maintain and preserve the integrity of biodiversity and ecosystems [27]. This will be beneficial for protecting natural water areas, islands, lands, and coastal areas, reducing waste, and increasing financial resources. In 2022, the Kunming–Montreal Global Biodiversity Framework was established to protect ecosystems. To accomplish the target of 2030, India has decided to increase its capacity for renewable energy and has jointly participated in the International Solar Alliance, France [27, 28]. Apart from that, India has already launched Jal Shakti Abhiyan, Clean India Mission for sanitation and hygiene, Namami Gange to reduce water pollution and save the natural water resources, Paramparagat Krishi Vikas Yogana for the promotion of organic farming and for a better agriculture system [12, 27].

To resolve existing challenges and fulfilment of Agenda 2030, the WHO has designed Triple Billion Targets based on the existing SDGs indicator's progress report [16, 27]. The Triple Billion Targets focus on more than one billion people receiving health protection through UHC [20, 28]; more than one billion people should receive health-related emergency services, and more than one billion people should enjoy healthy lives. The Triple Billion Target is based only on SDG 3, to provide better health facilities at three levels: 1. Preparation for emergency support, 2. prevention of health-related emergencies, and 3. Disease diagnosis in emergency patients with improved outcomes. In addition, there is an urgent need to develop vaccines against infectious diseases, antibiotics for drug-resistant tuberculosis, and calculative prescriptions for antibiotics. Further surveillance and infection control programs are required [29, 30].

CONCLUSION

The Sustainable Development Goals (SDGs) provide complex structural planning for sustainable environments and overall well-being. There are many challenges in planetary systems, including compact and interconnected forms of health, environment, and biodiversity. Many factors can hamper the system, such as climate change, biodiversity loss, uncontrollable consumption or production of goods, and an increase in waste generated from agriculture, households, and industries. The Sustainable Development Goal (SDG) approach targets every sector, which is necessary to address all global issues. It is a multifaceted approach that requires strategies and joint efforts from all countries. To achieve this goal, it is important to strengthen policies, promote the use of renewable energy sources, conduct more research on green technology and new therapeutic technology development, restore ecosystems, develop advanced technologies for waste management, build systems against climate actions, collaborate, support partnerships related to SDGs, and increase educational awareness.

LIST OF ABBREVIATIONS

SDGs	Sustainable Development Goals
WHO	World Health Organization
NTD	Neglected Tropical Diseases
WASH	Water Sanitation, and Hygiene
GDP	Gross Domestic Product
CVD	Cardiovascular Disease
UHC	Universal Health Coverage
PM	Particulate Matter
HIV	Human Immunodeficiency Virus
DTP	Diphtheria, Tetanus, and Pertussis Vaccine
MDGs	Millennium Development Goals
TFA	Trans Fatty acids
NCDs	Non-communicable Disease
APC	Alcohol Per Capita Consumption
HVAC	Heating Ventilation and SC system
GLO	Ground Level Oxygen
CO	Carbon Monoxide
NO	Nitrous Oxide
PAH	Polycyclic hydrocarbons
NTDs	Neglected Tropical Diseases
ART	Antiretroviral Therapy
BCG	Bacillus Calmette-Guerin
LF	Lymphatic Filariasis
MoHUA	Ministry of Housing and Urban Affairs
PHC	Primary Health Care
JJM	Jal Jeevan Mission
NGO	Nongovernment Organization
AMRUT	Atal Mission for Rejuvenation and Urban Transformation

KVK	Krishi Vigyan Kendra
PKVY	Pramparagat Krishi Vikas Yogna
FHTC	Function Household Tap connection
NRDWP	National Rural Drinking Water Programme
HGNSJ	Har Ghar Nal Se Jal

ACKNOWLEDGMENTS

The authors thank Atmiya University for their support in collecting the cited references.

REFERENCES

[1] United Nations. Transforming Our World: The 2030 Agenda for Sustainable Development. New York: United Nations; 2015. Available from: https://sustainabledevelopment.un.org/post2015/transformingourworld

[2] United Nations Development Programme. Sustainable Development Goals (SDGs). Geneva: UNDP; 2015. Available from: https://www.undp.org/sustainable-development-goals

[3] World Health Organization (WHO). Constitution of the World Health Organization. Geneva: WHO; 1948. Available from: https://apps.who.int/gb/bd/PDF/bd47/EN/constitution-en.pdf

[4] World Health Organization. Progress in the Implementation of the 2030 Agenda for Sustainable Development. Report by the Secretariat. Seventieth World Health Assembly. Geneva, 2017. Available from: http://apps.who.int/gb/ebwha/pdf_files/WHA70/A70_35-en.pdf

[5] World Health Organization Western Pacific Regional Office. People at the Center of Care. Manila: WHO; 2017. Available from: http://www.wpro.who.int/health_services/people_at_the_centre_of_care/definition/en/

[6] World Health Organization. Leaving No One Behind: Equity, Gender and Human Rights Policy to Practice. Geneva: WHO; 2017. Available from: https://iris.who.int/server/api/core/bitstreams/3cf1efa2-c12b-4ee2-8e65-78dba62e784f/content

[7] United Nations India. NITI Aayog – SDG India Index 2023–24: Towards Viksit Bharat, Sustainable Progress, Inclusive Growth. Available from: https://www.niti.gov.in/sites/default/files/2024-07/SDG_India_Index_2023-24.pdf

[8] Sachs JD, Lafortune G, Fuller. The SDGs and the UN Summit of the Future. Sustainable Development Report 2024. Dublin: Dublin University Press; 2024.

[9] United Nations Department of Global Communications. Sustainable Development Goals: Guidelines for the use of the SDG logo, including the colon wheel and 17 icons. 2024. Available from: https://www.un.org/sustainabledevelopment/news/communications-material/

[10] United Nations Department of Global Communications. Sustainable Development Goals – Guidelines for the Use of the SDG Logo Including the Colour Wheel and 17 Icons. September 2023. Available from: https://www.un.org/sustainabledevelopment/wp-content/uploads/2019/01/SDG_Guidelines_AUG_2019_Final.pdf

[11] World Health Organization. Technical Note: Developing an Index for the Coverage of Essential Health Services. Geneva: WHO; 2016. Available from: https://www.who.int/docs/default-source/gh-documents/world-health-statistic-reports/world-heatlth-statistics-2016.pdf

[12] United Nations Millennium Project, 2015. Available from: http://www.un.org/millenniumgoals/bkgd.shtml

[13] United Nations. Millennium Development Goals and Indicators. New York: United Nations. 2008. Available from: https://www.un.org/millenniumgoals/2008highlevel/pdf/newsroom/mdg%20reports/MDG_Report_2008_ENGLISH.pdf

[14] World Health Organization. Health in 2015: From MDGs to SDGs. Geneva: WHO; 2015. Available from: http://www.who.int/gho/publications/mdgs-sdgs/en/

[15] World Health Organization. World Health Statistics 2024: Monitoring Health for the SDGs. Geneva: WHO; 2024.

[16] Ghorani-Azam A, Riahi-Zanjani B, Balali-Mood M. Effects of air pollution on human health and practical measures for prevention in Iran. J Res Med Sci 2016; 21(1): 65.
[http://dx.doi.org/10.4103/1735-1995.189646] [PMID: 27904610]

[17] World Health Organization. Burden of disease due to ambient and household air pollution. Available from: https://www.who.int/data/gho/data/themes/air-pollution/total-burden-of-disease-from-household-and-ambient-air-pollution

[18] World Health Organization. Air Pollution [Internet]. Geneva: World Health Organization; 2019. Available from: http://www.who.int/airpollution/en/

[19] World Health Organization. Universal Health Coverage. Geneva: WHO; 2017. Available from: www.who.int/universal_health_coverage/en/

[20] Patra JK, Das G, Fraceto LF, et al. Nano based drug delivery systems: recent developments and future prospects. J Nanobiotechnology 2018; 16(1): 71.
[http://dx.doi.org/10.1186/s12951-018-0392-8] [PMID: 30231877]

[21] Abdallah NA, Prakash CS, McHughen AG. Genome editing for crop improvement: Challenges and opportunities. GM Crops Food 2015; 6(4): 183-205.
[http://dx.doi.org/10.1080/21645698.2015.1129937] [PMID: 26930114]

[22] Gao C. Genome engineering for crop improvement and future agriculture. Cell 2021; 184(6): 1621-35.
[http://dx.doi.org/10.1016/j.cell.2021.01.005] [PMID: 33581057]

[23] Sharma I, Murillo-Tovar MA, Saldarriaga-Noreña H, Saeid A. Bioremediation Techniques for Polluted Environment: Concept, Advantages, Limitations, and Prospects. In: IntechOpen; 2020.
[http://dx.doi.org/10.5772/intechopen.90453]

[24] World Health Organization. Declaration of the Sixth Ministerial Conference on Environment and Health. Ostrava, Czech Republic, June 13–15, 2017. Regional Office for Europe (EURO). Available from: https://iris.who.int/items/3ea5a252-eefa-454e-afe2-fd5576465613

[25] World Health Organization Regional Office for Europe. Health 2020: A European policy framework and strategy for the 21st century: WHO; 2012. Available from: https://iris.who.int/server/api/core/bitstreams/2731a5d0-ec46-48ca-b828-973bb9b50919/content

[26] United Nations. Resolution A/RES/66/288. The Future We Want. Sixty-sixth United Nations General Assembly, New York, 2012. Available from: https://www.un.org/en/development/desa/population/migration/generalassembly/docs/globalcompact/A_RES_66_288.pdf

[27] United Nations. Transforming Our World: The 2030 Agenda for Sustainable Development. New York: United Nations; 2015. https://sdgs.un.org/2030agenda

[28] World Health Organization. Minimum requirements for infection prevention and control programs. Geneva: WHO; 2019. Available from: https://www.who.int/publications/i/item/9789241516945

[29] World Health Organization. Declaration of the Sixth Ministerial Conference on Environment and Health: Compendium of Possible Actions to Advance the Implementation of the Ostrava Declaration. WHO Regional Office for Europe; 2017. Available from: https://www.who.int/europe/publications/i/item/WHO-EURO-2017-3898-43657-61366

[30] Dave R. Application of Nanotechnology for Drug and Vaccine Development. BioRender.com (2025). Available from: https://app.biorender.com/biorender-templates

CHAPTER 11

Transforming Wastewater into Renewable Energy: A Pathway to Achieve Sustainability and the Circular Economy

V.V. Vaishnavi[1], M. Mounica[1] and M. Vijay Pradhap Singh[1,*]

[1] *Department of Biotechnology, Vivekanandha College of Engineering for Women (Autonomous), Elayampalayam, Tamil Nadu, India*

Abstract: The growing demand for clean water, coupled with increasing wastewater generation due to urbanization and industrialization, necessitates innovative solutions for sustainable resource management. Energy recovery from wastewater, particularly through anaerobic digestion, presents a promising approach to achieving sustainability. This process converts organic matter in wastewater into biogas, a renewable energy source, reducing reliance on fossil fuels while addressing global energy demands. Biogas production from wastewater promotes resource recovery by minimizing waste, reducing greenhouse gas emissions, and reducing environmental pollution. The energy generated can be used to power wastewater treatment plants, decreasing reliance on fossil fuels and lowering operational costs. In addition to biogas production, advanced oxidation processes can be integrated into wastewater treatment to break down persistent organic pollutants, resulting in cleaner effluents and promoting resource recovery. The combination of these processes supports a circular economy by minimizing waste, recovering valuable by-products such as fertilizers, and reducing the carbon footprint of wastewater treatment plants. By adopting energy recovery technologies like biogas production and sludge incineration, wastewater treatment facilities can become self-sufficient, reducing operational costs and enhancing environmental sustainability. This chapter will address the sustainable production of biogas from wastewater, focusing on energy recovery methods and their role in promoting a circular economy.

Keywords: Biogas, Clean water, Circular economy, Energy recovery, Sustainability.

---

* **Corresponding author Pradhap Singh:** Department of Biotechnology, Vivekanandha College of Engineering for Women (Autonomous), Elayampalayam, Tamil Nadu, India; E-mail: vijaypradhapsingh@gmail.com

Rajneesh Kumar, Ram Sharan Singh & Maulin P. Shah. (Eds.)
All rights reserved-© 2026 Bentham Science Publishers

INTRODUCTION

The energy recovery process from waste via anaerobic digestion results in the generation of biogas along with some essential components. These strategies involved in the wastewater treatment process benefit by reducing the electricity consumption in wastewater treatment plants, reducing the reliance on fossil fuels, and helping to represent the possible areas for sustainable energy policy implementation [1]. Considering wastewater, there is a large amount of thermal energy present in it. When compared to the chemical energy present in the wastewater, the thermal energy seems to be neglected. Usually, the temperature of the waste is below 30 degrees Celsius. The recovered energy is further used in various sectors for the circular economy [2]. The energy recovery method is adapted to solve the arising energy and environmental problems simultaneously [3]. The energy recovered shows an increased interest in the demand for primary energy and reduces the reliance on fossil fuels. It also creates a concern about climate change and greenhouse gas emissions. These conditions are satisfied by bringing biofuels into use that are better alternatives for non-renewable energy sources [4]. The large energy consumption in the plant where the wastewater is being treated is satisfied by the use of recovered energy from wastewater, which has economic and environmental implications. The generation of wastewater is being increased by rapidly extending population, enhanced lifestyles, urbanization, and economic development [5].

This review involves techniques and strategies involved in the energy recovery process from wastewater in the development of the environment, which deals with the reduction of environmental pollution. This chapter further discusses the methodology and ideology adapted for the recovery process in detail. This study addresses the pathway to achieve sustainability and a circular economy.

WASTEWATER

The wastewater generated from various sectors contains components that pollute our environment. Wastewater can also be defined as the combination of solid waste carried by water from areas like residences, institutions, industries, and commercial establishments. It contains toxic substances that are harmful. Nowadays, the river and stream pollution with chemical contaminants is mentioned as a serious issue in our environment [6]. The increasing volume of urban waste is the result of allocating less fresh water to the agriculture sectors and allocating more fresh water to non-agriculture sectors. This is the main reason for the production of wastewater [7]. A decrease in fresh water creates an impact on the lifestyle and developing opportunities in the regions with insufficient water supply [8]. It is estimated that by 2025, 60% world's population may suffer from water insufficiency [9]. (Fig. **1**) represents the classification of wastewater.

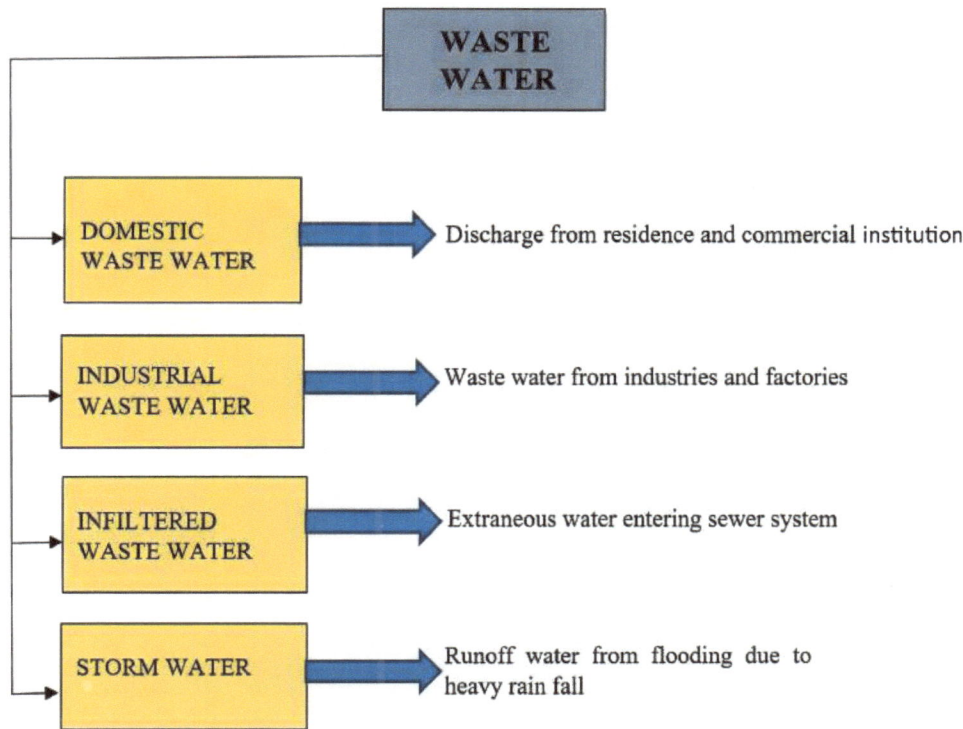

Fig. (1). Classification of wastewater [6]. .

Wastewater Generation

Wastewater is being generated by various sectors that include irrigation and industries. It is estimated that around 80% of the overall water that is supplied for domestic purposes is generated as wastewater. The estimation according to the CBCP states that the total wastewater generated from class I (cities) is around 35,000 MLD, and in class II (towns) is around 2700 MLD. Around 13470 MLD of wastewater is being generated by industries, where only 60-70% is treated [10]. The amount and types of wastewater produced from domestic or household use depend on the behaviour, lifestyle, and standard of living of people [11]. (Fig. **2**) tells about the water demand required by various sectors.

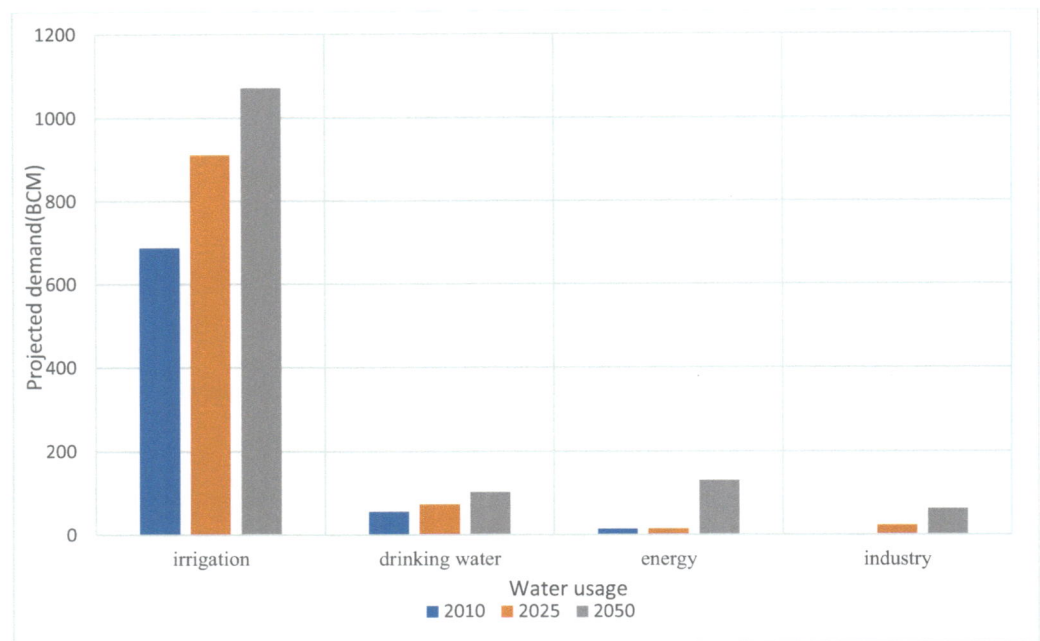

Fig. (2). Water demand by various sectors.

Composition

The composition of wastewater depends on several factors, such as the origin of generation [12]. The styles and technologies followed in society are reflected in the composition of wastewater. The effluent contains a combination of biological and chemical compositions. The effluent consists of chemical compositions, such as inorganic nitrogen and phosphorus, that cause eutrophication [13]. The inorganic constituents include some concentration of sodium, potassium, magnesium, calcium, bicarbonate, ammonia salts, and heavy metals [14]. The major N-containing compounds are indole, pyrrole derivatives, aminophenol, 3-amino-phenol, N-ethyl acetamide, pyrazine, methyl pyrazine, and their derivatives. Small acids like acetic acid and propanoic acid are also found. The effluent consists of pathogenic microorganisms that cause infectious diseases [15]. The wastewater also contains some organic materials like fibres (20.65%), proteins (12.40%), and sugars (10.70%) [16]. Fig. (**1**) illustrates the classification and origin of waste from various sectors.

ENERGY RECOVERY FROM WASTEWATER

Renewable energy from various sources has created a very big interest globally due to the rising demand. This minimizes the usage of fossils and preserves them. One example of such energy is biofuel, which has concerns due to greenhouse emissions and irregular climate change [4]. The extraction of such energy will simultaneously reduce the problems raised in our environment [17].

According to the Paris Climate Agreement, it is said that the energy recovered from the wastewater via. Wastewater treatment plants are being used urgently in many areas. The wastewater contains some organic substances (COD), which represent the amount of chemical energy present in it. Recovering this chemical energy serves as a sustainable method rather than destroying it through biological oxidation. This recovered energy is used for various functions that include heating or cooling buildings or for drying the dewatered sludge [18]. Energy recovery from wastewater is done by both aerobic and anaerobic processes. Aerobic organisms require oxygen in pure form for the degradation of bacteria and infectious agents by microorganisms. But the anaerobic process takes place during the lack of oxygen in the reactor, where biogas and other components are formed, and it contains methane in it [3].

Techniques used

There are many techniques available that are implemented in the energy recovery process from wastewater. This study includes the advantages and disadvantages of the various techniques used.

There are four main methods: anaerobic digestion, incineration, gasification, and pyrolysis.

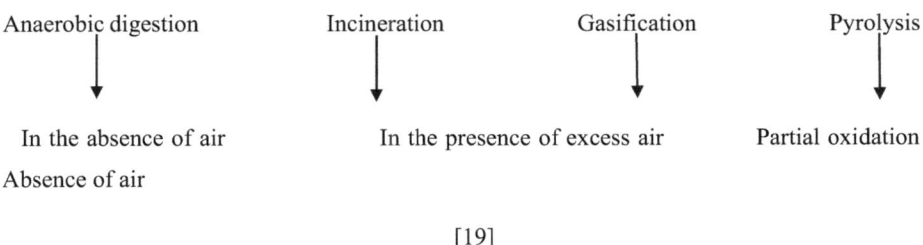

[19]

Anaerobic digestion is a biochemical process that involves the digestion of organic matter in absence of oxygen. This process is cost-effective. But it also has some disadvantages, like time consumption and low conversion efficiency [20].

To overcome these disadvantages, thermal process like pyrolysis, combustion, and gasification was invented. Even though these techniques consume less time, they require sludge with low moisture content. To achieve this, the sludge is being dried out, which requires high energy inputs. These thermal techniques are also costly [21]. Table **1** explains the advantages and disadvantages of the wastewater treatment techniques.

Table 1. Advantages and disadvantages of various techniques.

Techniques	Advantages	Disadvantages
ANAEROBIC DIGESTION	Cost-effective and used for wet sludge	Time-consuming and low conversion efficiency
COMBUSTION	Waste volume is reduced up to 90%.	Release pollutants that destroy air, and ash is generated as a product
PYROLYSIS	Waste like plastics can also be processed through this technique. Low emission of harmful gases. High conversion efficiency.	This process needs a high temperature for processing.
GASIFICATION	Produce less ash and low emissions of gases. This process consumes less time.	High operating cost. Should be monitored under controlled conditions.

Anaerobic Digestion

When compared to the methods mentioned above, the anaerobic process is used widely due to its advantages. In the Anaerobic process, the microorganism is broken down in the absence of oxygen. Anaerobic digestion is often used for decreasing the amount of organic solids and energy recovery. In this system, the organic substances present in the wastewater are converted into methane and carbon dioxide [22].

The bacteria that are sensitive to pH should be maintained in an optimum range between 6.5 and 7.5. To ensure this, the bicarbonate is used to maintain the pH [23].

Various reactors and filters are used to enhance the digestion of organic waste in wastewater. They are second generation anaerobic digestors [anaerobic filters, the downflow stationary fixed film reactor, The up flow anaerobic sludge blanket reactor, Anaerobic fluidized-bed reactor, Anaerobic sequencing batch reactor], Third generation high-rate digesters [The expanded granular sludge blanket (EGSB) reactor, Internal circulation (IC) reactor, Anaerobic migrating blanket reactor, Anaerobic hybrid reactors, Reactors with phase separation, Integrated anaerobic-aerobic reactors] [23]. Table **2** represents the design and flow pattern of reactors with its applications.

Table 2. Various types of anaerobic reactors [23].

Types	Design	Flow pattern	Advantages	Limitation	Applications
ANAEROBIC FILTERS	Made of plastics or stones	Either downflow or upflow	Simple design and cost-effective	Regular clogging	Food processing
DSFFR	Consists of a stationary bed covered by media	downflow	Shows stable performance	Not applicable for a highly fluctuating waste stream	Industrial wastewater
USAB	Consists of a suspended sludge bed	Up flow	High efficiency and easy operation	Long start-up period	Municipal wastewater treatment
AFBR	Consists of bio-plastics	Up flow	Good mixing and designed compactly	Complex designing	High-strength wastewater
ASBR	Uses an alternating feed system	Follow the discontinuous batch method	High efficiency	High control complexity	Variable wastewater streams
EGSB	Expanded sludge bed	Up flow	Highly efficient mass transfer	High energy requirement	Breweries and dairies
IC	Consists of an internal recirculation mechanism	Up flow along with internal mixing	High efficiency	High operational cost	Large-scale industrial wastewater
AMBR	Consists of various compartments	Horizontal flow	High efficiency	More complex mechanism	highly concentrated wastewater
HYBRID REACTORS	A combination of USAB and a fixed film system	Up flow	Easy operation	Expensive system	Both low and highly concentrated wastewater
PHASE SEPARATED REACTORS	Separated compartments for different phases	Multi stage	Highly stable	High surface area	Complex waste stream
INTEGRATED REACTORS	A combination of aerobic and anaerobic systems	sequential	High efficiency	Increased operational complexity	Municipal sewage

The diagram illustrated above Fig. (**3**) describes the workflow of basic anaerobic reactors. Here, the wastewater is treated in a bioreactor where the microorganisms are digested into simpler units like methane and carbon dioxide in the absence of oxygen. The obtained biogas is then transferred to the compressor, where it is compressed. Later, the obtained gas is stored and used for future purposes. The waste sludge obtained in the process is used as fertilizer in agricultural sectors and for other purposes.

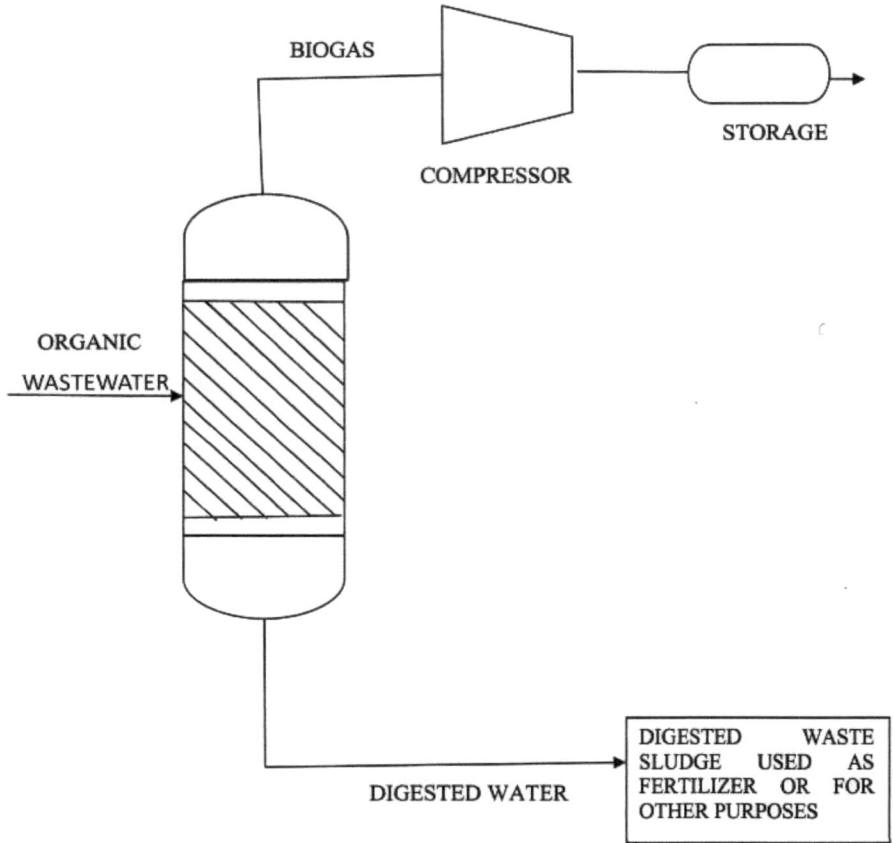

Fig. (3). Anaerobic reactor [22].

BIOGAS

Biogas is a renewable fuel that is used as an eco-friendly fuel due to its less polluting properties. The wastewater treatment plants help to digest the sludge to produce biogas (Fig. **4**) [24]. The fermentation of methane in organic waste results in the production of biogas [25]. The sludge developed in the wastewater treatment process could yield highly efficient energy in the form of biogas [21]. Usage of biogas can reduce the production of CH_4, CO, and CO_2. It also helps in the promotion of carbon sequestration. The usage of biogas results in the reduction of contamination in surface water and groundwater [24].

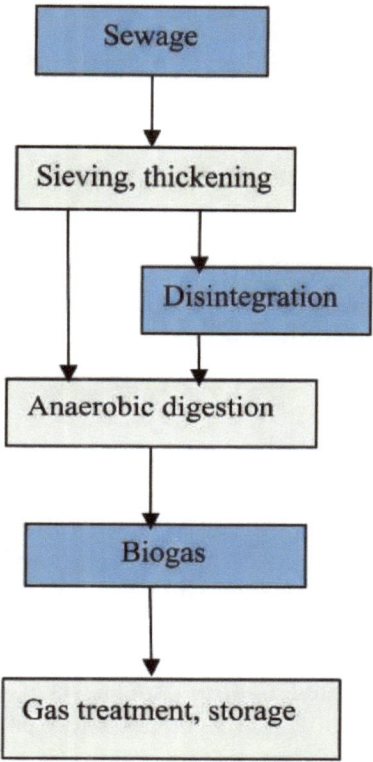

Fig. (4). Production of biogas [30].

Biogas fermentation is considered the best waste treatment alternative due to the formation of a renewable energy source, such as methane, by removing the organic waste simultaneously [26]. Two major groups of microorganisms are responsible for this formation. One is the acid-forming bacteria that break down the organic compounds into simpler units, resulting in the formation of primary products such as hydrogen, alcohol, and carbon dioxide. While the other is the methane-producing bacteria that obtains a food source from primary products for the conversion of organic acids into methane [27]. Some applications of biogas are listed below:

- Biogas is directly used in gas engines
- Used in the generation of steam or hot water
- Supply electricity for aeration tanks
- Used for domestic purposes

These are some applications of biogas that make it more beneficial and important [28, 29].

Other than the benefits mentioned above, there are several other benefits that include the usage of biogas as organic fertilizers for agricultural purposes. It decreases the dependency on chemical fertilizers and pesticides. This automatically helps in the reduction of greenhouse gases [31, 32].

The methane produced by the anaerobic digestion of waste is similar to the natural gas that is extracted from the well head and has less caloric value. Biogas is seen not only as a renewable source, but it is also discovered as a promising solution to the upcoming environmental problems [33].

CIRCULAR ECONOMY

The circular economy is defined as the concept where the raw material and product remain in the economy, and the obtained waste is considered as secondary raw material, which is further converted into value-added products [34]. The circular economy concept has an objective which is to eliminate waste throughout the life cycles and to utilize the byproducts [35].

This concept of circular economy is also applied in the field of wastewater treatment. The sludge produced in this process is used as fertilizer and is also further treated to produce biogas [36]. The waste treatment plant plays a crucial role in the circular economy due to the integration of energy production and recovery of resources [37, 38]. Here, the waste is reused and recycled as per the concept of circular economy [39]. The recovery potential of energy in the wastewater always depends on the various factors, including physical and chemical parameters [40, 41].

EMERGING TRENDS

This modern-day world has seen various improvements in industrial sectors, which have led to the production of large amounts of waste consisting of highly complex components. These wastes, when directly disposed of, pollute our environment. The present technologies in wastewater treatment are not enough to degrade or decompose all the contaminants. So various methodologies are emerging to show high efficiency.

Various techniques like ion exchange, adsorption, reverse osmosis, electrodialysis, flotation, coagulation, etc., show highly efficient output [42]. Among all other methods, the sorption technique by using different adsorbent materials is said to be the simplest one, which is manageable and cost-effective [43, 44]. This method does not cause any secondary pollution due to the production of other byproducts [45, 46]. Further research has stated that the modification of these adsorbent materials by treating them with other chemicals can enhance their capacity to adsorb pollutants. In further developments, nano adsorbents are used in the process [47, 48].

CONCLUSION

In a developing country, wastewater is being generated via various sectors like industries, agriculture, textiles, pharmaceuticals, and domestic. The waste present in the water can destroy our environment. Also, the demand for the use of renewable resources has increased. Considering the increasing production of wastewater and the need for renewable resources, wastewater treatment techniques can offer us a long-term solution to meet all our needs and to maintain a circular economy, leading to a sustainable environment. In this situation, the anaerobic digestion method has been an effective and sustainable method that offers an economic benefit. This wastewater treatment helps in reducing the contaminants in the wastewater, and the obtained sludge is considered a secondary raw material. This is treated to produce biogas and other value-added products. Here, biogas is produced by the fermentation of methane, which is a renewable energy source. This helps to reduce the reliability of non-renewable sources like fossil fuels, which leads to the production of greenhouse gases. This serves as a reason to satisfy the concept of the circular economy. Various research and studies are being conducted to invent methods for treating the wastewater containing more complex and highly toxic substances, which are more effective and eco-friendlier.

REFERENCES

[1] Stillwell AS, King CW, Webber ME, Duncan IJ, Hardberger A. Energy-Water Nexus in Texas; Environmental Defense Fund, University of Texas at Austin:2009 Energy Conservation in Wastewater Treatment Facilities Manual of Practice; Water Environment Federation: Alexandria. Austin, TX, USA VA, USA 1997; pp. 1-142.

[2] Cipolla SS, Maglionico M. Heat recovery from urban wastewater: analysis of the variability of flow rate and temperature. Energy Build. 2014 Feb; 69: 122–130.
[http://dx.doi.org/10.1016/j.enbuild.2013.10.017]

[3] Liu XW, Li WW, Yu HQ. Cathodic catalysts in bioelectrochemical systems for energy recovery from wastewater. Chem Soc Rev 2014; 43(22): 7718-45.

[http://dx.doi.org/10.1039/C3CS60130G] [PMID: 23959403]

[4] Nazari L, Sarathy S, Santoro D, Ho D, Ray MB, Xu CC. Recent advances in energy recovery from wastewater sludge. In: Xu CC, Ray MB, eds. Direct Thermochemical Liquefaction for Energy Applications. Cambridge (UK): Woodhead Publishing; 2018. p. 43-63.
[http://dx.doi.org/10.1016/B978-0-08-101029-7.00011-4]

[5] Garfi. M. Biological Treatment of Organic Waste in Wastewater—Towards a Circular and Bio-Based Economy. Water 2022; 14(3): 360.
[http://dx.doi.org/10.3390/w14030360]

[6] Sonune A, Ghate R. Developments in wastewater treatment methods. Desalination 2004; 167: 55-63.
[http://dx.doi.org/10.1016/j.desal.2004.06.113]

[7] Raschid-Sally L, Jayakody P. Drivers and Characteristics of Wastewater Agri culture in Developing Countries: Results from a Global Assessment. Research Report 127. International Water Management Institute. Colombo 2008.

[8] Qadir M, Sharma BR, Bruggeman A, Choukr-Allah R, Karajeh F. Non-conventional water resources and opportunities for water augmentation to achieve food security in water scarce countries. Agric Water Manage 2007; 87(1): 2-22.
[http://dx.doi.org/10.1016/j.agwat.2006.03.018]

[9] Rijsberman FR. Water scarcity: Fact or fiction? Agric Water Manage 2006; 80(1-3): 5-22.
[http://dx.doi.org/10.1016/j.agwat.2005.07.001]

[10] Koul B, Yadav D, Singh S, Kumar M, Song M. Insights into the domestic wastewater treatment (DWWT) regimes: a review. Water. 2022 Oct 28; 14(21): 3542.
[http://dx.doi.org/10.3390/w14213542]

[11] Henze M, van Loosdrecht MCM, Ekama GA, Brdjanovic D. Biological Wastewater Treatment: Principles, Modelling and Design. London: IWA Publishing; 2008.
[http://dx.doi.org/10.2166/9781780401867]

[12] Bisschops I, Spanjers H. Literature review on textile wastewater characterisation. Environ Technol 2003; 24(11): 1399–1411.
[http://dx.doi.org/10.1080/09593330309385684]

[13] Abdel-Raouf N, Al-Homaidan AA, Ibraheem IBM. Microalgae and wastewater treatment. Saudi J Biol Sci 2012; 19(3): 257-75.
[http://dx.doi.org/10.1016/j.sjbs.2012.04.005] [PMID: 24936135]

[14] Lim SL, Chu WL, Phang SM. Use of *Chlorella vulgaris* for bioremediation of textile wastewater. Bioresour Technol 2010; 101(19): 7314-22.
[http://dx.doi.org/10.1016/j.biortech.2010.04.092] [PMID: 20547057]

[15] Naidoo S, Olaniran AO. Treated wastewater effluent as a source of microbial pollution of surface water resources. Int J Environ Res Public Health. 2014; 11(1): 249–270.
[http://dx.doi.org/10.3390/ijerph110100249]

[16] Huang M, Li Y, Gu G. Chemical composition of organic matters in domestic wastewater. Desalination 2010; 262(1-3): 36-42.
[http://dx.doi.org/10.1016/j.desal.2010.05.037]

[17] Hao X, Li J, van Loosdrecht MCM, Jiang H, Liu R. Energy recovery from wastewater: Heat over organics. Water Res 2019; 161: 74-7.
[http://dx.doi.org/10.1016/j.watres.2019.05.106] [PMID: 31181448]

[18] McCarty PL, Bae J, Kim J. Domestic wastewater treatment as a net energy producer--can this be achieved? Environ Sci Technol 2011; 45(17): 7100-6.
[http://dx.doi.org/10.1021/es2014264] [PMID: 21749111]

[19] Oladejo J, Shi K, Luo X, Yang G, Wu T. A review of sludge-to-energy recovery methods. Energies

2018; 12(1): 60.
[http://dx.doi.org/10.3390/en12010060]

[20] Syed-Hassan SSA, Wang Y, Hu S, Su S, Xiang J. Thermochemical processing of sewage sludge to energy and fuel: Fundamentals, challenges and considerations. Renew Sustain Energy Rev 2017; 80: 888-913.
[http://dx.doi.org/10.1016/j.rser.2017.05.262]

[21] Campo G, Cerutti A, Zanetti M, Scibilia G, Lorenzi E, Ruffino B. Enhancement of waste activated sludge (WAS) anaerobic digestion by means of pre- and intermediate treatments. Technical and economic analysis at a full-scale WWTP. J Environ Manage 2018 Jun 15; 216: 372–382.
[http://dx.doi.org/10.1016/j.jenvman.2017.05.025]

[22] Nishio N, Nakashimada Y. Recent development of anaerobic digestion processes for energy recovery from wastes. J Biosci Bioeng 2007; 103(2): 105-12.
[http://dx.doi.org/10.1263/jbb.103.105] [PMID: 17368391]

[23] Tauseef SM, Abbasi T, Abbasi SA. Energy recovery from wastewaters with high-rate anaerobic digesters. Renew Sustain Energy Rev 2013; 19: 704-41.
[http://dx.doi.org/10.1016/j.rser.2012.11.056]

[24] Shen Y, Linville JL, Urgun-Demirtas M, Mintz MM, Snyder SW. An overview of biogas production and utilization at full-scale wastewater treatment plants (WWTPs) in the United States: Challenges and opportunities towards energy-neutral WWTPs. Renew Sustain Energy Rev 2015; 50: 346-62.
[http://dx.doi.org/10.1016/j.rser.2015.04.129]

[25] Makisha N, Semenova D. Production of biogas at wastewater treatment plants and its further application. MATEC Web of Conferences. 144: 1-7.
[http://dx.doi.org/10.1051/matecconf/201714404016]

[26] Deng L, Liu Y, Zheng D, et al. Application and development of biogas technology for the treatment of waste in China Renewable and Sustainable Energy Reviews 2017; 70: 845-51.
[http://dx.doi.org/10.1016/j.rser.2016.11.265]

[27] Volkov A, Kuzina O. Complementary assets in the methodology of implementation unified information model of the city environment project life cycle. Procedia Eng. 2016; 153: 838–43.
[http://dx.doi.org/10.1016/j.proeng.2016.08.252]

[28] Kulakov A, Makisha N. Removal of nitrogen and phosphorous from domestic wastewater. MATEC Web Conf 2017; 112: 10019.
[http://dx.doi.org/10.1051/matecconf/201711210019]

[29] Gogina E, Gulshin I. The Single-Sludge Denitri-Nitrification System in Reconstruction of Wastewater Treatment Plants in the Russian Federation. Appl Mech Mater 2014; 580-583: 2367-9.
[http://dx.doi.org/10.4028/www.scientific.net/AMM.580-583.2367]

[30] Lee ME, Steiman MW, St. Angelo SK. Biogas digestate as a renewable fertilizer: effects of digestate application on crop growth and nutrient composition. Renew Agric Food Syst. 2021 Apr; 36(2): 173–181.
[http://dx.doi.org/10.1017/S1742170520000186]

[31] Wang G, Liu W, Wang X, Gao DY, He DX, Chen W. Current status and prospect of biogas technology in China. Appl Energy Technol 2007; 12: 33-5. [In Chinese].

[32] Jiang X, Sommer SG, Christensen KV. A review of the biogas industry in China. Energy Policy 2011; 39(10): 6073-81.
[http://dx.doi.org/10.1016/j.enpol.2011.07.007]

[33] Berktay A, Nas B. Biogas production and utilization potential of wastewater treatment sludge. Energy Sources A Recovery Util Environ Effects 2007; 30(2): 179-88.
[http://dx.doi.org/10.1080/00908310600712489]

[34] Neczaj E, Grosser A. Circular Economy in Wastewater Treatment Plant–Challenges and Barriers

2018; 614: 614.
[http://dx.doi.org/10.3390/proceedings2110614]

[35] Casiano Flores C, Bressers H, Gutierrez C, de Boer C. Towards circular economy – a wastewater treatment perspective, the Presa Guadalupe case. Manag Res Rev 2018; 41(5): 554-71.
[http://dx.doi.org/10.1108/MRR-02-2018-0056]

[36] Guerra-Rodríguez S, Oulego P, Rodríguez E, Singh DN, Rodríguez-Chueca J. Towards the implementation of circular economy in the wastewater sector: Challenges and opportunities. Water 2020; 12(5): 1431.
[http://dx.doi.org/10.3390/w12051431]

[37] Rashidi H, GhaffarianHoseini A, GhaffarianHoseini A, Nik Sulaiman NM, Tookey J, Hashim NA. Application of wastewater treatment in sustainable design of green built environments: A review. Renew Sustain Energy Rev 2015; 49: 845-56.
[http://dx.doi.org/10.1016/j.rser.2015.04.104]

[38] Mo W, Zhang Q. Energy–nutrients–water nexus: Integrated resource recovery in municipal wastewater treatment plants. J Environ Manage 2013; 127: 255-67.
[http://dx.doi.org/10.1016/j.jenvman.2013.05.007] [PMID: 23764477]

[39] Salgot M, Folch M. Wastewater treatment and water reuse. Curr Opin Environ Sci Health 2018; 2: 64-74.
[http://dx.doi.org/10.1016/j.coesh.2018.03.005]

[40] Laureni M, Falås P, Robin O, et al. Mainstream partial nitritation and anammox: long-term process stability and effluent quality at low temperatures. Water Res 2016; 101: 628-39.
[http://dx.doi.org/10.1016/j.watres.2016.05.005] [PMID: 27348722]

[41] Sarpong G, Gude VG, Magbanua BS, Truax DD. Evaluation of energy recovery potential in wastewater treatment based on codigestion and combined heat and power schemes. Energy Convers Manage 2020; 222: 113147.
[http://dx.doi.org/10.1016/j.enconman.2020.113147]

[42] Baig U, Rao RAK, Khan AA, Sanagi MM, Gondal MA. Removal of carcinogenic hexavalent chromium from aqueous solutions using newly synthesized and characterized polypyrrole–titanium(IV)phosphate nanocomposite. Chem Eng J. 2015 Nov 15; 280: 494–504.
[http://dx.doi.org/10.1016/j.cej.2015.06.031]

[43] Jiryaei Sharahi F, Shahbazi A. Melamine-based dendrimer amine-modified magnetic nanoparticles as an efficient Pb(II) adsorbent for wastewater treatment: Adsorption optimization by response surface methodology. Chemosphere 2017; 189: 291-300.
[http://dx.doi.org/10.1016/j.chemosphere.2017.09.050] [PMID: 28942255]

[44] De Gisi S, Lofrano G, Grassi M, Notarnicola M. Characteristics and adsorption capacities of low-cost sorbents for wastewater treatment: A review. Sustain Mater Technol. 2016; 9: 10-40.
[http://dx.doi.org/10.1016/j.susmat.2016.06.002]

[45] Dubey SP, Gopal K, Bersillon JL. Utility of adsorbents in the purification of drinking water: a review of characterization, efficiency and safety evaluation of various adsorbents. J Environ Biol 2009; 30(3): 327-32.
[PMID: 20120453]

[46] Ersan G, Apul OG, Perreault F, Karanfil T. Adsorption of organic contaminants by graphene nanosheets: A review. Water Res 2017; 126: 385-98.
[http://dx.doi.org/10.1016/j.watres.2017.08.010] [PMID: 28987890]

[47] Kunduru KR, Nazarkovsky M, Farah S, Pawar RP, Basu A, Domb AJ. Nanotechnology for water purification: applications of nanotechnology methods in wastewater treatment. In: Domb AJ, ed. Water Purification. London: Academic Press; 2017. p. 33-74.
[http://dx.doi.org/10.1016/B978-0-12-804300-4.00002-2]

[48] Chaturvedi VK, Kushwaha A, Maurya S, Tabassum N, Chaurasia H, Singh M. Wastewater Treatment Through Nanotechnology: Role and Prospects. In: Singh A, Singh P, eds. Restoration of Wetland Ecosystem: A Trajectory Towards a Sustainable Environment. Berlin/Heidelberg: Springer; 2020. p. 227-47.
[http://dx.doi.org/10.1007/978-981-13-7665-8_14]

CHAPTER 12

Community-based Environmental Management

Preeti Mehta Kakkar[1,*], **Meet Sharma**[1], **Shivani Singh**[1], **Yashna Tiwari**[1] and **Ruchi Jakhmola Mani**[2]

[1] *Department of Biotechnology, Amity Institute of Biotechnology, Amity University, Noida, Uttar Pradesh, India*

[2] *Proteomics and Translational Research Lab, Centre for Medical Biotechnology, Amity Institute of Biotechnology, Amity University, Noida, Uttar Pradesh, India*

Abstract: Community-based Environmental Management (CBEM) is emerging as an important approach towards dealing with environmental issues and providing community control over local natural resource management and its protection. There are successful examples in many different types of geographies: inclusion of traditional knowledge, the incorporation of the most modern available technologies, and the involvement of all community groups through suitable policies are seen as some factors that facilitate success in their implementations. However, despite the successes, significant gaps persist in the implementation of CBEM. These gaps include the lack of long-term sustainability planning, insufficient integration of scientific research with traditional practices, underrepresentation of marginalized communities, lack of empowerment, socio-economic barriers, cultural and political resistance, and inadequate funding, which continue to hinder the widespread adoption of CBEM. Scientists have been working on a solution to these challenges by reviewing different case studies and existing literature for the identification of knowledge gaps, governance structures, and characteristics in the current CBEM practices. This chapter discusses evolution, contemporary trends, and challenges in CBEM, with specific emphasis on its importance to foster sustainable environmental practices and resilience to climate change. The chapter also identifies the innovative approaches that are now emerging in the field, for example, the integration of Artificial Intelligence (AI) and community-led monitoring systems. This can be used to identify the challenges and leverage new opportunities to achieve global environmental sustainability and resilience.

Keywords: Artificial Intelligence in environmental monitoring, Community-based environmental management, Climate resilience, Community empowerment, Natural resource governance, Sustainable practices, Traditional knowledge integration.

---

*Corresponding author **Preeti Mehta Kakkar**: Department of Biotechnology, Amity Institute of Biotechnology, Amity University, Noida, Uttar Pradesh, India; Email: pmkakkar@amity.edu

Rajneesh Kumar, Ram Sharan Singh & Maulin P. Shah. (Eds.)
All rights reserved-© 2026 Bentham Science Publishers

INTRODUCTION

Natural resource-based communities are communities that depend on land, water, forests, and other environmental resources for survival and economic stability at the interface of human society and environmental systems [1, 2]. Often, rural and indigenous populations around the world have a deep connection with the local ecosystems and rely on them for essentials like food, water, and shelter [3]. Given their close connection to natural resources, these communities are significantly affected by environmental changes, whether through climate variations, resource depletion, or biodiversity loss [4]. This dependency brings unique challenges, as traditional practices may sometimes clash with broader conservation efforts or government policies aimed at protecting the environment [5, 6]. However, recent research emphasizes the need to include these communities in environmental decision-making because they have intricate knowledge of their surroundings that can greatly benefit conservation efforts [4, 7].

CBEM has emerged as a promising framework that prioritizes community involvement in natural resource management. By incorporating local knowledge, CBEM aims to balance conservation with the sustainable use of resources, promoting both ecosystem health and community welfare [8]. The main idea behind CBEM is simple but powerful: by giving local people a say in managing their resources, we can better protect the environment and improve their quality of life.

Historically, many resource management policies ignored the human side of environmental issues, focusing mainly on ecosystem preservation. Traditional environmental policies often assumed that conservation and community needs could not coexist, especially for rural or indigenous populations who rely directly on the land for their livelihoods [2, 9]. This viewpoint led to policies that, although well-intentioned, sometimes restricted these communities from accessing their resources. However, with a more recent trend in environmental management over the past decades, researchers argue that traditional knowledge is irreplaceable in sustainable management as it gives insights that scientific data cannot always provide. CBEM, in this case, is then recognized for its potential in bridging the conservation to local needs by valuing the knowledge and lived experience of people directly connected to those resources [4].

The study of CBEM has attracted attention across various academic disciplines, including sociology, ecology, economics, and political science [1, 6, 10]. This broad interest has resulted in multiple perspectives, each contributing unique frameworks, methods, and goals to the field of CBEM [2, 11]. For instance, sociologists might examine how community structures affect resource-sharing practices, while ecologists may focus on biodiversity impacts. This interdisciplinary interest, while enriching the field, has also led to diverse definitions and approaches. There is still a lack of cohesive understanding of best practices across disciplines, highlighting a need for frameworks that integrate these perspectives.

One of the defining features of CBEM is its participatory approach [8, 9]. Rather than imposing top-down decisions, CBEM encourages local communities to be actively involved in every stage of environmental planning. This participatory model includes activities such as impact assessment, decision-making, and even long-term monitoring, all of which allow the community to shape policies directly affecting their environment [12, 13]. For example, local communities can contribute to impact assessments by identifying which species are most valuable to them or by suggesting practical conservation strategies that align with their traditions. This inclusive approach not only makes policies more effective but also increases the community's sense of ownership and responsibility toward the environment.

Another essential component of CBEM is decentralization, where power and responsibility for natural resource management are transferred from national authorities to local

governments and community organizations. This shift enables policies to be tailored specifically to local environmental and social contexts [14, 15]. It explains that when local authorities have control, they can implement measures that are more sensitive to the unique environmental dynamics of their region, making conservation efforts more effective.

Academic studies have further contributed to the development of CBEM by analyzing trends, identifying best practices, and suggesting areas for improvement [16, 17]. Researchers have utilized methods such as bibliometric analysis, which involves examining patterns in scientific publications, and meta-analysis to review existing literature on CBEM [18]. These studies often focus on specific branches within CBEM, like community forestry, which apply CBEM principles to sustainable forest management [2]. The study on Canadian community forestry highlighted shifts in research focus, moving from broad, general topics to more localized issues. Such research underscores the importance of combining traditional practices with scientific knowledge to improve CBEM's effectiveness [14]. CBEM offers numerous advantages to both communities and ecosystems [2, 3, 15]. First, it encourages dialogue among community members, which can help resolve conflicts over resource use. By fostering communication and cooperation, CBEM strengthens community adaptability, helping residents become more proactive in addressing environmental challenges. Additionally, CBEM encourages a better understanding of the human-ecology interface, enabling communities to balance their own needs with conservation goals. This balance is critical for long-term sustainability, as it allows communities to protect their environment while also meeting their basic needs [8]. Looking to the future, environmental assessments (EA) within CBEM are evolving to meet new sustainability challenges. The next-generation EA frameworks should strive for fairness and inclusivity, ensuring that benefits are equitably distributed among environmental, social, and economic spheres. These frameworks aim to minimize trade-offs and encourage communities to be actively involved in environmental governance. This inclusive approach will be essential as CBEM continues to expand globally.

In summary, CBEM provides a holistic framework for managing natural resources, one that acknowledges the invaluable role of local communities in conservation. By focusing on governance, scientific integration, and community participation, CBEM fosters sustainable resource use while empowering local populations. As technology advances, tools like AI and community-led monitoring are expected to enhance CBEM's capacity for global sustainability and resilience [19].

HISTORICAL EVOLUTION OF CBEM

CBEM has emerged as a response to the growing demand for sustainable and participatory ways of managing natural resources and ecosystems. Its roots lie in traditional and indigenous practices, where local communities established governance systems that would sustainably manage their environmental resources [3]. These governance frameworks frequently integrate extensive cultural and ecological insights, striving to harmonize human requirements with environmental conservation. For generations, indigenous groups practiced collective activities such as rotational agriculture, sacred grove preservation, and customary laws to ensure that their resource utilization remained within the ecological limits of their environments [11].

However, with the advent of industrialization, urban expansion, and centralized governance in the 19th and 20th centuries, indigenous peoples were marginalized by top-down approaches, promoting state authority and economic growth over community welfare, and these systems. This type of approach commonly led to alienation of local populations; resource conflict, inefficiencies, and environmental degradation [3, 20]. From the 1960s to the 1970s, the new, global environmental movement gained immense growth, driven by an increasing recognition of ecological deterioration and the repercussions of industrial growth

[4]. Pivotal publications, such as Rachel Carson's Silent Spring, underscored the ecological and social implications of abandoning environmental sustainability [1, 7, 20]. During this time, criticisms of fortress conservation- a model characterized by exclusion of local communities from access to forests, wildlife reserves, and other natural spaces emerged, showing that such strategies often resulted in conflict with local populations, especially in developing countries [5, 6].

International organizations, scholars, and activists started to advocate for participatory approaches that acknowledged the role of local communities as guardians of their natural environments [21]. In development, integrated rural development and community-based approaches have become a powerful approach to addressing poverty and environmental problems together, pointing to the intrinsic relationship between ecological integrity and human well-being. The 1980s and 1990s marked a watershed period in the development of CBEM as a global paradigm [4]. The 1987 Brundtland Report first presented the concept of sustainable development, emphasizing the need to balance economic growth, environmental protection, and social equity [4]. This was followed by the 1992 Rio Earth Summit, in which Agenda 21 clearly recognized the role of local communities in meeting international sustainability goals [10]. In response, governments and international entities began to decentralize power, transferring resource management duties from national bodies to local authorities, non-governmental organizations, and community collectives. Initiatives such as community forestry in Nepal, joint forest management in India, and community-based fisheries in the Philippines illustrated how empowering local populations could yield both conservation benefits and enhancements in livelihoods [10]. These approaches prioritized participatory planning, local stewardship, and the amalgamation of indigenous knowledge with contemporary scientific methodologies. As CBEM entered the 2000s, it faced complex issues such as climate change, biodiversity loss, and environmental justice. The international climate agreements that came into being, along with the increasing number of natural disasters, underscored the importance of ecosystems in providing resilience to climate-related stressors [7, 20]. As a result, community-based approaches assumed center stage in efforts such as ecosystem-based adaptation, watershed management, and mangrove restoration, which act as a safeguard ecosystem while the adaptive capacities of vulnerable communities are thereby strengthened [5, 10]. Furthermore, technological advancements, including Geographic Information Systems (GIS), satellite imagery, and mobile applications, have equipped communities with the tools necessary to monitor natural resources, report unlawful activities, and engage in informed decision-making [22]. This integration of technology has effectively connected local initiatives with broader global environmental objectives [23].

CBEM, despite its achievements, encounters considerable obstacles, such as disputes regarding resource rights, inadequate funding, a shortage of technical expertise, and fragile governance frameworks [24]. Disparities in power, especially within marginalized communities, may also impede fair participation and equitable distribution of benefits [4]. Nevertheless, CBEM presents substantial opportunities for the development of sustainable and resilient systems by cultivating collaborations among governmental bodies, civil society, and the private sector, enhancing local governance, and merging traditional knowledge with scientific advancements [5]. CBEM paves the way for tackling urgent environmental issues while fostering social and economic progress [4, 10]. Its progression signifies an increasing acknowledgment that effective resource management must be inclusive, adaptable, and driven by local initiatives, thereby safeguarding both ecological integrity and human welfare for future generations [4, 24].

CBEM has developed into a multidisciplinary hub that tackles global issues by integrating computational modeling, nanotechnology, and sustainable engineering practices. This integration aims to create renewable energy systems, combat climate change, and maintain the quality of water and air. These achievements highlight the progression of CBEM as an essential domain that operates at the convergence of technology, biology, and environmental conservation.

SUCCESS FACTORS IN CBEM

Strong Community Participation

Role of Community Involvement

Community participation is essential to CBEM as it ensures that management practices align with local needs, values, and knowledge systems [9]. Empowering community members brings ownership, sustainability, and accountability within environmental initiatives further into light [2, 15].

Methods for Effective Engagement

1. Workshops and Participatory Planning: Community workshops enable stakeholders to voice their concerns and co-design solutions [19]. Participatory Rural Appraisal (PRA) and other interactive methods help tap into local knowledge (Fig. **1**).
2. Incorporation of Traditional Knowledge: Integrating indigenous and traditional ecological knowledge can enhance the relevance and success of environmental strategies [9]

3. Capacity Building and Education: Training and educating communities on environmental issues and management practices can boost engagement and long-term commitment [2].

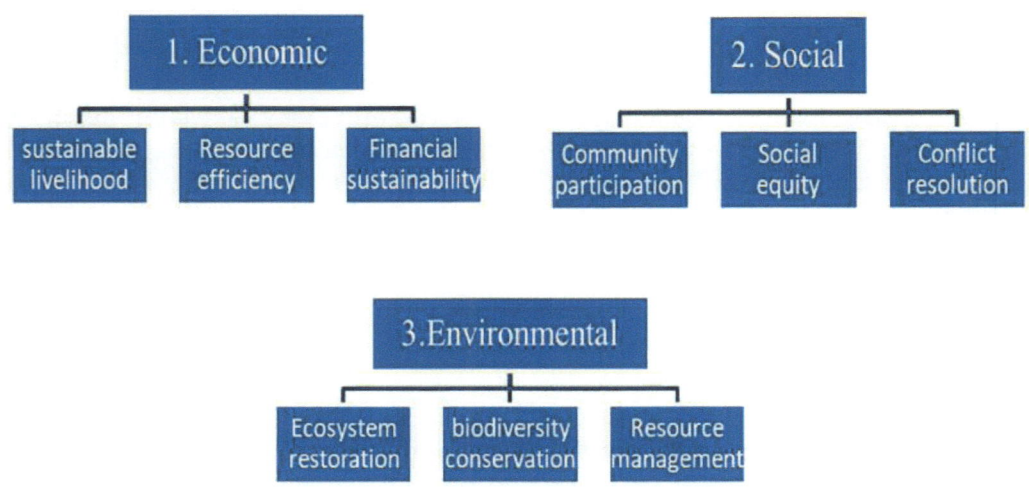

Fig. (1). The three pillars of sustainability.

Case Study

Successful stories of how community-led coastal management effectively conserved reefs can be drawn from countries such as the Philippines [25]

With its rich marine biodiversity and vast coastline in the Philippines, this country has supported various community-based initiatives intended to protect its coral reefs from overfishing and destructive fishing practices, as well as degradation of habitat.

A major example is 'The Apo Island Marine Reserve'. The Apo Island marine reserve is a global international model for marine resource sustainability management. Founded in the early 1980s, this initiative was the brainchild of the local community, coupled with scientists plus conservationists. As the fish populations started to dwindle and reefs began to be damaged, largely due to practices of unsustainability. As a result, the residents decided to ban fishing entirely over a specific area known as a "no-take" zone. This decision was based on traditional ecological knowledge and research [25].

The establishment of a "no-take zone" gave this ecosystem of coral reef a chance to regenerate, and it responded well with a marked increase in the populations of fish. The increase did not only occur within the protected area but also outside the no-take zones in the nearby fishing areas due to spillover. In fact, the local community experienced even greater economic benefits. When Apo Island emerged as a popular spot for divers and marine enthusiasts, all these activities created an influx of foreign tourism, generating cash for the local residents other than fishing [2, 25]. Regular involvement of the local community, strict rules implementation, and ongoing skill development programs supported by partnerships between academic institutions and NGOs played a significant role. This example illustrates how community-based coastal management can balance the environmental conservation aspect with socio-economic development. It points out the importance of integrating local knowledge, promoting cooperation among stakeholders, and building a solid policy framework for lasting success in reef preservation.

Supportive Policy Frameworks

Policy Enablers for CBEM

Effective CBEM requires policies that support decentralized decision-making, protect community rights, and facilitate financial resources for local projects [17]. Effective CBEM also implements policies that require all these important factors to cooperate.

Features of a Supportive Framework

1. Legal Recognition of Community Rights: Policies that legally recognize land, resource, and management rights are crucial to empowering communities [16].
2. Financial Incentives and Subsidies: Government support through grants or subsidies for conservation practices incentivizes communities.
3. Clear Governance Structures: Policies should establish transparent structures that clarify roles between government and community stakeholders [11, 26].

Examples of Successful Policy Frameworks

1. Costa Rica's Payment for Environmental Services (PES) program encourages communities to engage in conservation activities by providing financial rewards throughout the process [16, 17].
2. In Canada, the Aboriginal Forest Program provides policy support and funding to First Nations communities for sustainable forest management [27], which is yet another example of such.

Collaboration between Local and Global Stakeholders

Importance of Multi-Stakeholder Partnerships

Partnerships between local communities, NGOs, academic institutions, and international organizations enrich CBEM with diverse resources, expertise, and global perspectives [8, 25].

Models of Collaboration

1. Public-Private Partnerships (PPP): PPPs allow for shared funding and expertise in CBEM projects [12, 13].
2. International NGO Support: NGOs can bridge local efforts with global resources, bringing in technical knowledge, funding, and advocacy on international platforms [18].
3. Research Collaborations: Academic partnerships can provide communities with data and analysis tools, while field data from communities enhances scientific research [12].

Case Studies

Brazil: Sustainable Fisheries in the Amazon

In Brazil, partnerships among local fishers, government organizations, and the World Wide Fund for Nature (WWF) demonstrated the potential of partnerships to be used in the Amazon for achieving sustainable fisheries management. For many years, this region has been known for its richness of biodiversity and reliance on aquatic resources, but has been plagued by issues such as overfishing, destruction of habitats, and unregulated fishing activities. These issues not only put at risk the fish populations but also pose a threat to the livelihoods of the communities dependent on these resources [28].

To address these challenges, a participatory management plan was initiated. The local fishermen engaged in the formulation of plans for management, and their activities included fixing fishing quotas, identifying zones allotted for conservation, and observing closed seasons to rebuild the fish stock [29]. Technical assistance provided by agencies and supported government enforcement of the set rules and regulations, whereas WWF provided facilitation of community participation, undertaking research, and building up awareness of sustainable practices [27].

This approach has generated tangible fruits. Fish populations have regained their natural levels, whereas local people have enjoyed improvements and even more stable revenues from better access to fisheries. Second, the initiative built local capacity and instilled a culture of responsibility among fishers to preserve their natural resources. This is a case study exemplifying the benefits of combining the scientific and local knowledge on the sustainability of resource management systems while strengthening the resilience of communities in the area [30].

Africa: The Great Green Wall Project

This project is the largest effort to fight desertification and restore degraded lands in the Sahel, which covers Senegal to Djibouti. Backed by the African Union, working in tandem with organizations worldwide, national governments, and local communities. The aim of this project is to create a gigantic green belt to improve the livelihoods of the affected population [27].

Participation of local communities was a crucial component for the success of this project. They engage in activities such as tree plantation, cultivating drought-resistant crops, and using land management techniques sustainably. International agencies are instrumental in providing funding, technical support, and policies to ensure that the initiatives at the local level contribute to broader environmental and socio-economic objectives [28]. It has already borne fruit in several countries, with millions of planted trees. For instance, in Senegal, millions of trees have led to soil fertility enhancement, greater water retention, and even agricultural productivity improvement. Similarly, in Ethiopia, large-scale restorations of thousands of hectares of degraded land are being undertaken as a basis for food security and further poverty reduction [27].

The Great Green Wall also unlocked new opportunities in areas such as the economy, agriculture, forestry, renewable energy, and employment. It stands as one of the most visible examples of how partnerships between global organizations and local communities can build sustainable development and resilience to the impacts of climate change.

CHALLENGES AND GAPS IN CBEM

Lack of Long-term Sustainability

The lack of long-term sustainability is one of the most important issues in CBEM. Many efforts are intended to achieve immediate goals, such as reducing deforestation or increasing water quality, but they frequently fail to incorporate systems that assure continuity and adaptability in the face of shifting social, economic, and environmental situations [14]. In the lack of a long-term vision, projects usually lose pace once the early funding rounds have ended. Without consistent financial resources, many communities struggle to maintain or expand their management systems, resulting in resource deterioration or a return to unsustainable practices [1].

Another major issue is a failure to incorporate thorough resilience planning. Long-term sustainability in CBEM needs resilient systems that can adjust to unanticipated variables such as climate change, population increase, and economic changes [14, 31]. However, many projects lack the necessary tools or frameworks to predict these issues. For example, in areas where agriculture is highly dependent on local ecosystems (Fig. **2**), ignoring climate unpredictability or resource overexploitation might worsen vulnerabilities over time [12, 32].

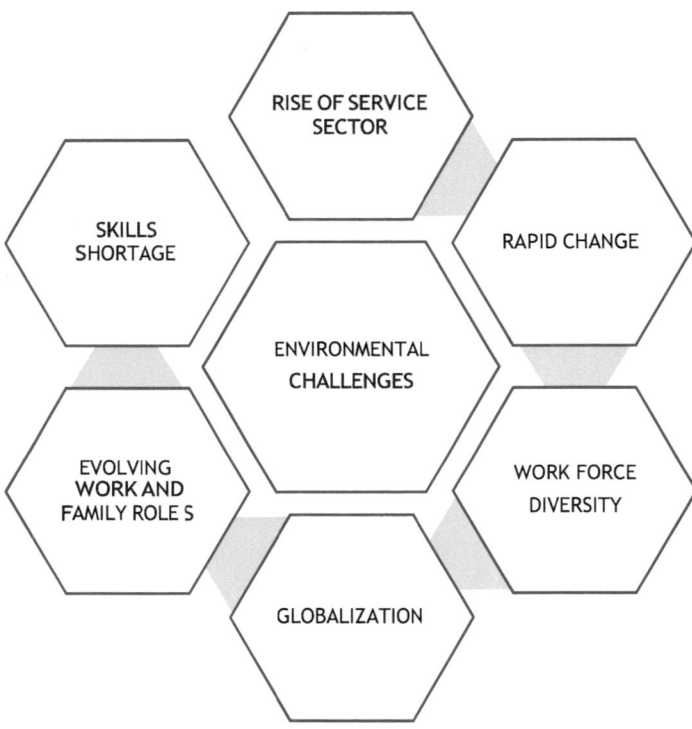

Fig. (2). Overview of challenges and gaps.

Intergenerational equity is frequently disregarded in CBEM. Sustainable development necessitates that current generations achieve their requirements without risking future generations' ability to do so [33]. However, CBEM initiatives generally lack explicit methods for addressing resource availability over time. This disparity is especially pronounced in circumstances when overutilization is motivated by short-term economic constraints, such as logging or fishing [14]. Balancing urgent livelihood demands with the necessity of conservation requires a sensitive, carefully planned approach, which is sometimes missing in practice (Fig. 2). Furthermore, the absence of monitoring and assessment methods worsens the difficulty of attaining long-term sustainability. Many CBEM programs do not build systematic feedback loops to evaluate outcomes, change strategy, or respond to changing community requirements. As a result, efforts may slow down, failing to learn from success and failure over time. This lack of dynamism impairs communities' ability to negotiate emerging risks, such as invasive species or changes in global trade patterns [34].

The inconsistency with which scientific knowledge is integrated into traditional practices contributes significantly to this difficulty. While traditional ecological knowledge provides invaluable insights into sustainable resource use, it is rarely complemented with rigorous scientific investigations that would improve its relevance to long-term management. For example, scientific modelling could assist communities in understanding how present resource use practices will influence ecosystems in the future. However, the disconnect between local practices and larger scientific frameworks frequently prevents CBEM initiatives from developing comprehensive sustainability plans. CBEM frameworks must undergo a fundamental change in order to meet these problems. Initiatives aimed at promoting resilience, adaptation, and equity must make long-term sustainability a top priority [14]. Over time, initiatives to build capacity can enable communities to efficiently manage resources, while financial tools like revenue-sharing plans or trust funds can offer steady assistance. Strategies can be kept current and successful by combining scientific and traditional knowledge with dynamic monitoring methods. CBEM can develop into a genuinely sustainable natural resource management paradigm by filling in these systemic shortcomings.

Insufficient Integration of Scientific Research and Traditional Practices

One major issue in CBEM is the inadequate fusion of traditional practices and scientific research. While scientific research gives tools like climate assessments, sophisticated monitoring systems, and predictive modelling that can supplement local practices, Traditional Ecological Knowledge (TEK) offers priceless insights derived from years of observation. Unfortunately, the efficacy of CBEM is frequently constrained by the disparity between these two strategies (Fig. 2). Many times, linguistic, infrastructure, or educational barriers prevent communities from accessing scientific results. On the other hand, scientists can ignore local knowledge, writing it off as informal or unstructured, which would limit prospects for collaboration [14].

Intentional strategies are needed to close this gap. First, context-sensitive solutions can be developed through co-creation of knowledge, in which local people and scientists work together to plan, carry out, and interpret research. For example, combining TEK with remote sensing technology can enhance biodiversity and land-use change monitoring [14]. Furthermore, by integrating community observations into scientific investigation, participatory research frameworks can strengthen communities and promote trust and respect for one another [18].

The lack of forums that encourage communication between scientists and traditional practitioners is another obstacle. Differences in terminology, priorities, and approaches continue to exist in the absence of efficient communication routes. Knowledge liberalisation can be aided by funding community-based training and education initiatives as well as the conversion of scientific data into easily comprehensible formats [34]. Mutual understanding can be fostered through workshops and community forums where stakeholders exchange ideas and work together to develop strategies.

Lastly, incorporating TEK and scientific research should go beyond project execution to include policymaking. The importance of both knowledge systems in developing comprehensive environmental strategies must be acknowledged by policymakers [32, 35]. For instance, localised vulnerabilities can be more successfully addressed by including TEK in climate adaptation programs in addition to scientific data (CBEM programs can guarantee that traditional and scientific knowledge make equal contributions to sustainable resource management by encouraging these linkages. Developing robust, flexible, and inclusive environmental initiatives requires closing this gap [36].

Socio-economic Barriers and Cultural Resistance in CBEM

Cultural opposition and socioeconomic hurdles provide significant obstacles to CBEM. The ability of many communities involved in CBEM to properly participate is limited by structural injustices like poverty, lack of access to education, and inadequate infrastructure [32, 33]. Adoption of sustainable practices is restricted by a lack of funding, and economic reliance on unsustainable resource exploitation, such as mining or logging, creates a contradiction between the demands of livelihood and conservation.

These issues are made worse by gaps in awareness and education. Since immediate economic concerns take precedence over long-term environmental advantages, CBEM efforts frequently encounter resistance due to a lack of knowledge (Fig. **2**). Communities might object to conservation efforts that limit their access to resources that they depend on for everyday survival, for instance [34]. Targeted interventions are needed to remove these obstacles, such as offering alternate sources of income, financial incentives, and capacity-building initiatives that strengthen communities and lessen reliance on exploitative practices.

The misalignment of CBEM tactics with conventional ideas or behaviours is frequently the cause of cultural resistance [31]. Communities run the risk of being alienated by conservation initiatives that disregard cultural values [33]. For example, confidence and collaboration may break down if externally devised resource management rules are imposed without first involving local authorities. Gender roles and cultural norms can also impede participation, particularly for women and marginalised groups, which limits the inclusiveness of CBEM programs.

A participatory approach that takes socioeconomic realities and cultural settings into account when designing and implementing CBEM programs is necessary to overcome these obstacles. These divides can be closed by offering forums for discussion, encouraging trust, and working with community members to jointly develop solutions. The success of CBEM projects can be increased by addressing economic inequality, reducing opposition, and providing culturally relevant participation, as well as by offering microfinance programs and opening markets for sustainable products [14].

Inadequate Funding and Resources in CBEM

CBEM implementation and sustainability are severely hampered by a lack of funds and resources [8, 12]. Financial resources are necessary for CBEM efforts to be effective in carrying out tasks, including infrastructure construction, community participation, monitoring, capacity-building, and the adoption of sustainable practices [13, 25]. However, many communities are unable to maintain long-term programs because of financial constraints brought either by their reliance on outside donors or the lack of adequate government support [8, 12, 18]. For instance, a shortage of funding may make it impossible to organise seminars for community training (Fig. 2) or purchase necessary equipment for environmental monitoring [13]. Furthermore, rural or marginalised areas frequently lack basic equipment, transportation, and communication networks, which makes it difficult for local actors to carry out conservation efforts [8]. These restrictions undermine the impact of CBEM initiatives and lower community motivation, which feeds environmental deterioration cycles.

The short-term nature of many funding possibilities, which results in fragmented project deadlines, is another significant obstacle [37]. It can be challenging for communities to plan or execute initiatives that call for consistent work since they frequently rely on grants with set objectives across short time frames [33]. The accessibility and efficacy of CBEM activities are further limited by administrative errors in cash allocation or resource misuse [18]. Innovative funding strategies like microfinance for sustainable businesses, public-private partnerships, and using carbon credits or ecotourism opportunities are required to address these problems [8]. Co-financing long-term initiatives with governments, corporations, and non-profits can also aid in closing the financial gap. To guarantee effective use of funds, capacity-building initiatives should concentrate on enhancing community members' financial literacy and resource management abilities [18]. To promote resource adequacy and guarantee the sustainability of CBEM projects, transparent and inclusive financial frameworks and consistent donor engagement are essential [8, 13].

EMERGING SOLUTIONS AND INNOVATIONS

New innovations and solutions in community-based CBEM revitalize the communities' perspective on how to attack environmental matters by integrating knowledge from locales with modern tools and instruments of governance [37, 38]. One popular trend is making use of digital tools and equipment, for instance, drones and satellite imagery; Geographic Information Systems (GIS), enabling real-time monitoring, and that's how management of natural resources can empower communities to take well-informed decisions which can be data-driven.

Eco-enterprise models, including sustainable livelihoods based on eco-tourism and green businesses, can provide the required economic incentives for environmental preservation, while circular economy strategies encourage waste minimization and resource reutilization [39, 40]. Environmental education and capacity building programs are provided through community workshops and mobile applications, which empower local populations, complemented by Citizen science projects, which include engaging communities in monitoring biodiversity and collecting climate data [41]. Collaborative government frameworks, including co-management systems, enhance inclusive decision-making processes [42].

Re-vegetation, such as reforestation and restoration of wetlands, has become a popular strategy in building up ecosystem resilience against climate change impacts. In addition, blockchain technology is improving transparency in resource management [42]. On the other hand, community-based adaptation strategies are promoting adaptability towards climate change through initiatives such as flood-resistant infrastructure and early warning systems. Lastly, the use of Indigenous knowledge and the establishment of green infrastructure, like community gardens and urban green spaces, is revolutionizing traditional practices and advancing sustainability in both rural and urban settings [40]. These innovations make a shift towards more inclusive, technology-driven, and sustainable strategies for managing the environment [39 - 41].

ROLE OF AI IN CBEM

Artificial Intelligence (AI) is increasingly integral to CBEM, thereby greatly improving decision-making, resource management, and monitoring of the environment [43]. Technologies like machine learning, computer vision, and data analytics are enhancing the ways in which communities oversee and manage their natural resources, thereby increasing the effectiveness and scalability of environmental management [21].

The key applications of AI in CBEM are through data analysis and predictive modeling. AI algorithms can handle vast environmental data sets such as satellite imagery, climate information, and sensor data to establish patterns, predict future environmental changes, and monitor in real-time [44, 45]. For example, AI may analyze satellite images to spot cases of deforestation or land degradation, through which communities are able to respond in a timely manner to the needs for conservation. Also, machine learning models can predict climate change impact on the local ecosystems, equipping the community with better preparation for disasters or the shift in biodiversity [44].

Other key contributions to CBEM are AI decision support systems that integrate the environmental data with community feedback to form solutions that resonate with local priorities [46]. For example, AI can help communities in designing sustainable use of land strategies or water management frameworks by analyzing data on soil conditions, water resources, and agricultural practices. By integrating AI with GIS, communities can visualize environmental data, thus making informed decisions on resource distribution and conservation initiatives [47].

In addition, AI supports handling and reporting tasks, and AI-enabled remote sensing technologies can autonomously gather environmental data from regions with less accessibility, such as forests or coastal areas, without requiring much human intervention [39]. This capability enhances the ability of communities to monitor most of the locations, thereby improving the overall effectiveness of environmental management.

COMMUNITY LED MONITORING SYSTEM

Community-Led Monitoring Systems (CLMS) also play an important role in CBEM, empowering the grassroots population to play an active role in observation and resource management, including water, which, through these systems, collect all the data across environmental aspects [38]. Biodiversity and land use, for instance, often utilize technologies like mobile applications, drones, and sensors. CLMS integrates indigenous knowledge with scientific methodologies to yield a holistic and accurate representation of environmental dynamics [37].

They also foster participation in data analysis, giving the opportunity for community members to work together with experts on the interpretation of data, thus leading to better management strategies that respond to the contexts of communities. CLMS also supports local governance in terms of capacity building, as it offers training and ensures that communities have the tools and skills for environmental monitoring and reporting. As such, this empowers the communities to advocate for policies that conserve their ecosystems. It involves residents, scientists, and governmental bodies; hence, it creates a group approach to solving environmental problems [8].

This community-led monitoring not only increases resilience in the environment but also builds accountability in addition to supporting the sustainable management of resources, making significant progress in CBEM [33]. Innovative funding mechanisms play an essential role in the sustainability and effectiveness of CBEM initiatives. Payment for Ecosystem Services is an excellent example whereby communities are paid for conserving ecosystems that deliver essential services for human livelihoods, such as clean water, among others [8]. The crowdfunding phenomenon has also promoted the aggregation of small resources of local communities toward funding targeted environmental projects [32, 35].

Blended finance approaches merged public and private investments to support larger environmental initiatives, while carbon markets enable communities to generate income through forest conservation and activities that help reduce the amount of greenhouse gases in the atmosphere [14]. SIBs and green bonds offer early financing for environmental projects, with returns to investors based on successful achievement of specific outcomes, such as improved water quality or habitat restoration. Philanthropic donations and impact investing also offer grants or equity investments to strengthen projects that address both environmental and social goals [8]. Grants and other subsidies from local and central governments, as well as the establishment of Environmental Trust Funds ETFs, further strengthen financial resource mobilization for community-led environmental activity [41]. These new funding structures, apart from providing indispensable funds for environmental management activities, also promote collaboration among communities, bodies of government, businesses, and other international organizations to sustain CBEM projects over time for effective implementation.

FUTURE DIRECTIONS IN CBEM

Adopting Advanced Technologies

The future of CBEM revolves around the integration of Geographic Information Systems (GIS), remote sensing, and big data analytics. These advances can improve resource monitoring (Fig. 3) while allowing for informed decision-making on the community level. Important to this end are platforms supporting data sharing to empower local stakeholders [8]. Encouraging Collaborative Approaches: Strengthening relationships between governments, NGOs, and community members is vital for CBEM success [7]. Establishing participatory governance systems helps align policies with the unique needs of each community, fostering trust and cooperation [8].

Fig. (3). A directional future view of CBEM.

1. Defining Sustainability Benchmarks: To guide improvement, it is necessary to set long-term sustainability goals and measure CBEM's effectiveness through clear metrics. Some of the examples include biodiversity indicators and resource use efficiency rates.
2. Incorporation of Climate Adaptation Strategies: Future CBEM programs must focus on climate adaptation, which includes practices like agroforestry, wetland restoration, and the incorporation of renewable energy systems in order to avoid climate-related risks and enhance resilience.

Addressing Challenges

Resolving Conflicts: Disputes over resource access and benefits often hinder CBEM progress. Establishing platforms for mediation and training community leaders in conflict resolution techniques can alleviate these challenges [44]. Overcoming Funding Constraints: Insufficient funding is a common barrier for CBEM initiatives (Fig. **3**). Innovative solutions such as leveraging eco-tourism revenue, microfinance, and forming partnerships with the private sector can help bridge financial gaps.

1. Improving Policy Frameworks: Current policies may not entirely support CBEM. For sustainable progress, there is a need to promote a robust legal framework that supports community rights and the powers of local stakeholders.
2. Strengthening Community Involvement: Low participation can be attributed to socio-economic barriers or ignorance. Some effective strategies include workshops and outreach programs that are all-inclusive to reach every member of the community [45].

Leveraging Technological Advancements

Advanced technologies, like AI-based predictive models and mobile applications, can facilitate resource management in a more efficient manner while making CBEM even more transparent (Fig. **3**).

1. Involving Marginalized Sectors: Involving youths and women in the initiatives of CBEM will add more significance to the cause, utilizing their latent potential towards resource stewardship.
2. Link with Global Sustainability Objectives: Integrating CBEM projects with global frameworks, such as the SDGs, could bring in funding and create an international platform.

Strategies for Strengthening CBEM

1. Building Community Capacity: Training local participants in sustainable practices and Leadership ensures that communities can independently manage and sustain their projects over time [8].
2. Adopting Flexible Management Models: Flexible, adaptive management approaches allow CBEM programs to evolve based on lessons learned and changing environmental conditions.

3. Ensuring Transparency and Accountability: Transparent systems for decision-making and resource allocation build trust and encourage long-term commitment from community members [43].

Promoting Knowledge Exchange

The establishment of networks would promote innovation and cross-learning of successful models, tools, and experiences among CBEM practitioners. Incorporation of Traditional Knowledge: Combining traditional ecological knowledge with modern practices will provide culturally relevant solutions and thereby enhance the effectiveness of any initiatives of CBEM [2].

Through addressing challenges and capitalizing on emerging opportunities, CBEM programs can become better systems for community-driven environmental conservation and sustainable development (Fig. **3**).

CONCLUSION

CBEM stands as a transformative approach to addressing environmental challenges, uniting ecological preservation with social and economic progress of the society using measurable processes [2, 35]. By prioritizing local community involvement, CBEM fosters an inclusive framework that integrates, expands, and works with indigenous knowledge along with scientific advancements, enabling tailored solutions for sustainable resource management [7, 8]. The historical evolution of CBEM underscores the importance of participatory and decentralized governance systems, which have demonstrated significant success in fostering ecosystem resilience and addressing the needs of local populations of places [37, 38]. Specific examples of successful CBEM projects are, e.g., Joint Forest management in India. Joint forest management in India has empowered local communities to conserve forests while improving their livelihoods, showcasing the potential of integrating indigenous practices with scientific forestry for enhancements [2].

Despite its promising outcomes, CBEM faces notable challenges, such as inadequate funding for fulfilling the requirements of desired goals, socio-economic barriers like resource inequities, gender disparities, or cultural resistance, and gaps in the integration of traditional and scientific practices. Addressing socio-economic disparities, such as unequal access to resources and decision-making power, particularly for marginalized groups, requires targeted capacity-building programs and inclusive policy frameworks [33]. Overcoming these hurdles requires innovative funding mechanisms to be boosted, robust policy support with consistency, and capacity-building initiatives to ensure the long-term sustainability of CBEM projects [2]. Emerging technologies, such as AI and GIS, coupled with community-led monitoring systems, provide exciting opportunities for enhancing the efficiency and scalability of these initiatives [44]. Using these advancements, AI can be used for predictive modelling, bringing positive or negative approximate predictions for climate-related risks, if not the accurate results. While AI helps with these processes, emerging technologies, such as blockchain, offer a potential for enhancing transparency in resource allocation. By adopting these major advancements, CBEM can explain the complexities of modern environmental issues with modern solutions, including climate change and loss of biodiversity, in a better way [21, 45].

The future of CBEM lies in embracing collaborative and equal partnerships among governments, NGOs, the private sector, and local communities of the desired area or on an international level. The strength of these relationships is directly proportional to the cocreation of strategies that adopt both inclusive and adaptable habits to these changing environmental conditions [2, 8]. Moreover, the integration of global sustainability goals with local practices and the community's support will enhance CBEM's role in achieving and promoting equitable and long-lasting environmental solutions. With the challenges mentioned, recommendations for policy changes or actions that can bolster CBEM include focus on community strengthening rights for the communities to pair up and take a step towards the goal, providing financial incentives for conservation practices, and fostering collaborations among stakeholders to ensure CBEM initiatives achieve their full potential with these. In conclusion, CBEM offers a dynamic and inclusive paradigm for managing natural resources, emphasizing the interconnectedness of ecological integrity, social equity, and economic development as explained using diagrams and flowcharts [2, 8]. CBEM works with global sustainability frameworks like the UN Sustainable Development Goals alongside better systems and examples. As the years move ahead of us, the evolution of CBEM, directed by innovation, guided by collaboration, and committed to sustainability, will be crucial in addressing a lot of major global environmental challenges and promoting resilience in the face of future uncertainties.

REFERENCES

[1] Rodgers P, Vershinina N, Williams CC, Theodorakopoulos N. Leveraging symbolic capital: the use of *blat* networks across transnational spaces. Glob Netw 2019; 19(1): 119-36.
[http://dx.doi.org/10.1111/glob.12188]

[2] Surjeet Kumar, Vyas N. Socio-economic empowerment of local communities through ecotourism: A review analysis. EPRA Int J Soc-Econ Environ Outlook 2022; 9(3): 10-8. [SEEO].

[3] Shepherd DA, Parida V, Wincent J. The surprising duality of jugaad: Low firm growth and high inclusive growth. J Manage Stud 2020; 57(1): 87-128.
[http://dx.doi.org/10.1111/joms.12309]

[4] Magni G. Indigenous knowledge and implications for the sustainable development agenda. Eur J Educ

2017; 52(4): 437-47.
[http://dx.doi.org/10.1111/ejed.12238]

[5] Aria M, Cuccurullo C. *bibliometrix*: An R-tool for comprehensive science mapping analysis. J Informetrics 2017; 11(4): 959-75.
[http://dx.doi.org/10.1016/j.joi.2017.08.007]

[6] Fordham AE, Robinson GM. Robinson GM. Mechanisms of change: Stakeholder engagement in the Australian resource sector through CSR. Corp Soc Resp Environ Manag 2018; 25(4): 674-89.
[http://dx.doi.org/10.1002/csr.1485]

[7] Kraus S, Breier M, Dasí-Rodríguez S. The art of crafting a systematic literature review in entrepreneurship research. Int Entrep Manage J 2020; 16(3): 1023-42.
[http://dx.doi.org/10.1007/s11365-020-00635-4]

[8] Mistry J, Berardi A. The challenges and opportunities of participatory video in geographical research: exploring collaboration with indigenous communities in the North Rupununi, Guyana. Area 2012; 44(1): 110-6.
[http://dx.doi.org/10.1111/j.1475-4762.2011.01064.x]

[9] Das M, Chatterjee. B. Community empowerment and conservation through ecotourism: A case of Bhitarkanika Wildlife Sanctuary, Odisha, India. Tour Rev Int 2020; 24(4): 215-31.
[http://dx.doi.org/10.3727/154427220X15990732245655]

[10] Nnamani C, Ajayi S, Oselebe H, Atkinson C, Igboabuchi A, Ezigbo E. *Sphenostylis stenocarpa* (Ex. A. Rich.) harms., a fading genetic resource in a changing climate: Prerequisite for conservation and sustainability. Plants 2017; 6(3): 30.
[http://dx.doi.org/10.3390/plants6030030] [PMID: 28704944]

[11] Singh A, Mishra R. Role of civil society in environmental governance: Case studies from India. Env Polit 2018; 27(4): 700-18.

[12] Berardi A, Mistry J, Tschirhart C, *et al.* Applying the system viability framework for cross-scalar governance of nested social-ecological systems in the Guiana Shield, South America. Ecol Soc 2015; 20(3): 42.
[http://dx.doi.org/10.5751/ES-07865-200342]

[13] Ruiz-Mallén I, Schunko C, Corbera E, Rös M, Reyes-García V. Meanings, drivers, and motivations for community-based conservation in Latin America. Ecol Soc 2015; 20(3): art33.
[http://dx.doi.org/10.5751/ES-07733-200333]

[14] Beausoleil D, Munkittrick K, Dubé MG, Wyatt F. Essential components and pathways for developing Indigenous community-based monitoring: examples from the Canadian oil sands region. Integr Environ Assess Manag. 2021 Nov; 17(6): 1139–1152.
[http://dx.doi.org/10.1002/ieam.4485]

[15] Demkova M, Sharma S, Mishra PK, *et al.* Potential for sustainable development of rural communities by community-based ecotourism: A case study of rural village Pastanga, Sikkim Himalaya, India. Geo J Tour Geosites 2022; 43(3): 964-75.
[http://dx.doi.org/10.30892/gtg.43316-910]

[16] Gupta M, Sharma A. Effectiveness of environmental laws in addressing pollution: Empirical evidence from India. J Environ Manage 2021; 290: 112712.

[17] Beg N, Morlot JC, Davidson O, et al. Linkages between climate change and sustainable development. Climate Policy. 2002 Sep; 2(2–3): 129–144.
[http://dx.doi.org/10.3763/cpol.2002.0216]

[18] Franco S, Mandla VR, Rao KRM. Urbanization, energy consumption and emissions in the Indian context: a review. Renew Sustain Energy Rev. 2017 May; 71: 898–907.
[http://dx.doi.org/10.1016/j.rser.2016.12.117]

[19] Kunjuraman V. Community-based ecotourism managing to fuel community empowerment? An

evidence from Malaysian Borneo. Tour Recreat Res 2022; 47(4): 384-99.
[http://dx.doi.org/10.1080/02508281.2020.1841378]

[20] Donthu N, Kumar S, Mukherjee D, Pandey N, Lim WM. How to conduct a bibliometric analysis: An overview and guidelines. J Bus Res 2021; 133: 285-96.
[http://dx.doi.org/10.1016/j.jbusres.2021.04.070]

[21] Birner R, Daum T, Pray C. Who drives the digital revolution in agriculture? A review of supply-side trends, players and challenges. Appl Econ Perspect Policy 2021; 43(4): 1260-85.
[http://dx.doi.org/10.1002/aepp.13145]

[22] Addison J, Friedel M, Brown C, Davies J, Waldron S. A critical review of degradation assumptions applied to Mongolia's Gobi Desert. Rangeland J. 2012; 34(2): 125–137.
[http://dx.doi.org/10.1071/rj11013]

[23] Priyadarshini P, Abhilash PC. Promoting tribal communities and indigenous knowledge as potential solutions for the sustainable development of India. Environ Dev 2019; 32(4): 100459.
[http://dx.doi.org/10.1016/j.envdev.2019.100459]

[24] Segger MCC, Phillips FK. Indigenous traditional knowledge for sustainable development: the biodiversity convention and plant treaty regimes. J For Res 2015; 20(5): 430-7.
[http://dx.doi.org/10.1007/s10310-015-0498-x]

[25] Mudge L. Use of community perceptions to evaluate and adapt coastal resource management practices in the Philippines. Ocean Coast Manage 2018; 163: 304-22.
[http://dx.doi.org/10.1016/j.jaridenv.2010.12.019]

[26] Cheng Y, Tsubo M, Ito TY, Nishihara E, Shinoda M. Impact of rainfall variability and grazing pressure on plant diversity in Mongolian grasslands. J Arid Environ. 2011 May; 75(5): 471–476.
[http://dx.doi.org/10.1016/j.jaridenv.2010.12.019]

[27] Bernard E, Penna LAO, Araújo E. Downgrading, downsizing, degazettement, and reclassification of protected areas in Brazil. Conserv Biol 2014; 28(4): 939-50.
[http://dx.doi.org/10.1111/cobi.12298] [PMID: 24724978]

[28] Hallwass G, Lopes PF, Juras AA, Silvano RAM. Fishers' knowledge identifies environmental changes and fish abundance trends in impounded tropical rivers. Ecol Appl 2013; 23(2): 392-407.
[http://dx.doi.org/10.1890/12-0429.1] [PMID: 23634590]

[29] Sá-Oliveira JC, Ferrari SF, Vasconcelos HCG, et al. Restoration effects of the riparian forest on the intertidal fish fauna in an urban area of the Amazon River. Sci World J 2016; 2016: 1-9.
[http://dx.doi.org/10.1155/2016/2810136] [PMID: 27699201]

[30] McGrath DG, Castello L, Almeida OT, Estupiñán GMB. Market formalization, governance, and the integration of community fisheries in the Brazilian Amazon. Soc Nat Resour 2015; 28(5): 513-29.
[http://dx.doi.org/10.1080/08941920.2015.1014607]

[31] Coppock DL, Fernández-Giménez ME, Hiernaux P, et al. Rangeland systems in developing nations: Conceptual advances and societal implications. In: Briske DD, Ed. Rangeland systems: Processes, management and challenges. Cham: Springer International Publishing 2017; pp. 569-641.
[http://dx.doi.org/10.1007/978-3-319-46709-2_17]

[32] Moritz M, Gardiner E, Hubbe M, Johnson A. Comparative study of pastoral property regimes in Africa offers no support for economic defensibility model. Curr Anthropol 2019; 60(5): 609-36.
[http://dx.doi.org/10.1086/705240]

[33] Jacobs MJ, Schloeder JA, Tanimoto PD. Dryland agriculture and rangeland restoration priorities in Afghanistan. J Arid Land. 2015; 7(3): 412–420.
[http://dx.doi.org/10.1007/s40333-015-0002-7]

[34] Baggio JA, Barnett AJ, Perez-Ibarra I, et al. Explaining success and failure in the commons: the configural nature of Ostrom's institutional design principles. Int J Commons 2016; 10(2): 417-39.
[http://dx.doi.org/10.18352/ijc.634]

[35] Keshkamat SS, Tsendbazar NE, Zuidgeest MHP, Shiirev-Adiya S, van der Veen A, van Maarseveen MFAM. Understanding transportation-caused rangeland damage in Mongolia. J Environ Manage. 2013 Jan 15; 114: 433–444.
[http://dx.doi.org/10.1016/j.jenvman.2012.10.043]

[36] Leisher C, Hess S, Boucher TM, van Beukering P, Sanjayan M. Measuring the impacts of community-based grasslands management in Mongolia's Gobi. PLoS One. 2012 Feb 1; 7(2): e30991.
[http://dx.doi.org/10.1371/journal.pone.0030991]

[37] Dell'Angelo J, McCord PF, Gower D, Carpenter S, Caylor KK, Evans TP. Community water governance on Mount Kenya: An assessment based on Ostrom's design principles of natural resource management. Mt Res Dev 2016; 36(1): 102-15.
[http://dx.doi.org/10.1659/MRD-JOURNAL-D-15-00040.1]

[38] Luintel H, Bluffstone RA, Scheller RM, Adhikari B. The effect of the Nepal Community Forestry Program on equity in benefit sharing. J Environ Dev 2017; 26(3): 297-321.
[http://dx.doi.org/10.1177/1070496517707305]

[39] Mössner S, Miller B. Sustainability in one place? Dilemmas of sustainability governance in the Freiburg Metropolitan Region. Regions Magazine 2015; 300(1): 18-20.
[http://dx.doi.org/10.1080/13673882.2015.11668692]

[40] Nicholson S. The West African Sahel: a review of recent studies on the rainfall regime and its interannual variability. ISRN Meteorol. 2013; 2013: 453521.
[http://dx.doi.org/10.1155/2013/453521]

[41] Sen PK. Sri Aurobindo: His Life and Yoga. HarperCollins Publishers India 2018.

[42] Kapur A. Better to Have Gone: Love, Death, and the Quest for Utopia in Auroville. Scribner Book Company. 2021.

[43] Cohen-Shacham E, Andrade A, Dalton J, *et al.* Core principles for successfully implementing and upscaling Nature-based Solutions. Environ Sci Policy 2019; 98: 20-9.
[http://dx.doi.org/10.1016/j.envsci.2019.04.014]

[44] Dwivedi YK, Kshetri N, Hughes L, *et al.* Opinion Paper: "So what if ChatGPT wrote it?" Multidisciplinary perspectives on opportunities, challenges and implications of generative conversational AI for research, practice and policy. Int J Inf Manage 2023; 71: 102642.
[http://dx.doi.org/10.1016/j.ijinfomgt.2023.102642]

[45] Kitamura FC. ChatGPT is shaping the future of medical writing but still requires human judgment. Radiology 2023; 307(2): e230171.
[http://dx.doi.org/10.1148/radiol.230171] [PMID: 36728749]

[46] Pérez-Peña MC, Jiménez-García M, Ruiz-Chico J, Peña-Sánchez AR. Analysis of research on the SDGs: The relationship between climate change, poverty, and inequality. Appl Sci (Basel) 2021; 11(19): 8947.
[http://dx.doi.org/10.3390/app11198947]

[47] Mocuta DN. Influence of the climate changes on the human life quality in rural areas. Revista de Chimie 2017; 68(6): 1392-6.
[http://dx.doi.org/10.37358/RC.17.6.5680]

CHAPTER 13

Reduced Graphene Oxide-based Solutions for Water Purification: Advances in Sustainable Nanocomposites

Deepak Dahiya[1], Ashish Sharma[2], Sweety Dahiya[3], Pooja Yadav[1], Kiran Kaushik[1] and Sudesh Chaudhary[1,*]

[1] *Center of Excellence for Energy and Environmental Studies, Deenbandhu Chhotu Ram University of Science and Technology, Murthal, Sonipat, Haryana, India*

[2] *Department of Chemical Engineering, Deenbandhu Chhotu Ram University of Science and Technology, Murthal, Sonipat, Haryana, India*

[3] *Department of Environmental Sciences, School of Basic and Applied Sciences, SGT University, Gurugram, Haryana, India*

Abstract: Environmental pollution, particularly water contamination, is a major hazard to world ecosystems and public health. With increased concerns about the inefficiencies and ecological risks connected with traditional cleanup procedures, there is a renewed emphasis on sustainable alternatives. This work provides a complete overview of reduced Graphene Oxide (rGO)-based nanocomposites, including synthesis, rGO characteristics, pollutant removal processes, environmental benefits, and obstacles. rGO-based nanocomposites have gained popularity due to their distinct physical, chemical, and adsorption capabilities, making them appropriate for the removal of a wide range of contaminants, including heavy metals, organic dyes, and medicines. Their ability to provide long-term, scalable, and environmentally friendly remedial solutions is highlighted.

Keywords: Environmental benefits, Pollutant removal processes, Reduced Graphene Oxide (rGO), rGO-based nanocomposites, Water contamination.

INTRODUCTION

The pollution of aquatic ecosystems, and thus environmental degradation, has become a critical global concern in the 21st century. Aquatic ecosystems frequently suffer contamination from various heavy metals, organic pollutants, pharmaceuticals, and toxic industrial chemicals, predominantly sourced from

---

[*] **Corresponding author Sudesh Chaudhary:** Center of Excellence for Energy and Environmental Studies, Deenbandhu Chhotu Ram University of Science and Technology, Murthal, Sonipat, Haryana, India; E-mail: sudesh.energy@dcrustm.org

agricultural runoff, industrial discharge, and urban emissions. Waterborne pollutants present considerable risks to aquatic life and adversely affect human health *via* bioaccumulation and biomagnification. The growing apprehensions over the detrimental impacts of these pollutants on the ecosystem's sustainability and public health underscore the imperative for effective wastewater treatment solutions [1]. Traditionally, standard wastewater treatment techniques such as coagulation, sedimentation, adsorption, and chemical oxidation have been extensively employed to alleviate water contamination. Nonetheless, these methodologies frequently exhibit constraints. Chemical treatments may produce dangerous by-products, whereas coagulation and sedimentation procedures often generate significant amounts of sludge, leading to secondary pollution. The elevated operational and maintenance expenses of conventional technologies render them less viable for broad, long-term environmental management [2].

Recognizing the constraints of traditional methods, academics and environmental engineers are currently concentrating on creating innovative, eco-friendly, and economically viable remediation technologies. Nanoparticles, due to their distinctive physicochemical characteristics, have demonstrated significant potential in enhancing pollutant removal efficacy, rendering nanotechnology an effective instrument in addressing environmental contamination [3]. Among the several nanomaterials investigated, nanocomposites utilizing rGO have attracted considerable attention. rGO, a derivative of graphene, preserves numerous beneficial properties of graphene, including a high specific surface area, exceptional electrical conductivity, and enhanced mechanical strength, while also providing additional advantages from the presence of oxygen-containing functional groups. These functional groups facilitate the interaction and amalgamation of rGO with diverse materials, such as metals, metal oxides, and polymers, leading to rGO-based nanocomposites with improved pollutant removal efficacy [3]. The distinctive characteristics of rGO-based nanocomposites render them exceptionally effective in adsorbing and removing various pollutants, including heavy metals such as lead, cadmium, and mercury, along with organic compounds, including colors, medicines, and pesticides. The large surface area of rGO offers multiple active sites for pollutant adsorption, and its remarkable chemical stability and ease of functionalization facilitate the creation of nanocomposites designed to target specific pollutants. Moreover, the superior recyclability and reusability of rGO render it a cost-efficient and ecologically friendly choice for enduring pollution cleanup [4]. This review is to thoroughly examine the function of rGO-based nanocomposites in environmental remediation, in light of the growing focus on sustainability and the pressing necessity to tackle water contamination [5]. This study will investigate their use in removing diverse contaminants, assess their efficacy relative to traditional materials, and analyze the challenges and future potential for their incorporation into extensive water treatment systems.

SYNTHESIS AND CHARACTERISTICS

By lowering the oxygen-containing functional groups in GO, rGO is a type of GO with a partially recovered graphitic structure. While maintaining certain oxygen functions that are necessary for surface interactions with contaminants, this partial decrease enhances the material's mechanical and conductivity qualities [6].

RGO Synthesis

Chemical reduction techniques—such as the use of reducing agents like hydrazine, sodium borohydride, or ascorbic acid—are commonly employed in the synthesis of rGO. In order to recover the conductivity and structural integrity of GO, thermal and electrochemical reduction techniques are also utilised [7]. In an effort to reduce the negative environmental effects of GO synthesis, recent trends emphasise the use of sustainable and \environmentally friendly reducing agents, such as plant extracts.

As shown in Fig. (1), the different methods of preparing rGO usually involve the reduction of GO, which is obtained from graphite using chemical, thermal, or electrochemical processes [8]. The reduction approach aims to improve the electrical, thermal, and mechanical properties of GO by building the conjugated carbon network through the removal of oxygen-containing functional groups. Various methods are available for synthesising rGO, each with unique advantages and limitations regarding reduction efficiency, environmental impact, and scalability [7].

Chemical Reduction

Chemical reduction is a widely used and extensively utilized technique for the synthesis of rGO. Under this procedure, reducing agents are introduced into suspensions of GO to eliminate oxygen-containing groups, hence producing rGO.

Commonly used reducing agents: Hydrazine hydrate is a well-established and widely used reducing agent. Hydrazine hydrate efficiently decreases Gallic Oxide, yet it is both poisonous and not ecologically benign [9].

Natrium Borohydride ($NaBH_4$): An extensively used reducing agent that achieves a moderate reduction of Graphene Oxide (GO) but necessitates elevated temperatures for optimal effectiveness. Aside from its toxicity, it has a restricted capacity for reduction [10].

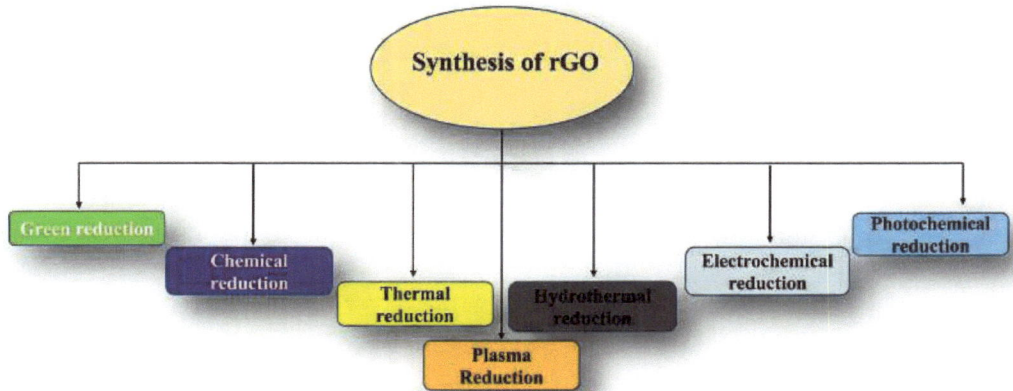

Fig. (1). Different methods of preparing (rGO).

Ascorbic Acid, often known as Vitamin C: Ascorbic acid, a more environmentally friendly and non-toxic alternative to hydrazine, is highly effective under modest treatment settings. It has become prominent for the sustainable synthesis of rGO [11].

Benefits

Simplicated procedure, possible to be executed under moderate circumstances. Enables manipulation of the extent of reduction by changes in reagent concentration and reaction duration.

Drawbacks

The probable outcome is a partial decrease. Utilization of hazardous substances in specific techniques, the possibility of aggregation of rGO sheets.

Thermal reduction reaction

Thermal reduction is the process of subjecting GO to elevated temperatures in order to eliminate oxygen-containing functional groups. The aforementioned procedure reinstates the graphitic arrangement of GO and yields rGO with enhanced electrical and thermal characteristics.

Conventional methodologies: Direct heating refers to the process of heating GO in an inert or reducing environment, such as argon or hydrogen, at temperatures between 200°C and 1000°C. Typically, the process has a duration ranging from a few minutes to several hours.

Microwave heating is a fast technique that utilises microwave irradiation to heat GO. The speed and energy efficiency of this approach surpass those of traditional heating methods [7].

Benefits

Yields rGO with superior conductivity in comparison to chemical reduction techniques. Excludes the use of recuing agents, therefore enhancing cleanliness. Attains a significant level of reduction.

Drawbacks

Demands elevated temperatures and substantial energy input. Thermal stress-induced potential structural damage to graphene. Forms rGO with the remaining oxygen groups.

Reduction using electrochemical means

Electrochemical reduction is the process of reducing GO inside an electrolyte solution by the application of an electric current. This method entails the application of a negative voltage to electrodes coated with GO in order to stimulate the reduction of oxygen-containing groups.

Typical Procedures

In electrochemical cells, GO is commonly applied as a coating over a conductive substrate, such as glassy carbon, and then immersed in an aqueous electrolyte solution, such as potassium chloride or sulphuric acid. A constant voltage is used to convert GO to rGO [12]. In flow cells, GO dispersions are subjected to an electric current in order to decrease the concentration of GO.

Benefits

Sanitary and ecologically sustainable: It enables meticulous regulation of the level of decrease by manual adjustment of the applied potential. Suitable for integration with other processes in real-time applications.

Drawbacks

Considerably time-consuming in comparison to chemical and thermal techniques. It has restricted scalability and has comparatively low conductivity of rGO in comparison to thermal techniques.

Reduction by Photochemical Processes

Photochemical reduction is the exposure of GO to light, often ultraviolet (UV) or visible light, in the presence of a reducing agent. Photons stimulate the surface of GO, hence promoting the reduction of oxygen groups and reinstating the graphitic crystal structure [13].

Conventional Techniques

GO is subjected to UV radiation when reducing agents such as hydrogen gas or alcohol are present. UV-induced photoreduction disrupts the carbon-oxygen bonds, resulting in the creation of rGO. Visible light has been employed in certain investigations to decrease GO under gentle circumstances, in conjunction with photocatalysts such as TiO_2.

Benefits

Capable of being executed at moderate temperatures and pressures, enabling precise spatial reduction, such as the patterning of GO sheets. Eschews the usage of hazardous substances.

Drawbacks

Necessitates a light source and meticulous regulation of irradiation. Generally yields reduced conductivity compared to thermal techniques. Complete decrease varies depending on the intensity and duration of light.

Biological (Green) Analytical Reduction Biological or green reduction is an environmentally benign technique that employs natural reducing catalysts such as plant extracts, microbes, or proteins to convert GO into rGO. This method obviates the need for hazardous substances, therefore establishing it as a viable choice for mass manufacturing [2].

Prevalent agents:

Plant Extracts: Extracts derived from plants such as tea leaves, eucalyptus, or lemon have been employed for the purpose of downregulating GO. The intrinsic antioxidants present in these extracts enhance the process of reduction [14]. Microbial species, including bacteria and fungi, have the ability to decrease GO by enzymatically metabolising oxygen groups.

Benefits

Ecologically sound and enduring, free from toxicity and appropriate for mass manufacture, susceptible to biological functionalization of rGO.

Drawbacks

Extended reaction durations in comparison to chemical and thermal techniques constrained the ability to regulate the extent of decline, which may yield rGO with reduced conductivity.

Plasma Reduction Procedure

Plasma reduction refers to the utilisation of ionised gases, known as plasma, to rGO. This technique enables rapid reduction and is applicable at rather low temperatures. Plasma treatment may be conducted in alternative atmospheres, including hydrogen, argon, or ammonia.

Typical methodologies

GO is subjected to hydrogen plasma, a process that induces the reduction of oxygen-containing groups without the need for elevated temperatures.

Argon Plasma: Grey Oxide (GO) can be reduced by argon plasma, resulting in the elimination of oxygen groups while maintaining the structural integrity of rGO.

Benefits

Rapid reductive procedure can be performed at reduced temperatures, with minimal chemical residues present.

Drawbacks

Specialised equipment, namely plasma reactors, is necessary. High-cost setup, limited-scale manufacturing.

Reduction by Hydrothermal Process

Hydrothermal reduction is a highly pressurised process in which GO is heated in an aqueous solution within a sealed autoclave. Heat and pressure synergistically promote the reduction of GO into rGO.

Typical Procedures

Autoclave Procedure: GO is dispersed in water or an aqueous reducing agent and subjected to high-pressure heating at temperatures ranging from 120 to 200°C for a duration of several hours. The solvothermal method is analogous to the hydrothermal method, save that it utilises non-aqueous solvents such as ethanol.

Benefits

Straightforward and economically efficient formulation of rGO with excellent solubility in solvents. Operationally scalable for industrial manufacturing.

Drawbacks

Necessitates elevated pressure and extended reaction durations, some residual oxygen groups may persist.

All synthesis methods have their own advantages and disadvantages. Table **1** presents a comparison of different rGO synthesis methods.

Table 1. Different methods of rGO synthesis.

Method	Description	Reducing agent/conditions	Advantages	Disadvantages
Chemical reduction	Chemical-reducing compounds are used to reduce GO.	Hydrazine, NaBH$_4$, Ascorbic acid	Simple, effective under mild conditions, scalable	Toxic chemicals (hydrazine, NaBH$_4$) Incomplete reduction
Thermal Reduction	High temperature to reduce GO.	Inert gases(argon,nitrogen), temperatures up to 1000 °C	High conductivity of rGO, no reducing agents needed	High energy consumption, possible damage to graphene structure
Electrochemical reduction	Application of electric current to reduce GO in an electrolyte solution.	Electrolytes (*e.g.*, KCl, H$_2$SO$_4$), constant potential	Clean and environmentally friendly, precise control of reduction	Time-consuming, difficult scalability
Photochemical Reduction	Use of UV or visible light to reduce GO, often with a reducing agent or catalyst.	UV/Visible light, hydrogen gas, alcohol	Low temperature, spatial control of reduction	Requires a light source, incomplete reduction

(Table 1) cont.....

Method	Description	Reducing agent/conditions	Advantages	Disadvantages
Biological (Green) Reduction	Use of natural reducing agents like plant extracts or microorganisms.	Plant extracts (tea, eucalyptus), bacteria, fungi	Eco-friendly, non-toxic, sustainable for large-scale production	Long reaction times, less control over reduction
Plasma Reduction	Reduction of GO using ionized gases (plasma).	Plasma gases (hydrogen, argon)	Fast, lower temperature required, minimal chemical residues	Expensive equipment, limited scalability
Hydrothermal Reduction	Reduction of GO under high temperature and pressure in an autoclave.	Water or solvents (ethanol) under pressure	Simple, scalable, produces rGO with good dispersibility	High pressure and long reaction time, residual oxygen groups

rGO's Characteristics

The exceptional physical and chemical characteristics of rGO make it preferred for environmental remediation applications. Various approaches for synthesizing nanoparticles are summarized in Fig. (**2**), illustrating chemical, biological, and physical methods that complement rGO-based systems. Let us analyze the fundamental attributes that render rGO a highly desirable material for such applications:

Large Specific Surface Area Adsorption Capacity rGO has a remarkably large surface area because of its stratified structure, which offers a plentiful number of active sites for adsorption. This characteristic renders it very efficient in the capture of pollutants, including heavy metals, organic contaminants, and even chemical gases. Furthermore, the large surface area of the material promotes enhanced contact with pollutants, leading to more effective elimination of pollutants [15].

Multifunctionality: The significant specific surface area is crucial in several applications, including water purification, air filtration, and soil remediation, since it may retain substantial quantities of pollutants in proportion to its dimensions.

Enhanced Electrical Conductivity

The exceptional electrical conductivity of rGO is a crucial determinant in catalytic and photocatalytic applications. By enabling the transfer of electrons, rGO enhances the effectiveness of redox processes, which are essential for the degradation or conversion of pollutants. In photocatalysis, rGO can facilitate the process of electrons being transferred to reactive oxygen species, including hydroxyl radicals, which break down organic pollutants in water. rGO can be employed in electrocatalysis to accelerate the rate of electrochemical processes, including the reduction of heavy metal ions.

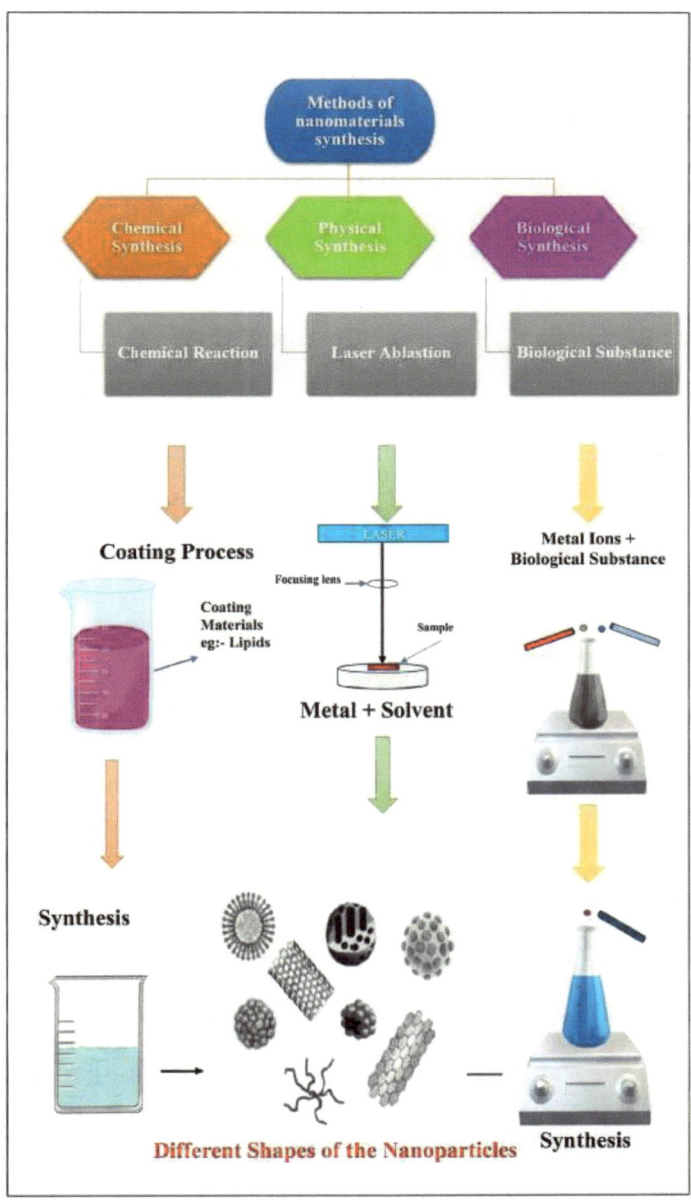

Fig. (2). Methods of nanoparticle synthesis.

The electrical characteristics of rGO enable its integration into microbial fuel cells or its usage in conjunction with other conductive materials to improve the breakdown of pollutants using electrochemical methodologies.

Chemical Functional Capability

Classification of functional groups: Although rGO has undergone significant oxygen removal compared to GO, it still maintains certain oxygen-containing groups, including hydroxyl, carboxyl, and epoxy groups. These specific functional groups have a crucial effect in the binding of contaminants, such as metal ions, and facilitate subsequent chemical alterations.

These oxygen groups enable further chemical changes to customize the surface chemistry of rGO for particular environmental remediation purposes. For instance, rGO can be modified with organic compounds or nanoparticles to increase its attraction to specific contaminants or to provide catalytic characteristics.

Metal ion capture: Functionalized rGO is highly efficient in dissociating heavy metals such as lead, cadmium, or mercury from water.

Optimization of Photocatalytic Efficiency: By incorporating specific functional groups or combining rGO with metal oxides (such as TiO_2), the photocatalytic efficiency of the material can be greatly enhanced.

Mechanical robustness and pliability

RGO maintains the mechanical characteristics of graphene, exhibiting robustness and longevity. This is crucial for environmental remediation materials, as they must endure diverse physical conditions (such as pressure and temperature) while being applied.

The versatility of rGO enables its multifunctionality in many forms, such as thin films, membranes, or composites, therefore rendering it suitable for application in filters, coatings, or absorbents.

Ecological Sustainability

RGO exhibits overall stability throughout various environmental circumstances, including diverse pH levels and temperatures. This characteristic is crucial for its sustained effectiveness in remediation procedures.

This material's resistance to degradation enables it to retain its functioning for long periods, making it a cost-efficient choice for environmental applications.

These properties render rGO a versatile and very effective material for eliminating a broad spectrum of pollutants, such as heavy metals, organic contaminants, and even gases, from various environmental matrices. As shown in Fig. (**3**), the Benefits of using nanomaterials in environmental remediation [16].

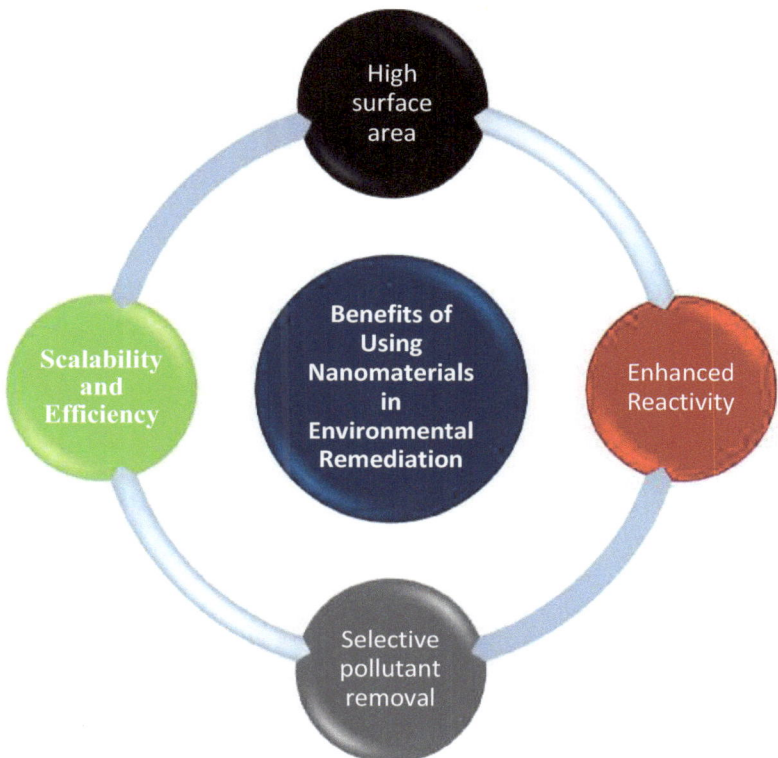

Fig. (3). Benefits of using nanomaterial in environmental remediation.

METHODS OF REMOVING POLLUTANTS USING RGO-BASED NANOCOMPOSITES

The kind of pollution and the composition of the composite material have an impact on how well rGO-based nanocomposites remove it. Adsorption, ion exchange, photocatalysis, and catalysis are the primary mechanisms.

Mechanism of Adsorption

Adsorption is a highly important method by which nanocomposites based on rGO eliminate pollutants from the environment. This method is extensively used to remove several contaminants, including heavy metals (lead, cadmium, arsenic) and organic dyes (methylene blue, rhodamine B), from water and other aquatic settings. The exceptional adsorption capability of rGO is attributed to its extensive surface area, porosity, and the existence of oxygen-containing functional groups.

The adsorption of heavy metals onto rGO nanocomposites is facilitated through multiple mechanisms, including ion exchange, electrostatic attraction, surface complexation, π–π interactions, and precipitation, as illustrated in Fig. (**4**). These interactions enhance the affinity of rGO for heavy metal ions, resulting in efficient removal from aqueous solution.

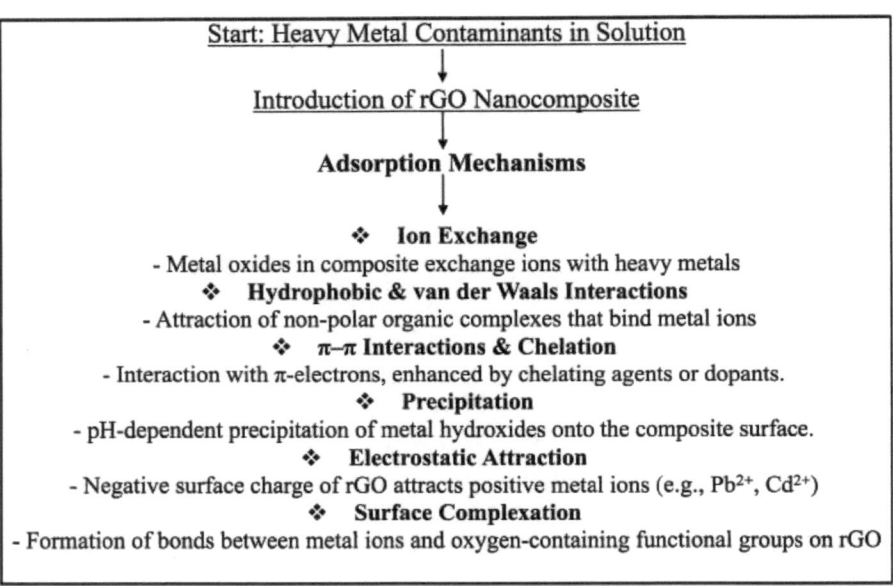

Fig. (**4**). Mechanism of heavy metals adsorption on rGO nanocomposite

Significance of Surface Area and Porosity

The extensive specific surface area of rGO offers a multitude of active sites for pollutants to attach to, therefore facilitating the adsorption of substantial quantities of contaminants. The permeable structure of the material enables the penetration of contaminants into its internal layers, therefore optimizing the interaction between the adsorbate (pollutant) and the adsorbent (rGO surface) [17].

Absorption of heavy metals, including lead (Pb) and cadmium (Cd), occurs on the surface of rGO through interactions with its oxygen-containing functional groups, namely hydroxyl (-OH), carboxyl (-COOH), and epoxide (-O-). Through processes such as ion exchange, electrostatic attraction, and complexation, these functional groups serve as binding sites for metal ions.

Lead ions (Pb^{2+}) have a robust interaction with the carboxyl and hydroxyl groups on rGO, resulting in the formation of stable complexes that are firmly attached to the surface of rGO [5].

The adsorption capability of rGO for the removal of hazardous metals is enhanced by the capture of cadmium ions (Cd^{2+}) through electrostatic interactions and coordination bonds [17].

Organic Dye Adsorption: rGO is very efficient at adsorbing organic dyes, which are frequently present in industrial wastewater. Adsorption of dyes like methylene blue and rhodamine B occurs *via* π-π interactions, which are aromatic ring contacts between the dye molecules and the surface of rGO, as well as electrostatic forces between the functional groups of rGO and the dye molecules [4].

Actuation of Functional Groups Containing Oxygen

The oxygen-containing functional groups present on rGO are essential for the adsorption of pollutants, particularly metal ions and organic compounds. These functional groups augment the interaction between rGO and pollutants by:

Electrostatic attraction originates from the negative charge on oxygen-containing groups, particularly carboxyl and hydroxyl groups, which strongly attract positively charged metal ions.

Coordination bonds: The stability of the adsorption process is greatly enhanced by the formation of coordination complexes between metal ions, namely Pb^{2+} and Cd^{2+}, and the oxygen atoms in rGO [18].

Hydrogen bonding: Furthermore, the adsorption mechanism can be enhanced by the formation of hydrogen bonds between organic dyes and polar organic pollutants with hydroxyl and carboxyl groups on rGO.

Augmenting Adsorption *via* Functionalization

Functionalization of the surface of rGO with other nanomaterials, including iron oxide (Fe_3O_4) or polymers, can significantly improve its adsorption performance. Functionalization enhances the reactivity of rGO, thereby increasing its efficacy in collecting pollutants.

The functionalization of rGO with Fe_3O_4 nanoparticles results in the formation of a magnetic nanocomposite that demonstrates enhanced adsorption capabilities for heavy metals and organic contaminants. Ferromagnetic characteristics of Fe_3O_4 facilitate the separation and retrieval of the nanocomposite from water-based solutions upon adsorption. In addition to boosting the overall adsorption effectiveness, Fe_3O_4 nanoparticles also provide supplementary active sites for metal ion binding.

The Fe_3O_4 catalyst is particularly advantageous for the elimination of heavy metals, as it may also engage in redox processes that decrease the toxicity of specific metals, such as the conversion of Cr(VI) to the less detrimental Cr(III) [19]. Polymer Functionalization: rGO can be modified with different polymers to improve its effective adsorption capacity. Grafting rGO with polyethylenimine (PEI) enhances the surface content of amine groups, enabling them to engage in ion exchange and electrostatic interactions with negatively charged pollutants such as phosphate or nitrate ions, as well as organic dyes [20]. The polymer-coated rGO exhibits selective adsorption for particular contaminants, rendering it very versatile for various environmental remediation requirements.

The Kinetics and Isotherms of Adsorption

The adsorption process of nanocomposites based on rGO can be elucidated by kinetic and isothermal models. Typically, the kinetics of adsorption are regulated by pseudo-second-order models, which suggest that the rate of adsorption is directly proportional to the quantity of active sites present on rGO. Adsorption isotherms, such as the Langmuir and the Freundlich models, are employed to characterize the state of equilibrium in the adsorption process. The Langmuir isotherm postulates the occurrence of a single layer of adsorption on a surface that is uniform in nature, whereas the Freundlich model proposes the occurrence of adsorption on a surface that is heterogeneous and has different values of affinity for the adsorbate [21].

Catalysis and Photocatalysis Catalytic Pathways in Photocatalytic Degradation of Organic Pollutants Using rGO Nanocomposites Explanation of the Flow:

1. Light Irradiation: Initiates electron-hole pair generation in the semiconductor.
2. Charge Separation and Migration: rGO helps separate and transport electrons and holes to prevent recombination.
3. Pathways for Pollutant Degradation:
    ○ Enhanced charge transfer through rGO.
    ○ Reactive Oxygen Species (ROS) formation, including hydroxyl and superoxide radicals.
    ○ Direct electron transfer to pollutants.
4. Decomposition and Mineralization:
    ○ Degradation of pollutants through oxidative decomposition.
    ○ Sequential oxidation or reduction to final products (CO_2 and H_2O).

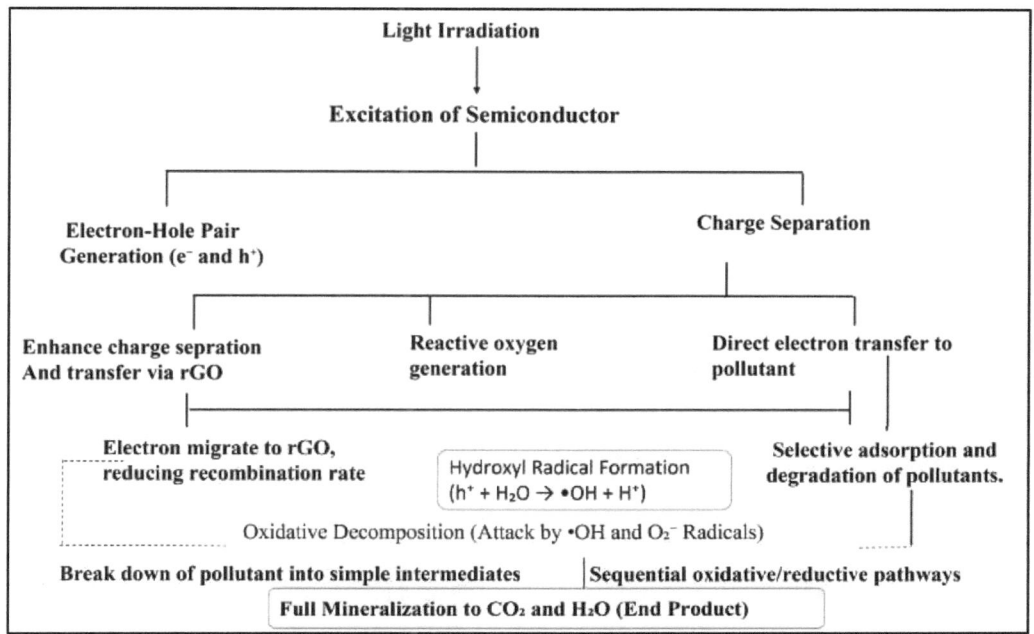

Fig. (5). A flowchart simplifies the complex catalytic pathways into actionable steps.

The combination of rGO with photocatalytic materials like ZnO (Zinc Oxide) or (Titanium Dioxide) TiO_2 significantly impacts the enhancement of both photocatalytic and catalytic activity. As (Fig. 5) presents a flowchart that simplifies the complex catalytic pathways into clear, actionable steps for understanding the mechanism. The integration of reduced graphene oxide (rGO) with photocatalytic materials such as zinc oxide (ZnO) and titanium dioxide (TiO_2) significantly enhances both their catalytic and photocatalytic efficiencies.

The improvement can be mostly ascribed to the capacity of rGO to promote electron transit and ameliorate charge separation, both of which are critical aspects in photocatalytic reactions. When combined with substances such as ZnO and TiO2, rGO functions as an electron sink, efficiently absorbing photo-excited electrons and decreasing the rate at which electron-hole pairs recombine. The enhanced separation of charges greatly enhances the total efficiency of the system in photocatalysis [22]. Photocatalytic processes involve the degradation of organic contaminants by the generation of ROS, such as hydroxyl radicals, when materials like TiO2 or ZnO are exposed to visible light. The addition of rGO facilitates faster electron transfer to the ROS, therefore improving the degradation of hazardous organic pollutants such as dyes, insecticides, and industrial chemicals into less toxic compounds, such as carbon dioxide (CO_2) and water (H_2O). This process has shown remarkable efficacy in the breakdown of organic dyes, which are prevalent contaminants in water [23].

Moreover, rGO-metal composites, which consist of rGO bonded with noble metals such as silver (Ag), gold (Au), or palladium (Pd), have exceptional catalytic characteristics in the chemical reduction of various pollutants, including organic dyes and nitro compounds(Liu *et al.*, 2020)demonstrated that rGO-Au composites have exceptional catalytic activity in the reduction of 4-nitrophenol, a poisonous nitro compound, to 4-aminophenol, a less hazardous substrate. The high electrical conductivity of rGO facilitates this catalytic process by enhancing electron transport in redox reactions, hence accelerating the reduction of potentially hazardous compounds. rGO-based composites improve the degradation of organic dye molecules into harmless byproducts in catalytic processes. This is crucial for the treatment of effluent generated by the textile and chemical industries. The inclusion of rGO in these composites serves the dual purpose of increasing the number of active sites for the reaction and helping to stabilize the metal nanoparticles, therefore preventing their aggregation and preserving their exceptional catalytic activity [24].

Therefore, rGO operates not only as a passive material in photocatalysis and catalysis, but also actively improves the efficiency of these processes, rendering it an indispensable material in endeavors to clean the environment.

RGO-BASED NANOCOMPOSITES FOR THE ELIMINATION OF POLLUTANTS

Nanocomposites based on rGO offer a flexible foundation for eliminating a variety of contaminants. When rGO is mixed with other nanomaterials, the adsorption, catalytic, and mechanical properties of the composites are enhanced, increasing their suitability for environmental applications. Table **2** presents the Mechanisms of different nanocomposites in pollutant removal.

Table 2. Different nanocomposite mechanisms in pollutant removal.

Type of Pollutant	rGO-Based Nanocomposite	Mechanism of Pollutant Removal	Efficiency	Reference
Industrial wastewater pollutants	rGO/CuO nanocomposite	Adsorption and catalytic oxidation	Up to 93% removal of phenolic compounds	[25]
Pesticides (*e.g.*, atrazine)	rGO/Ag nanocomposite	Adsorption and catalytic degradation *via* reduction reactions	Up to 87% removal in water treatment applications	[26]

(Table 2) cont.....

Type of Pollutant	rGO-Based Nanocomposite	Mechanism of Pollutant Removal	Efficiency	Reference
Pharmaceuticals (*e.g.*, antibiotics)	rGO/Fe_3O_4 nanocomposite	Adsorption through π-π interactions and hydrogen bonding	Removal efficiency up to 90% for common antibiotics	[27]
Heavy metals (*e.g.*, Pb^{2+}, Hg^{2+})	rGO/ZnO nanocomposite	Adsorption *via* electrostatic interactions and ion exchange	High adsorption capacity, up to 98% removal of Pb^{2+} (Gupta *et al.*, 2019)	[28]
Organic dyes (*e.g.*, methylene blue)	rGO/TiO_2 nanocomposite	Photocatalytic degradation under UV light	95% dye degradation under UV light in 60 minutes	[29]

Nanocomposites of Metal Oxides (rGO)

RGO nanocomposites composed of metal oxides, including TiO_2, magnetite (Fe_3O_4), and ZnO, are extensively employed in environmental remediation because of their improved photocatalytic and adsorptive characteristics. The combined effects of (rGO) and metal oxides enhance the overall efficacy of pollutant elimination, rendering these composites tremendously efficient for the breakdown of organic pollutants and the adsorption of heavy metals.

rGO-TiO_2 Nanocomposites

TiO_2 is renowned for its photocatalytic properties, namely in the breakdown of organic contaminants such as dyes and phenols when exposed to UV light. The photocatalytic efficacy of TiO_2 is much improved when it is coupled with rGO, as rGO can effectively promote electron transport and minimize pairings of electrons and holes recombining. The electron acceptor property of rGO extends the lifespan of excited electrons, hence enhancing the efficiency of photocatalysis. Multiple studies have demonstrated that rGO-TiO_2 nanocomposites exhibit superior photocatalytic activity in comparison to TiO_2 alone, particularly in the degradation of pollutants such as methylene blue and phenols [25].

Nanocomposites in rGO-Fe_3O_4

The combination of Fe_3O_4 and rGO results in a nanocomposite that exhibits exceptional adsorption capacity and magnetic characteristics. rGO-Fe_3O_4 composites exhibit high efficacy in eliminating heavy metals, including arsenic (As) and chromium (Cr), which are recognized to have significant environmental and health implications. Fe_3O_4's magnetic characteristics facilitate the easy separation of these nanocomposites from water by the application of an external magnetic field, therefore enabling their reuse. A study found that rGO-Fe_3O_4

composites exhibit enhanced adsorption capabilities for As and Cr [26]. This can be attributed to the rGO's large surface area and the great affinity of Fe_3O_4 for these heavy metals.

Furthermore, apart from the elimination of heavy metals, these nanocomposites have been actively investigated for their ability to break down organic contaminants. For example, the adsorption of dyes and phenols from wastewater has shown encouraging outcomes when utilizing rGO-Fe_3O_4 composites [27]. The nanocomposites have a double capability: they can immobilize the contaminants and, when exposed to light, break them down by photocatalytic mechanisms [28]. This attribute renders them efficient for use in wastewater treatment applications.

The rGO-ZnO Nanocomposites

Metal oxide ZnO is often mixed with rGO to create nanocomposites that exhibit improved photocatalytic and adsorptive properties [29]. ZnO has robust photocatalytic characteristics when exposed to UV radiation and is heavily employed for the degradation of organic contaminants. By combining rGO, the composite takes advantage of rGO's large surface area and electron mobility, resulting in improved charge separation and enhanced photocatalytic performance. rGO-ZnO nanocomposites have demonstrated efficacy in eliminating dyes, including Rhodamine B and methyl orange, from aqueous solutions [17].

The inclusion of rGO in these composites introduces supplementary chemical functional groups, which can either bind pollutants more effectively or enable alterations that improve the overall performance of the nanocomposite in water treatment. Furthermore, the nanocomposites' stability and recyclability render them a sustainable choice for extended environmental remediation applications [30].

Nanocomposites of rGO-polymers

The improved flexibility, adsorption capacity, and catalytic capabilities of nanocomposites of (rGO) have attracted considerable interest in the field of environmental remediation. Through the integration of (rGO) with polymers or other substances, the nanocomposites can be customized to efficiently and precisely target particular pollutants.

Iron-graphene Oxide-polyaniline Nanocomposites

One effective combination includes polyaniline (PANI), a conductive polymer renowned for its redox action and resistance to environmental degradation.

Nanocomposites comprising rGO and PANI have demonstrated exceptional efficacy in the adsorption and catalytic breakdown of industrial dyes, including methylene blue, commonly present in wastewater. These nanocomposites exploit the advantageous characteristics of rGO, such as its large surface area and electrical conductivity, together with the redox and adsorption capabilities of PANI.

Huang and colleagues (2020) showed that rGO-polyaniline nanocomposites have excellent adsorption efficiency and catalytic activity. The integration enables the nanocomposites to engage with the dye molecules, therefore facilitating their degradation into less detrimental constituents *via* oxidation and reduction biochemical processes. The composite's high conductivity facilitates electron transmission, a crucial factor in the catalytic degradation process.

rGO-Polyvinyl Alcohol Nanocomposites

The coupling of rGO with polyvinyl alcohol (PVA), a synthetic polymer renowned for its exceptional film-forming characteristics and environmental friendliness, is another effective approach. Water purification applications have been facilitated by the development of rGO-PVA nanocomposites, which possess notable mechanical strength, flexibility, and improved adsorption capability. Hydroxide groups in PVA give more active sites for binding pollutants, while rGO improves the overall structural integrity and conductivity, resulting in enhanced performance in adsorption and photocatalytic applications [26].

Specialization of Nanocomposites for Targeted Pollutants

Nanocomposites can be customized to selectively address particular contaminants by changing their surface or by modifying the proportion of (rGO) to the polymer matrix [27]. It has been demonstrated that rGO-metal oxide nanocomposites can be designed to enhance the adsorption of heavy metals or improve the degradation of organic pollutants by improved photocatalytic activity. Nanocomposite design's adaptability enables a deeply tailored method to environmental remediation, rendering them appropriate for many applications such as wastewater treatment, soil remediation, and air purification. These advancements emphasize the capacity of rGO-based nanocomposites as adaptable materials for utilization in environmental applications. Researchers have achieved an improvement in adsorption characteristics, catalytic activities, and durability by combining rGO with different polymers or functional materials. This development offers a sustainable approach for pollution reduction.

Biopolymer Composites using rGO

The efficient removal of pollutants, including organic contaminants and heavy metals, from wastewater has made biopolymer composites containing rGO a subject of growing attention in sustainable water treatment. A biodegradable cationic biopolymer containing amino and hydroxyl groups, chitosan, may efficiently bind negatively charged contaminants, while rGO increases the adsorption capacity by virtue of its extensive surface area and electrostatic interactions [28]. The aforementioned composites integrate the exceptional adsorption efficacy of GO with the environmentally favorable characteristics of biopolymers such as chitosan and alginate, therefore rendering them highly valuable for environmental purposes [30].

Recent research has demonstrated that rGO-chitosan composites possess exceptional mechanical characteristics and resilience, rendering them well-suited for extensive use in wastewater treatment [31]. GO-chitosan composites have shown great efficacy in eliminating 90% of methylene blue dye from wastewater. Additionally, they display robust affinity for heavy metals, including nickel and cobalt [32]. Nevertheless, the significant expenses associated with synthesis, the need for pH control, and the lack of effectiveness at low concentrations continue to hinder their extensive use [31, 32].

In brief, biopolymer composites based on rGO offer a potential opportunity for environmentally friendly water filtration technologies by combining the advantages of adsorption and biodegradability. However, additional study is required to overcome obstacles related to cost and efficiency [31].

ENVIRONMENTAL IMPACT AND SUSTAINABILITY

The environmental remediation potential of rGO-based nanocomposites is considerable; yet, there are concerns regarding their sustainability and environmental impact. Conventional rGO production predominantly depends on hazardous substances, such as hydrazine, that provide hazards to both the environment and human well-being. Nevertheless, green chemistry has made significant progress by introducing environmentally acceptable reducing agents, like plant extracts, and naturally existing antioxidants, which provide more sustainable options. Another study emphasizes the potential of plant-mediated green synthesis methods, which effectively decrease the generation of toxic waste and mitigate ecological damage [33]. The objective of these advancements is to tackle the sustainability issues linked to traditional synthesis methodology [34].

A crucial concern with rGO-based nanocomposites is their proper disposal following their use. Failure to handle these materials properly may result in the release of the pollutants they have absorbed back into the environment, therefore causing secondary contamination. To avoid this risk, stress the significance of implementing efficient regeneration and recycling procedures. Current research is advancing in the development of reusable materials based on rGO that may be regenerated using techniques such as heat treatment, chemical washing, or electrochemical processes. This approach aims to minimize waste and improve the environmental sustainability of these materials [35].

These recent developments emphasize continuous endeavours to enhance the environmental characteristics of (rGO) nanocomposites, by achieving a balance between their efficacy in eliminating pollutants and the long-term sustainability and safety factors [36].

CONCLUSION

Nanocomposites based on (rGO) display great potential in the field of environmental cleanup, namely in the domain of water treatment. The substantial surface area, adjustable functionalization, and adaptability of these materials render them exceptionally effective at getting rid of a wide range of contaminants, such as heavy metals, organic colors, and pharmaceutical contaminants. These characteristics emphasize their capacity to have a substantial influence on environmental remediation and the control of pollutants. Nevertheless, in order for rGO-based nanocomposites to fully realize their capabilities, it is necessary to tackle some crucial obstacles. Challenges such as the ecological consequences of synthesis techniques, the economic feasibility of manufacturing, and the durability of these materials must be addressed. The conventional methods of synthesis frequently include dangerous chemicals, which present hazards to both human health and the environment. Furthermore, there is ongoing worry regarding the efficient disposal or recycling of used nanocomposites, since they have the potential to release pollutants that have been adsorbed back into the environment. In order to surmount these obstacles, forthcoming research should prioritize the development of more environmentally friendly synthesis techniques that employ non-hazardous and renewable ingredients. Important advancements in composite design will include integrating biocompatible materials and improving the stability of rGO nanocomposites. Moreover, it will be crucial to develop effective recycling and regeneration procedures in order to preserve the long-term environmental advantages of these materials. Therefore, to fully harness the potential of rGO-based nanocomposites in addressing environmental pollution, it is necessary to adopt a comprehensive strategy that tackles both their environmental and practical obstacles. Given ongoing progress in synthesis methods, material design, and recycling approaches, nanocomposites based on rGO have the potential to play a crucial role in scalable, cost-effective, and environmentally friendly solutions for removing pollution and managing the environment.

REFERENCES

[1] Ali H, Khan E, Ilahi I. Environmental chemistry and ecotoxicology of hazardous heavy metals: environmental persistence, toxicity, and bioaccumulation. J Chem. 2019; 2019: 6730305.
[http://dx.doi.org/10.1155/2019/6730305]

[2] Akhavan O, Ghaderi E. *Escherichia coli* bacteria reduce graphene oxide to bactericidal graphene in a self-limiting manner. Carbon. 2012; 50(5): 1853–60.
[http://dx.doi.org/10.1016/j.carbon.2011.12.035]

[3] Kumar RA, Mary VD, Josephine GSA, Mohamed RA. Graphene/GO/rGO based nanocomposites: emerging energy and environmental application–review. Hybrid Adv. 2024; 5: 100168.
[http://dx.doi.org/10.1016/j.hybadv.2024.100168]

[4] Avani AV, Babu CR, Anila EI. rGO-MoO_3 nanocomposite for superior methylene blue removal by adsorption and photocatalysis. Mater Res Bull 2024; 180: 113002.
[http://dx.doi.org/10.1016/j.materresbull.2024.113002]

[5] Goyat R, Saharan Y, Sraw J, Umar A, Akbar SA. Synthesis of graphene-based nanocomposites for environmental remediation applications: a review. Molecules. 2022 Sep 29; 27(19): 6433.
[http://dx.doi.org/10.3390/molecules27196433]

[6] Das P, Deoghare AB, Maity SR. A novel approach to synthesize reduced graphene oxide (RGO) at low thermal conditions. Arab J Sci Eng. 2021; 46(6): 5467–75.
[http://dx.doi.org/10.1007/s13369-020-04956-y]

[7] Mondal A, Prabhakaran A, Gupta S, Subramanian VR. Boosting photocatalytic activity using reduced graphene oxide (RGO)/semiconductor nanocomposites: issues and future scope. ACS Omega. 2021; 6(13): 8734–43.
[http://dx.doi.org/10.1021/acsomega.0c06045]

[8] Jia F, Xiao X, Nashalian A, *et al*. Advances in graphene oxide membranes for water treatment. Nano Res. 2022 Jul; 15(7): 6636–6654.
[http://dx.doi.org/10.1007/s12274-022-4273-y]

[9] Loryuenyong V, Totepvimarn K, Eimburanapravat P, Boonchompoo W, Buasri A. Preparation and characterization of reduced graphene oxide sheets *via* water-based exfoliation and reduction methods. Adv Mater Sci Eng. 2013; 2013: 923403.
[http://dx.doi.org/10.1155/2013/923403]

[10] Patil VS, Thoravat SS, Kundale SS, Dongale TD, Patil PS, Jadhav SA. Synthesis and testing of polyaniline grafted functional magnetite (Fe_3O_4) nanoparticles and rGO based nanocomposites for supercapacitor application. Chem Phys Lett. 2023; 814: 140334.
[http://dx.doi.org/10.1016/j.cplett.2023.140334]

[11] Stankovich S, Dikin DA, Dommett GH, Kohlhaas KM, Zimney EJ, Stach EA, Piner RD, Nguyen ST, Ruoff RS. Graphene-based composite materials. Nature. 2006 Jul 20; 442(7100): 282-6.
[PMID: 16855586]

[12] Pham VH, Cuong TV, Hur SH, *et al*. Chemical reduction of graphene oxide by lemon juice. J Mater Chem 2013; 22(21): 10530-6.
[http://dx.doi.org/10.1039/c2jm30562c]

[13] Alharthi FA, Alsyahi AA, Alshammari SG, Al-Abdulkarim HA, AlFawaz A, Alsalme A. Synthesis and characterization of rGO@ZnO nanocomposites for esterification of acetic acid. ACS Omega. 2022; 7(3): 2786–97.
[http://dx.doi.org/10.1021/acsomega.1c05565] [PMID: 35097275]

[14] Yang L, Xiang Y, Jia F, Xia L, Gao C, Wu X, et al. Photo-thermal synergy for boosting photo-Fenton activity with rGO-ZnFe$_2$O$_4$: novel photo-activation process and mechanism toward environment remediation. Appl Catal B Environ. 2021; 292: 120198.
[http://dx.doi.org/10.1016/j.apcatb.2021.120198]

[15] Kolya H, Kang CW. Next-generation water treatment: exploring the potential of biopolymer-based nanocomposites in adsorption and membrane filtration. Polymers. 2023 Aug 16; 15(16): 3421.
[http://dx.doi.org/10.3390/polym15163421]

[16] Gopika G, Sathish A, Senthil Kumar P, Nithya K. A review on current progress of graphene-based ternary nanocomposites in the removal of anionic and cationic inorganic pollutants. Chemosphere. 2022 Dec; 309(Pt 1): 136617.
[http://dx.doi.org/10.1016/j.chemosphere.2022.136617]

[17] Joshi NC, Gururani P. Advances of graphene oxide based nanocomposite materials in the treatment of wastewater containing heavy metal ions and dyes. Curr Res Green Sustain Chem. 2022; 5: 100306.
[http://dx.doi.org/10.1016/j.crgsc.2022.100306]

[18] Benjwal P, Kumar M, Chamoli P, Kar KK. Enhanced photocatalytic degradation of methylene blue and adsorption of arsenic(III) by reduced graphene oxide (rGO)-metal oxide (TiO$_2$/Fe$_3$O$_4$) based nanocomposites. RSC Adv. 2015; 5(89): 73249–60.
[http://dx.doi.org/10.1039/C5RA13689J]

[19] Husein DZ, Hassanien R, Khamis M. Cadmium oxide nanoparticles/graphene composite: synthesis, theoretical insights into reactivity, and adsorption study. RSC Adv. 2021; 11(43): 27027-41.
[http://dx.doi.org/10.1039/D1RA04754J] [PMID: 35480026]

[20] Chandra V, Park J, Chun Y, Lee JW, Hwang IC, Kim KS. Water-dispersible magnetite-reduced graphene oxide composites for arsenic removal. ACS Nano. 2010 Jul 27;4(7):3979-86.
[http://dx.doi.org/10.1021/nn1008897] [PMID: 20552997]

[21] Kumar SRA, Mary DV, Josephine GAS, Ahamed MAR. Graphene/GO/rGO based nanocomposites: emerging energy and environmental application–review. Hybrid Adv. 2024 Apr; 5: 100168.
[http://dx.doi.org/10.1016/j.hybadv.2024.100168]

[22] Xu P, Wang P, Wang Q, et al. Facile synthesis of Ag$_2$O/ZnO/rGO heterojunction with enhanced photocatalytic activity under simulated solar light: kinetics and mechanism. J Hazard Mater. 2021 Feb 5; 403: 124011.
[http://dx.doi.org/10.1016/j.jhazmat.2020.124011]

[23] Liu X, Li X, Liu X, He S, Jin J, Meng H. Green preparation of Ag–ZnO–rGO nanoparticles for efficient adsorption and photodegradation activity. Colloids Surf A Physicochem Eng Asp. 2020 Jan 5; 584: 124011.
[http://dx.doi.org/10.1016/j.colsurfa.2019.124011]

[24] Stephen D, Sivasubramanian M, Ramakrishnan K, et al. Graphene oxide-based nanomaterials for the treatment of pollutants in the aquatic environment: recent trends and perspectives – a review. Environ Pollut. 2022 Aug 1; 306: 119377.
[http://dx.doi.org/10.1016/j.envpol.2022.119377]

[25] Kumar SRA, Mary DV, Josephine GAS, Ahamed MAR. Graphene/GO/rGO based nanocomposites: emerging energy and environmental application–review. Hybrid Adv. 2024 Apr; 5: 100168.
[http://dx.doi.org/10.1016/j.hybadv.2024.100168]

[26] Priyadharshini SD, Manikandan S, Kiruthiga R, et al. Graphene oxide-based nanomaterials for the treatment of pollutants in the aquatic environment: recent trends and perspectives – a review. Environ Pollut. 2022 Aug 1; 306: 119377.

[27] Sharma A, Lee BK. Structure and activity of TiO$_2$/FeO co-doped carbon spheres for adsorptive-photocatalytic performance of complete toluene removal from aquatic environment. Appl Catal A Gen. 2016 Aug 5; 523: 272–282.

[http://dx.doi.org/10.1016/j.apcata.2016.06.018]

[28] Safajou H, Ghanbari M, Amiri O, *et al.* Green synthesis and characterization of RGO/Cu nanocomposites as photocatalytic degradation of organic pollutants in wastewater. Int J Hydrog Energy 2021 Jun 4; 46(39): 20534–20546.
[http://dx.doi.org/10.1016/j.ijhydene.2021.03.175]

[29] Aher A, Thompson S, Nickerson T, Ormsbee L, Bhattacharyya D. Reduced graphene oxide–metal nanoparticle composite membranes for environmental separation and chloro-organic remediation. RSC Adv. 2019 Nov 22; 9(66): 38547–38557.
[http://dx.doi.org/10.1039/C9RA08178J]

[30] Agarwal S, Tyagi I, Gupta VK, *et al.* Applications of reduced graphene oxide-based materials in environmental cleanup. Mater Today Commun. 2020 Sep; 24: 101187.

[31] Ma Z, He J, Song H, *et al.* Reduced graphene oxide-based heterojunctions for enhanced photocatalytic removal of emerging pollutants: a comprehensive review. Chem Eng J. 2024 Mar 15; 484: 149607.

[32] Wu J, Li Q, Zhang R. Magnetic rGO-Fe_3O_4 nanocomposites for arsenic and chromium removal: synthesis, characterization, and adsorption mechanisms. Environ Sci Pollut Res Int. 2021 Mar; 28(12): 15632–15644.

[33] Chen Y, Li W, Zhang T. Machine learning approaches for optimizing the design of rGO-based nanocomposites in environmental applications. J Environ Nanotechnol. 2022; 14(3): 45-59.

[34] Dangi SB, Leel NS, Quraishi AM, *et al.* Poly(vinyl alcohol)/reduced graphene oxide (rGO) polymer nanocomposites: ecological preparation and application-oriented characterizations. Opt Mater. 2024 Feb; 148: 114965.
[http://dx.doi.org/10.1016/j.optmat.2024.114965]

[35] Sharma A, Lee BK. Structure and activity of TiO_2/FeO co-doped carbon spheres for adsorptive-photocatalytic performance of complete toluene removal from aquatic environment. Appl Catal A Gen. 2016 Aug 5; 523: 272–282.
[http://dx.doi.org/10.1016/j.apcata.2016.06.018]

[36] Saini R, Mishra RK, Kumar P. Green synthesis of reduced graphene oxide using the Tinospora cordifolia plant extract: exploring its potential for methylene blue dye degradation and antibacterial activity. ACS Omega. 2024 Apr 24; 9(18): 20304–20321.
[http://dx.doi.org/10.1021/acsomega.4c00748]

SUBJECT INDEX

A

Accountability 203, 218, 223, 273, 284, 287
Acetic acid 256
Acidophilic multifunctionality 193
Acoustic cavitation 193
Activated sludge processes 63
Activated sludge systems 47, 50, 54, 55, 56
Active remote sensing 205
Active sites, multiple 293
Adaptability, nanocomposite design 311
Adaptation, ecosystem-based 272
Adaptive policy measures 217
Adsorbate 304, 306
Adsorption 48, 293, 300, 303, 304, 306, 308, 309, 310, 311, 312
   selective adsorption 306
   isotherms 306
   catalytic degradation 308
   exceptional capacity 309
Advanced electrode materials 62
Advanced oxidation techniques 195
Advanced recycling technologies 199
Aeration configuration 55
Aerobic granular sludge 101
Agricultural by-products 5, 12
Agricultural productivity improvement 277
Agriculture, climate-smart 169
AI-based predictive models 286
AI-IoT integration 221
Air quality management 26, 36, 38
Algae-based textiles 6
Alginate beads 73
Algorithms, learning-based 103
Anaerobic digestion method 263
Anaerobic fluidized-bed reactor 258
Anaerobic reductive dechlorination 114
Analysis, time-series 206
Antibiotic resistance genes 193
AOPs (advanced oxidation processes) 177, 181, 192
APIs (active pharmaceutical ingredients) 180, 181, 195
Aquatic ecosystems 32, 33, 49, 77, 149, 153, 178, 180, 182, 183, 292
Arbuscular mycorrhizae 158
Artificial intelligence 26, 40, 49, 199, 212
Atmospheric carbon 166

B

Bacterial cellulose acetobacter 7
BBD (biodegradation database) 103, 131
Big data analytics 202, 207, 211, 285
Bio-adhesives 11
Bio-piling 101
Bioaugmentation strategies 74
Biochemical oxygen demand (BOD) 50, 120
   Biochar 61, 79, 81, 83, 84
  addition of 83, 84
Biodegradable packaging 5
Biodegradation 9, 53, 55, 70, 75, 77, 79, 80, 81, 83, 113, 116, 117, 124, 126, 131
  aerobic 83, 116
  dichlorprop 70
  mecoprop 77
   Biodiversity indicators 285
Bioelectrochemical systems 7
Biofuel production 28, 31, 125
Biogeochemical cycles 146, 148, 154
Biological nutrient removal (BNR) 37, 47, 56
Biomass concentrations 61
Bioremediation 1, 12, 13, 93, 94, 98, 102, 103, 110, 116, 126, 132, 165
  ex-situ 94, 116, 132
  in-situ 94, 98, 102
Biosensors monitoring 202
Biosparging 96, 104, 115
Biostimulation approach 83
Bioventing, aerobic 112
Blockchain 199, 202, 222
Brundtland commission 1, 228

C

Carbon capture and storage (CCS) 27, 214
Carbon footprint 19, 169, 253
Carbon sequestration 4, 61, 155, 172
Catalytic oxidation 308
Catalytic ozonation 187, 191
CBEM (community-based environmental management) 268, 272, 281
Circular bioeconomy 35, 132
Climate-smart agriculture 169
CLMS (community-led monitoring systems) 268, 283
Co-financing 282
COD (chemical oxygen demand) 50, 181, 257
Community empowerment 268
Composting environments 5
CRISPR-Cas9 technology 13, 33
Cyanobacteria 31, 152, 172

D

Data collection strategies 246
Digital environmental sustainability 200
Disaster risk reduction 211
Drone surveillance 211

E

EBPR (enhanced biological phosphorus removal) 57
Ecological integrity 272, 288
Effluent treatment plants 178
Electrochemical oxidation (EO) 186, 190
Energy recovery methods 253

F

Fenton reaction 186, 188
Financial resources 248, 275, 277, 281
  consistent 277
  increasing 248
Fossil fuels 9, 11, 13, 14, 15, 18, 26, 172, 253, 254, 263
Functional groups, oxygen-containing 294, 305

G

GEMs (genetically engineered microbes) 8, 95, 129
GIS and remote sensing 199, 212
Global biodiversity framework 248
Graphene oxide, reduced (rGO) 292, 307
Greenhouse gas mitigation 35, 168

H

Habitat restoration 284
Hazardous compounds 308
Heavy metals 13, 26, 27, 30, 31, 32, 34, 48, 62, 77, 92, 119, 120, 129, 130, 292, 304, 309, 310, 312
  absorbing 130
  eliminating 309
  removal of 48, 120
  removing 62
  toxic 30, 129
Hydrothermal reduction 298, 300
Hydroxyl radicals 60, 184, 191

I

Image recognition 206
Impact assessments 203, 270
  environmental 203
Incentives 21, 275, 280, 282
  financial 275, 280
  required economic 282
Industrial chemicals 48, 59, 77, 80, 292, 307
  toxic 292
IoT (internet of things) 199, 212, 221

J

Jal Jeevan Mission (JJM) 245

L

Land degradation 283
LCA (life cycle assessment) 19
Legal recognition 275
Light irradiation 306
Local governance 273, 284
  enhancing 273
Lyme disease 159, 162, 171

M

Machine learning algorithms 199, 214
Management decisions 216, 220
    environmental 220
Mass transfer 116, 187, 191, 259
    efficient 259
    enhanced 187
    improving 116
Mechanical strength 293, 311
    enhanced 293
Methane-producing settings 156
Microbial fuel cells (MFCs) 7, 49, 124

N

Nanocomposites magnetic 305
Nanoremediation 103
Nanotechnology 48, 104, 186, 238, 239, 240, 241, 273, 293
    rendering 293
Natural disasters 208, 211, 214, 216, 272
Natural fertilization 155
Nitrogen fixation 157, 164

O

Ocean acidification 146, 152, 172
Operational costs 51, 62, 195, 253
    lowering 253
    minimizing 51, 195
Organic pollutants, persistent 177, 194

P

Payment for environmental services (PES) 275
Phytoremediation processes 98
Public-private partnerships (PPPs) 222, 241

R

Radicals 62, 130, 180, 183, 184, 185, 186, 187, 189, 191, 194, 306
    free 130, 180, 184, 185, 189, 194
    hydroxy 183, 191
    oxidative 184
    reactive 183
    reactive oxidizing 187

superoxide 306
Radio frequencies 209
Resource recovery methods 47

S

SBRs (sequencing batch reactors) 47, 56, 110
Scalability 12, 16, 17, 20, 33, 49, 61, 186, 283, 288, 294, 296, 299, 300
    increasing 12
    limited 300
    production process's 17
    restricted 296
Scalable digital solutions 221
Scaling 21, 62, 192, 237
    industrial 21
SDGs (sustainable development goals) 202, 230, 240
Synthetic biology tools 26, 34, 39

T

Targeted Pollutants 311
Targeted resource allocation 217
Technological advancements 20, 40, 178, 220, 272
    industrial 178
TEK (traditional ecological knowledge) 274, 287

U

UAVs (unmanned aerial vehicles) 208, 209
Underutilisation 219
Universal health coverage (UHC) 237, 248

W

Waste-to-resource conversion 7
Waste reduction 10, 240
Wastewater treatment plants (WWTPs) 178, 181, 254
Wildlife reserves 272

www.ingramcontent.com/pod-product-compliance
Lightning Source LLC
Chambersburg PA
CBHW041456280526
45792CB00004B/1025